Collins

Student Book

CAMBRIDGE IGCSE® BIOLOGY

Sue Kearsey, Gareth Price,
Jackie Clegg and Mike Smith

William Collins' dream of knowledge for all began with the publication of his first book in 1819. A self-educated mill worker, he not only enriched millions of lives, but also founded a flourishing publishing house. Today, staying true to this spirit, Collins books are packed with inspiration, innovation and practical expertise. They place you at the centre of a world of possibility and give you exactly what you need to explore it.

Collins. Freedom to teach

Published by Collins
An imprint of HarperCollins *Publishers*
77 – 85 Fulham Palace Road
Hammersmith
London
W6 8JB

Browse the complete Collins catalogue at
www.collinseducation.com

© HarperCollins*Publishers* Limited 2012

10 9 8 7 6 5 4 3 2 1
ISBN 13 978 0 00 745442 6

The authors assert their moral rights to be identified as the authors of this work.

All rights reserved. No part of this publication may be reproduced, stored in a retrieval system, or transmitted in any form or by any means, electronic, mechanical, photocopying, recording or otherwise, without the prior written permission of the Publisher or a licence permitting restricted copying in the United Kingdom issued by the Copyright Licensing Agency Ltd., 90 Tottenham Court Road, London W1T 4LP.

British Library Cataloguing in Publication Data
A Catalogue record for this publication is available from the British Library

Commissioned by
Letitia Luff and Rebecca Richardson
Project edited by **Caroline Green**
Project managed by
Jim Newall and Lifelines Editorial Services
Edited by **Lauren Bourque**
Proofread by **Maggie Rumble**
Indexed by **Jane Henley**
Designed by **Jouve India Private Limited**
New illlustrations by **Jouve India Private Limited**
Picture research by
Caroline Green and Grace Glendinning
Concept design by **Anna Plucinska**
Cover design by **Angela English**
Production by **Rebecca Evans**
Printed and bound by **L.E.G.O. S.p.A. Italy**

Fully safety checked but not trialled by CLEAPSS

The syllabus content is reproduced by permission of Cambridge International Examinations.

Acknowledgements

The publishers wish to thank the following for permission to reproduce photographs. Every effort has been made to trace copyright holders and to obtain their permission for the use of copyright materials. The publishers will gladly receive any information enabling them to rectify any error or omission at the first opportunity:

(t = top, c = centre, b = bottom, r = right, l = left)

Cover & p1 Ribe/Dreamstime, p8-9 Stephen Coburn/Shutterstock, p10t Science Photo Library/Alamy, p10b formiktopus/Shutterstock, p12 Hazel Appleton, Centre for Infections/Health Protection Agency, p14 Aleexandru-Radu Borzea/Shutterstock, p15 Juan Camilo Bernal/Shutterstock, p17 Brisbane City Council, p20 Sergey Popov V/Shutterstock, p21t Brian Lasenby/Shutterstock, p21b Doug Ellis/Shutterstock, p22 Aleksandar Todorovic/Shutterstock, p23 Aleksei Verhovski/Shutterstock, p27 Martin Fowler/Shutterstock, p28 Lehrer/Shutterstock, p29t MindStorm/Shutterstock, p29b dmvphotos/Shutterstock, p30 Matthew Cole/Shutterstock, p31 John A. Anderson/Shutterstock, p35 Tracing Tea/Shutterstock, p44-45 Jubal Harshaw/Shutterstock, p46 PHOTOTAKE Inc./Alamy, p47t Ed Reschke/Getty Images, p47b Claude Nuridsany & Marie Perennou/Science Photo Library, p49 Melba Photo Agency/Alamy, p52 Tinydevil/Shutterstock, p55 Dr. Richard Kessel & Dr. Gene Shih/Getty Images, p58 Carol and Mike Werner/Alamy, p59 Dr. David M. Phillips/Getty Images, p63 Dr. Stanley Flegler, Visuals Unlimited/Science Photo Library, p64 Andrew Lambert Photography/Science Photo Library, p66 Picsfive/Shutterstock, p68 David Cook/blueshiftstudios/Alamy, p70 PHOTOTAKE Inc./Alamy, p76 stefanolunardi/Shutterstock, p84 photo provided courtesy of Joseph Lee of http://www.gratuitousfoodity.com/and used with permission, p85 shae cardenas/Shutterstock, p89 Mary Evans Picture Library/Alamy, p92 oksana.perkins/Shutterstock, p96t Martin Shields/Alamy, p96b Andrew Lambert Photography/Science Photo Library, p97 David Vincent, p100 sciencephotos/Alamy, p102 A.Krotov/Shutterstock, p105 Triff/Shutterstock, p106l Nigel Cattlin/Science Photo Library, p106r Nigel Cattlin/Alamy, p109 Jeff Rotman/Alamy, p110t Images of Africa Photobank/Alamy, p110b HLPhoto/Shutterstock, p113 Cate Turton/Department for International Development, p119 david n madden/Shutterstock, p126 Julien Harneis/WikiMedia Commons, p136 Alain Pol, ISM/Science Photo Library, p138 Power and Syred/Science Photo Library, p139t Dr Keith Wheeler/Science Photo Library, p139b Zastol`skiy Victor Leonidovich/Shutterstock, p140 Nigel Cattlin/Alamy, p141l Nigel Cattlin, Visuals Unlimited/Science Photo Library, p141r Adam Hart-Davis/Science Photo Library, p146 lolloj/Shutterstock, p149 Andrew Brookes, National Physical Laboratory/Science Photo Library, p151 LeventeGyori/Shutterstock, p158 National Cancer Institute/Science Photo Library, p160 Paul Gunning/Getty Images, p162 Dmitry Naumov/Shutterstock, p167 aureapterus/iStockphoto, p170 Nickolay Vinokurov/Shutterstock, p171 Jan Kratochvila/Shutterstock, p172 Maxisport/Shutterstock, p173 Adam Hart-Davis/Science Photo Library, p178 Andrew Gentry/Shutterstock, p185 Keith A Frith/Shutterstock, p190 Poco a poco/WikiMedia Commons, p195 Tudor Stanica/Shutterstock, p206 Anest/Shutterstock, p213 aGinger/Shutterstock, p215 Westend61 GmbH/Alamy, p231 Ed Reschke/Getty Images, p238-239 Pakhnyushcha/Shutterstock, p240 PHOTOTAKE Inc./Alamy, p242l Stephen Emerson/Alamy, p242r oksix/Shutterstock, p246 Dr Jeremy Burgess/Science Photo Library, p247 glyn/Shutterstock, p248t piyato/Shutterstock, p248c WILDLIFE GmbH/Alamy, p248b Dartmouth Electron Microscope Facility, p249 Tim Gainey/Alamy, p251 Valentyn Volkov/Shutterstock, p257 Francis Leroy, Biocosmos/Science Photo Library, p258 Nic Cleave Photography/Alamy, p260 Bork/Shutterstock, p266 mknobil/WikiMedia Commons, p270 Galyna Andrushko/Shutterstock, p272 worldswildlifewonders/Shutterstock, p274 Perry Mastrovito/Getty Images, p276 Pi-Lens/Shutterstock, p279 Andy Lim/Shutterstock, p281 CNRI/Science Photo Library, p282l Eric Isselée/Shutterstock, p282r Eric Isselée/Shutterstock, p290t Mary Evans Picture Library/Alamy, p290b Science Photo Library, p294 Denis Kuvaev/Shutterstock, p295t Sebastian Kaulitzki/Shutterstock, p295b ISM/Science Photo Library, p297t Photocrea/Shutterstock, p297b Francesco Tonelli/Alamy, p301 M. J. Mayo/Alamy, p316-317 NASA, p318 Colin Pickett/Alamy, p321 Anan Kaewkhammul/Shutterstock, p331 Rick Wylie/Shutterstock, p333 Anest/Shutterstock, p335 Jane McIlroy/Shutterstock, p336 Tropical Rain Forest Information Center (TRFIC)/Basic Science and Remote Sensing Initiative (BSRSI)/Landsat 7 Project Science Office/NASA Goddard Space Flight Center, p338 Frank Vincentz/WikiMedia Commons, p340 Fotokostic/Shutterstock, p344t nadirco/Shutterstock, p344b Peter Gudella/Shutterstock, p345 Jim Cartier/Science Photo Library, p353 Celso Diniz/Shutterstock, p356 Darrin Henry/Shutterstock, p358 Earth Observations Laboratory, Johnson Space Center, p360 NASA, p361 Roger Eritja/Alamy, p362t Julio Etchart/Alamy, p362b Jeremy Sutton-Hibbert/Alamy, p363 Martyn F. Chillmaid/Science Photo Library, p364 Sue Kearsey, p369 NASA/JPL, p371 Photoshot Holdings Ltd/Alamy, p375 cozyta/Shutterstock, p376 neelsky/Shutterstock, p378 Oxfam East Africa, p379 Andrey Kekyalyaynen/Shutterstock, p402 Ed Phillips/Shutterstock.

Contents

Getting the best from the book 4

Section 1
Characteristics and classification of living organisms 8
a) Characteristics of living organisms 10
b) Classification and diversity of living organisms 15
c) Simple keys 35
d) Exam-style questions 39

Section 2
Organisation and maintenance of the organism 44
a) Cell structure and organisation 46
b) Levels of organisation 52
c) Size of specimens 59
d) Movement in and out of cells 63
e) Enzymes 76
f) Nutrition 92
g) Transportation 136
h) Respiration 167
i) Excretion in humans 185
j) Coordination and response 195
k) Exam-style questions 220

Section 3
Development of the organism and the continuity of life 238
a) Reproduction 240
b) Growth and development 272
c) Inheritance 279
d) Exam-style questions 308

Section 4
Relationships of organisms with one another and with their environment ... 316
a) Energy flow, food chains and food webs 318
b) Nutrient cycles 333
c) Population size 345
d) Human influences on the ecosystem 356
e) Exam-style questions 385

Doing well in examinations 396
Introduction 396
Overview 396
Assessment objectives and weightings 396
Examination techniques 397
Answering questions 398

Developing experimental skills 400
Introduction 400
Using and organising techniques, apparatus and materials 400
Observing, measuring and recording 403
Handling experimental observations and data 406
Planning and evaluating investigations 412

Glossary 423
Answers 434
Index 455

Getting the best from the book

Welcome to Collins *Cambridge IGCSE Biology*.

This textbook has been designed to help you understand all of the requirements needed to succeed in the Cambridge IGCSE Biology course. Just as there are four sections in the Cambridge syllabus, there are four sections in the textbook: Characteristics and classification of living organisms, Organisation and maintenance of the organism, Development of the organism and the continuity of life and Relationships of organisms with one another and with their environment.

Each section is split into topics. Each topic in the textbook covers the essential knowledge and skills you need. The textbook also has some very useful features which have been designed to really help you understand all the aspects of Biology which you will need to know for this syllabus.

SAFETY IN THE SCIENCE LESSON

This book is a textbook, not a laboratory or practical manual. As such, you should not interpret any information in this book that related to practical work as including comprehensive safety instructions. Your teachers will provide full guidance for practical work and cover rules that are specific to your school.

A brief introduction to the section to give context to the science covered in the section.

Starting points will help you to revise previous learning and see what you already know about the ideas to be covered in the section.

The section contents shows the separate topics to be studied matching the syllabus order.

Knowledge check shows the ideas you should have already encountered in previous work before starting the topic.

Learning objectives cover what you need to learn in this topic.

Cell structure and organisation

INTRODUCTION

All 'complex' cells (those that contain a nucleus) in all animals, plants and protoctists on Earth have the same basic structure. Scientists say that this is because we have all evolved from a single complex cell. This cell evolved from a simple bacteria-like cell (without a nucleus) over 1600 million years ago. This is the origin of all the millions of different species of plants, animals and protoctists that live on Earth today.

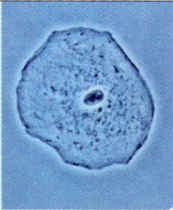

△ Fig. 2.1 All plant, protoctist and animal cells (apart from some very specialised cells), like this human cheek cell, have a cell nucleus like this.

KNOWLEDGE CHECK
✓ Know that most organisms are formed from many cells.
✓ Know that cells may be specialised in different ways to carry out different functions.

LEARNING OBJECTIVES
✓ Be able to state that living organisms are made of cells.
✓ Be able to identify and describe the structure of a plant cell (palisade cell) and an animal cell (liver cell), as seen under a light microscope.
✓ Be able to describe the differences in structure between typical animal and plant cells.
✓ EXTENDED Be able to relate the structures seen under the light microscope in the plant cell and in the animal cell to their functions.

PLANT AND ANIMAL CELLS

The diagrams below show a typical animal cell and typical plant cells. These cells all have a nucleus and cytoplasm.

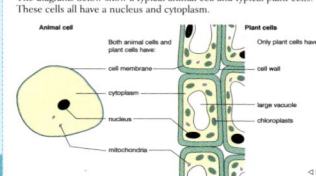

◁ Fig. 2.2 The basic structures of an animal cell and plant cells.

All living organisms are made of cells. Some, such as bacteria and some fungi, are formed from a single cell; others, such as most plants and animals, are **multicellular**, with a body made of many cells. All animal and plant cells have certain features in common:
- a **cell membrane** that surrounds the cell
- the **cytoplasm** inside the cell, in which all the other structures are found
- a large **nucleus**.

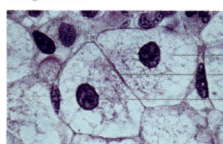

△ Fig. 2.3 A human liver cell is a typical animal cell.

Plant cells also have features that are not found in animal cells, such as:
- a **cell wall** surrounding the cell membrane
- a large central **vacuole**
- green **chloroplasts** found in some, but not all, plant cells.

Each structure in a cell has a particular role.

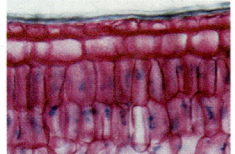

△ Fig. 2.4 A typical plant cell is a palisade cell in the upper part of a leaf. The chloroplasts are supported by the cytoplasm.

Osmosis in animal and plant cells

If an animal cell is placed in a solution that has a higher concentration of solute (and so a lower concentration of water molecules) than the cytoplasm inside the cell, water will leave the cell by osmosis and the cell will shrink (Fig. 2.21). In living cells this causes problems because many reactions inside the cell take place in solution.

Even bigger problems occur if an animal cell is placed in a solution that has a lower concentration of solute (and so a higher concentration of water molecules) than the cytoplasm inside the cell. Osmosis will cause water to enter the cell and increase its volume (Fig. 2.20). If too much water enters, the pressure inside the cell may become too much for the cell membrane to hold, and the cell may burst.

Plant cells are surrounded by cell walls outside the cell membrane. Cell walls are completely permeable, which means that water and solute molecules pass easily through them. In a solution of high concentration of solute (low concentration of water), water will leave a plant cell by osmosis and the cell will become **flaccid**. This can be seen in plant cells as the cytoplasm reduces in volume inside the cell, and so reduces the pressure on the cell wall. However, the cell doesn't shrink, unlike in animal cells, because the plant cell wall controls the structure of the cell. Instead, the whole plant wilts.

In a solution of low concentration of solute (high concentration of water), water will enter the plant cell by osmosis. However, the plant cell does not eventually burst. The strong cell wall provides strength when the cytoplasm is full of water and **turgid**, preventing the cell from expanding any further and bursting. Turgid cells also provide strength, making the plant stand upright with its leaves held out to catch the sunlight.

△ Fig. 2.24 When the cells of the plant are not full of water, the cell walls are not strong enough to support the plant, and the plant collapses (wilts). When the cells are fully turgid, the plant stands upright.

Developing investigative skills

Dandelions are weed plants in northern Europe that have a hollow fleshy stem. Strips of dandelion stem about 5 cm long and 3 mm wide were placed in sodium chloride solutions of different concentration. (Note: care should be taken with sharp instruments used for cutting; dandelion parts are very mildly poisonous and should not be eaten and hands should be washed after handling.) After 10 minutes, the strips looked as shown in the diagram. (Note that the outer layer of a dandelion stalk is 'waterproofed' with a waxy layer to protect them from water loss to the environment.)

△ Fig. 2.25 Investigating osmosis

Using and organising techniques, apparatus and materials
① Write a plan for an experiment to carry out this investigation. Your plan should include:
  a) Instructions on how to prepare the stem samples
  b) Instructions on how to keep the stem samples until the experiment starts.

Observing, measuring and recording
② Using the diagram, describe the results of this investigation.

Handling experimental observations and data
③ Explain as fully as you can the results of this investigation.
④ Use the results to suggest the normal concentration of cell cytoplasm. Explain your answer.

Examples of investigations are included with questions matched to the investigative skills you will need to learn.

Getting the best from the book *continued*

Science in context boxes put the ideas you are learning into real-life context. It is not necessary for you to learn the content of these boxes as they do not form part of the syllabus. However, they do provide interesting examples of scientific application that are designed to enhance your understanding.

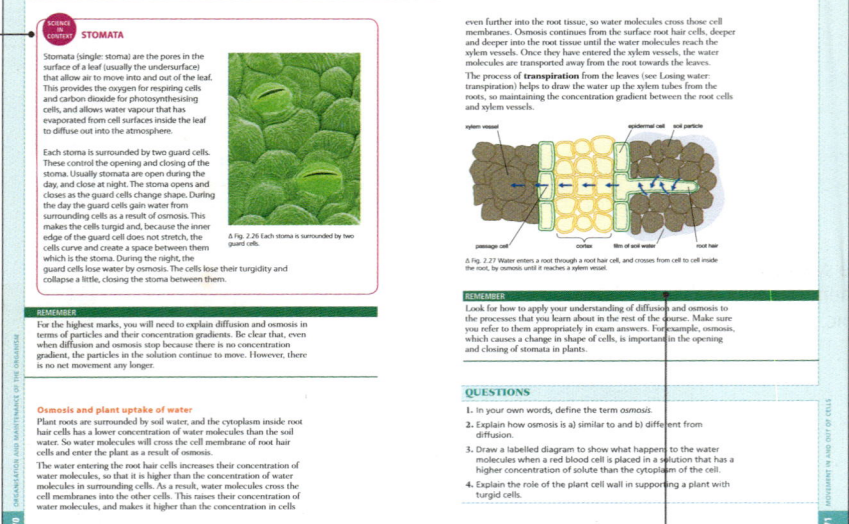

Remember boxes provide tips and guidance to help you during your course and to prepare for examination.

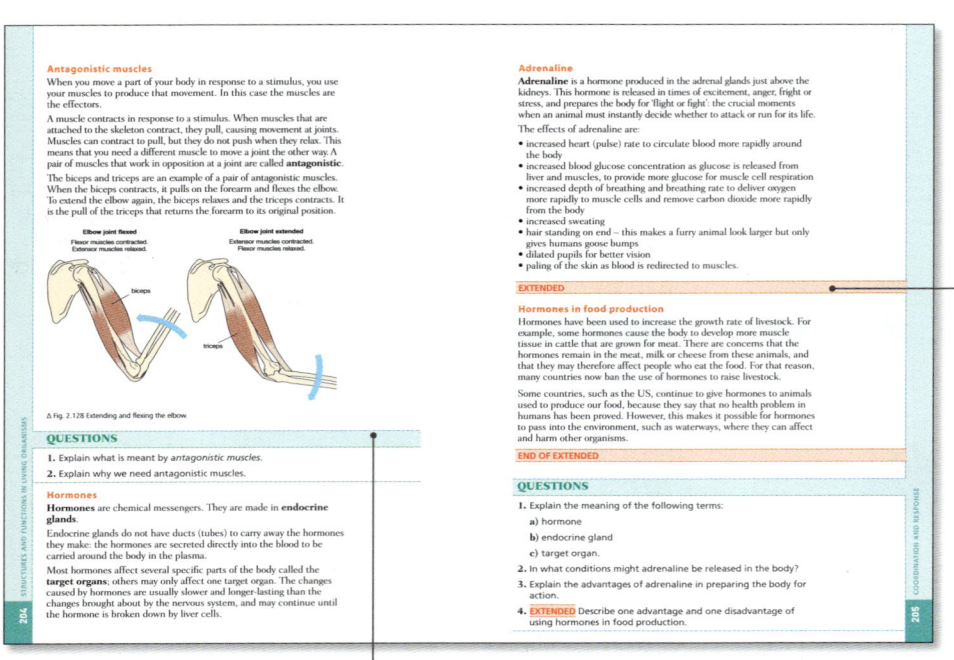

Clearly differentiated Extended material takes your learning even further.

Questions to check your understanding.

End of topic questions allow you to apply the knowledge and understanding you have learned in the topic to answer the questions.

A full checklist of all the information you need to cover the complete syllabus requirements for each topic.

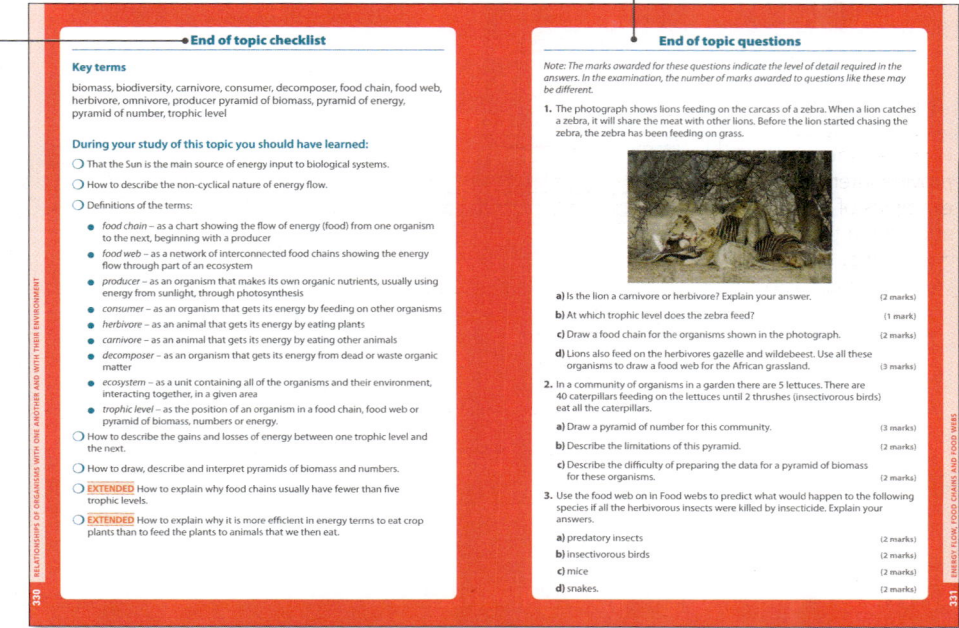

Each section includes exam-style questions to help you prepare for your exam in a focussed way and get the best results.

The first question is a student sample with teacher's comments to show best practice.

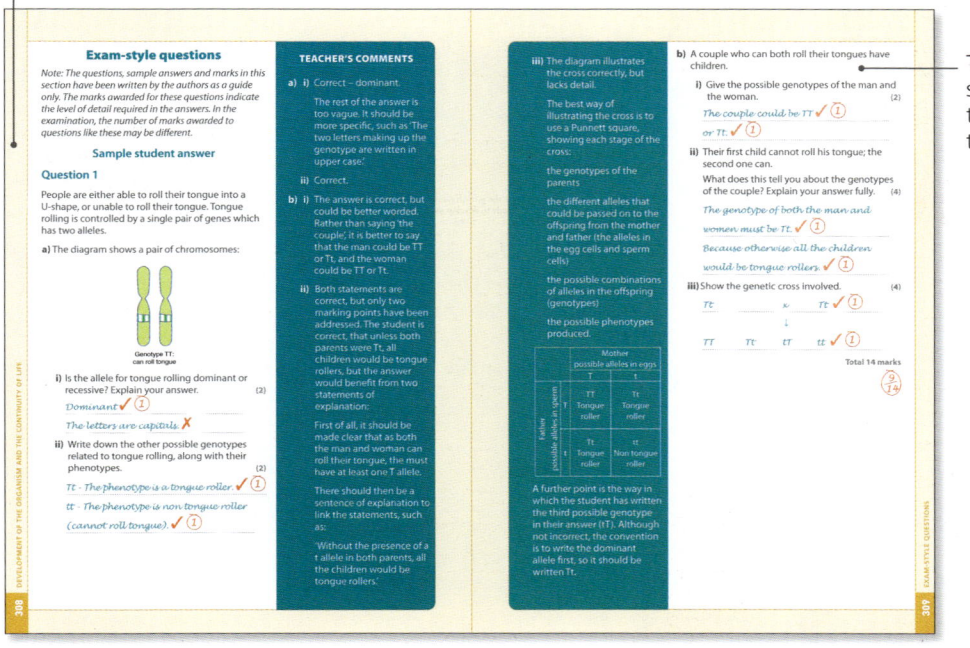

Around 1.74 million living species have been identified on Earth, not including bacteria. Over 320 000 of these species are classified as plants and around 1.36 million species are classified as animals. Over 62 000 of the animal species are vertebrates (animals with bony skeletons) and the rest are invertebrates (animals without backbones), of which the majority (around 1 million species) are insects.

It is difficult to know how many species are still to be discovered, although scientists believe they have discovered most living mammals, birds and coniferous (cone-bearing) trees. The smaller the organism, the greater the chance that there are species we don't yet know about. So although around 4000 species of bacteria have been identified, there could be more species of bacteria than of all the other kinds of organisms put together.

STARTING POINTS

1. What are the characteristics shared by living organisms?
2. Crystals can grow in size, but does that mean they are alive?
3. We talk of 'feeding' a fire when we add fuel, but does that mean fire is a living thing?
4. Why is it useful to group organisms?
5. What features are the most useful for grouping organisms?

SECTION CONTENTS

a) Characteristics of living organisms
b) Classification and diversity of living organisms
c) Simple keys
d) Exam-style questions

1 Characteristics and classification of living organisms

△ Many species of different kinds of organisms live on a coral reef.

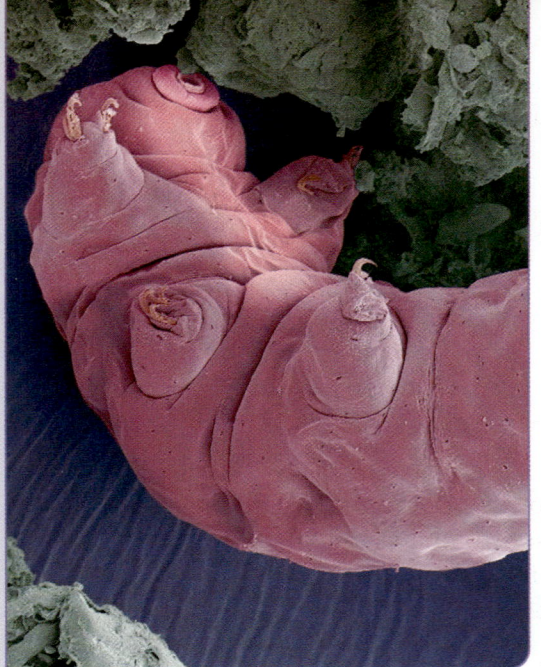

Fig.1.1 Tiny tardigrades (about 1 mm long) are the toughest organisms known. They can survive temperatures below −200 °C, 10 days in the vacuum of space and over 10 years without water!

Characteristics of living organisms

INTRODUCTION

Sometimes it is easy to tell when something dies – an animal stops moving around, a plant may wilt and all the green parts collapse. But does a tree in a seasonal region die in winter, when its leaves have dropped off? Are animals 'dead' when they hibernate underground for months to avoid winter cold? As technology gets increasingly sophisticated, and we can create machines with 'brains' and new organisms from basic molecules, distinguishing between living and dead could get even more difficult. We need a set of 'rules' that work for most organisms, most of the time.

KNOWLEDGE CHECK

✓ Know that living organisms show a range of characteristics that distinguish them from dead or non-living material.
✓ Know that the life processes are carried out by the cells, tissues, organs and systems of the body.

LEARNING OBJECTIVES

✓ Be able to list and describe the characteristics of living organisms.
✓ Define the terms: nutrition, excretion, respiration, sensitivity, reproduction, growth and movement.

SEVEN LIFE PROCESSES

There are seven life processes that most living organisms will show at some time during their life.

- Nutrition: This is the absorption of nutrients, such as organic substances and mineral ions, into the body and assimilating them (using them to make new substances). The

△ Fig 1.2 Sunflowers follow the sun as it moves across the sky through the day.

nutrients are the raw materials that cells need to release energy and to make more cells.
- **Excretion:** Living cells produce many products from the reactions that take place inside them. Some of these are waste products – materials that the body does not use. For example, animals cannot use the carbon dioxide produced during respiration. These waste products may also be toxic, so they must be removed from the body, by excretion. Organisms also excrete substances that are in excess, where there is more in the body than is needed.
- **Respiration:** This is a series of reactions that take place in living cells to release energy from nutrient molecules. This energy can then be used for all the chemical reactions that help to keep the body alive.
- **Sensitivity:** Living organisms are able to detect (or sense) changes in their external and internal conditions (stimuli) and respond to them.
- **Reproduction:** This includes the processes that result in making more individuals of that kind of organism (offspring).
- **Growth:** This is the permanent increase in the size and/or dry mass (mass without water content) of cells or the whole body of an organism. Your mass changes throughout the day, depending on how much you eat and drink, but your growth is the amount by which your body increases in size when you take nutrients into the cells to increase their number and size.
- **Movement:** This is an action by an organism that causes part, or all, of the organism to change position or place. For example, in all living cells, structures in the cytoplasm move; in more complex organisms the whole structure may move. Animals may move their entire bodies; plants may move parts of their body in response to external stimuli, such as light.

An easy way to remember all seven processes is to take the first letter from each process. This spells 'Mrs Gren'. Alternatively, you may wish to make up a sentence in which each word begins with the same letter as one of the processes: for example, My Revision System Gets Really Entertaining Now.

QUESTIONS

1. For each of the seven characteristics of life, give one example for:

   a) a human

   b) an animal of your choice

   c) a plant.

2. For each of the seven characteristics, explain why they are essential to a living organism.

EXTENDED

Viruses are very simple structures consisting of an outer protein coat that protects the genetic material inside. They have no cell structures or cytoplasm, so they do not respire or sense their surroundings. They also do not take in substances to build more cells or excrete anything. In many ways they behave like simple crystalline chemicals. However, when they infect a cell – whether a bacterial, plant or animal cell – they cause that cell to produce many copies of the virus. So they do reproduce.

1. Which characteristic of living organisms do viruses have?
2. List the other characteristics of living organisms. For each one, describe what viruses can and cannot do.
3. Using what you know about viruses, prepare an argument for classifying them as living organisms.
4. Using what you know about viruses, prepare an argument for *not* classifying them as living organisms.

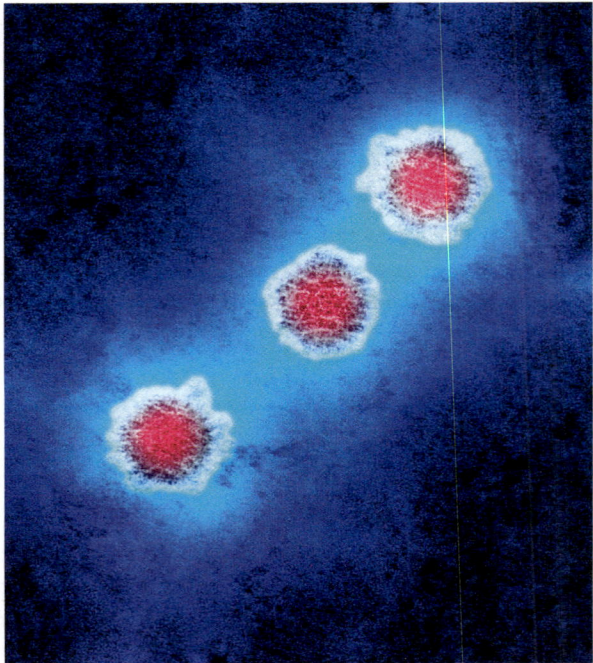

△ Fig 1.3 Not everyone agrees on whether viruses can be called *living* organisms.

REMEMBER

Be prepared to make a decision and use your knowledge to argue your point of view for difficult examples such as viruses.

END OF EXTENDED

End of topic checklist

Key terms

excretion, growth, respiration, sensitivity

During your study of this topic you should have learned:

○ How to list and describe the characteristics of living organisms.

○ Definitions of the terms:

- *nutrition* as taking in of nutrients, which are organic substances and mineral ions, containing raw materials or energy for growth and tissue repair, absorbing and assimilating them
- *excretion* as removal from organisms of toxic materials, the waste products of metabolism (chemical reactions in cells including respiration) and substances in excess of requirements
- *respiration* as the chemical reactions that break down nutrient molecules in living cells to release energy
- *sensitivity* as the ability to detect or sense changes in the environment (stimuli) and to make responses
- *reproduction* as the processes that make more of the same kind of organism
- *growth* as a permanent increase in size and dry mass by an increase in cell number or cell size or both
- *movement* as an action by an organism or part of an organism causing a change of position or place.

End of topic questions

Note: The marks awarded for these questions indicate the level of detail required in the answers. In the examination, the number of marks awarded to questions like these may be different.

1. Name the seven processes of life. Try making up your own sentence to help you remember them all. **(9 marks)**

2. Name two life processes necessary for an organism to release energy. **(2 marks)**

3. Explain why dry mass is used to measure growth. **(2 marks)**

4. When you place a crystal of copper(II) sulfate in a saturated solution of the same compound, the crystal will increase in size. Does this mean that the crystal is alive? Explain your answer. **(2 marks)**

5. Generally, plants cannot move about as animals can. Does that mean animals are more alive than plants? Explain your answer. **(2 marks)**

6. During winter, an oak tree in the UK will lose its leaves and not grow. Is the tree still living during this time? Explain your answer using all the characteristics of life. **(4 marks)**

Classification and diversity of living organisms

INTRODUCTION

The many different kinds of living organisms come in a confusing variety of forms. Classifying (grouping) organisms using their similar characteristics helps us to make sense of all the variation. This information can help us understand which organisms are most closely related to each other, which groups have evolved from other groups, and which groups play the most important roles in an ecosystem.

△ Fig. 1.4 The largest plants are giant redwood trees, capable of growing to over 90 m high.

KNOWLEDGE CHECK
✓ Know that living organisms show great variety.
✓ Know that organisms can be classified according to their characteristics.

LEARNING OBJECTIVES
✓ Be able to define and describe the *binomial* system used to name and classify organisms.
✓ Be able to list the main features of the following vertebrates: bony fish, amphibians, reptiles, birds and mammals.
✓ Be able to list the main features used to classify flowering plants and some invertebrates.
✓ **EXTENDED** Be able to describe other ways of classifying organisms, such as cladistics.
✓ **EXTENDED** Be able to list the main features and adaptations used in the classification of viruses, fungi and bacteria and their adaptation to the environment.

CONCEPT AND USE OF A CLASSIFICATORY SYSTEM
Classifying organisms

The process of classifying organisms is called taxonomy. Classifying plants and animals makes it easier to see patterns and relationships when studying organisms and the environment.

The first attempts to classify animals and plants grouped them according to a single characteristic, such as the ability to fly. So, for example, bats, wasps and owls could all be placed in the same group. This method of classification is called an artificial system. An artificial system makes it difficult to make predictions about the grouped organisms, as their other characteristics could be very different.

Modern scientists classify organisms using a natural system. This system groups organisms according to their most common characteristics, such as structure of flower, number of legs, body covering or skeleton. It may also include evolutionary relationships.

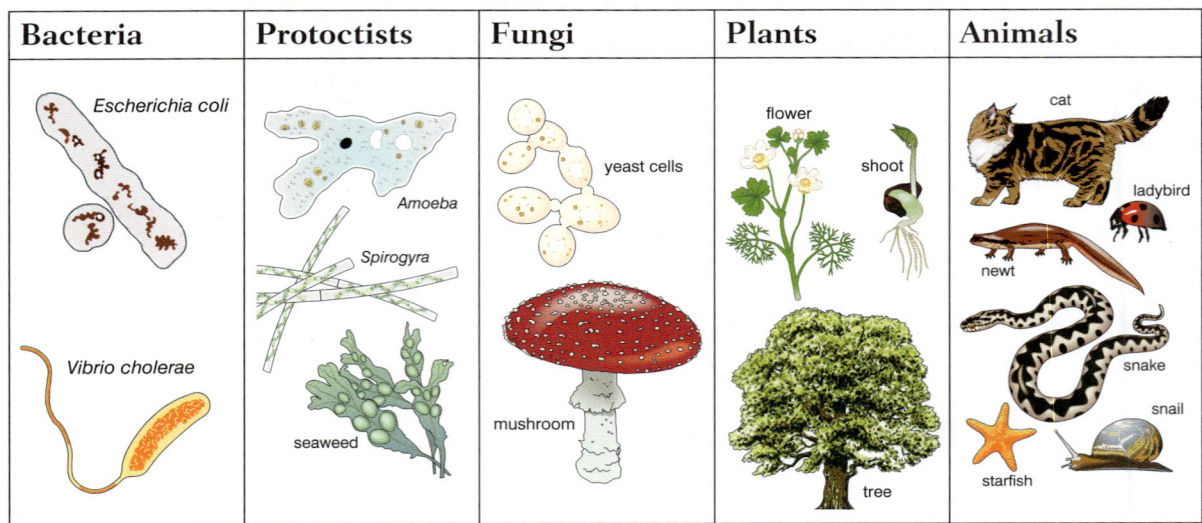

△ Fig. 1.5 Most scientists classify living organisms into one of five large groups, called kingdoms: bacteria, protoctista, fungi, plants and animals.

The kingdoms are subdivided into smaller and smaller groups, ending with **genus** and **species**.

The smallest classification group contains only one type of organism, called a **species**. A species is defined as a group of similar organisms that can breed with each other to produce **fertile offspring**.

The binomial system

Plants and animals are often called by different names in different parts of each country as well as across the world. The flower in the photo can be called a yellow wood violet, two-flower violet or arctic yellow violet. To avoid confusion, biologists use a standard international system for naming species. This is the **binomial system**.

This system of naming gives every species a name made up of two parts. Each name is in Latin and it is unique. The first part is the genus and the second part is the species. The correct biological name for a human is *Homo sapiens*. The full classification is as follows:

Kingdom	Animals	Cells with nucleus but no cell wall, organism able to move around
Phylum	Chordata	Backbone
Class	Mammalia	Hair, live young, mammary glands make milk for feeding young
Order	Primates	Five digits (fingers, toes) with opposable 'thumb', enlarged cerebrum in brain
Family	Hominidae	Great apes, no tail
Genus	*Homo*	Human-like apes, walk upright, use tools
Species	*Sapiens*	Flat-faced skull, large brain with areas for speech and reasoning, culture

Note that the genus and species names may be written in italic in a book, or underlined when written by hand.

SCIENCE IN CONTEXT: THE TROUBLE WITH GROUPING BY CHARACTERISTICS

Grouping organisms by characteristics can cause problems, because you have to use the right characteristics. For example, a duck-billed platypus has a beak and webbed feet, and it lays eggs; these are all characteristics more commonly associated with birds. However, the platypus is not a bird but a mammal, because it has fur and the mother produces milk from mammary glands to feed her young – two characteristics that are unique to mammals. The beak and feet are adaptations to its environment, and it lays eggs because it belongs to the oldest group of mammals.

△ Fig. 1.6 A duck-billed platypus has three characteristics commonly found in birds, but it is actually a mammal.

QUESTIONS

1. The binomial name of the lion is *Panthera leo*.

   a) Which part of the name is unique to the lion?

   b) What does the other part of the name indicate? Explain your answer.

2. Explain why the *binomial system* for naming organisms is useful.

EXTENDED

Other classificatory systems

In the past, scientists have only been able to identify similarities between organisms using their physical characteristics. Comparing these characteristics produced the Kingdom/…/Species structure described above, which treats different groups (kingdoms, phyla, etc. on down to species) as completely separate. However, many scientists realised that classification is strongly related to how all organisms evolved from other organisms in the past. They see the organisms that are alive now as the living part of an evolutionary tree that stretches back to the beginning of life on Earth.

To help explore the evolutionary process, these scientists used the fullest possible range of physical characteristics to calculate a value for the similarity or difference between organisms. They used this information to construct diagrams, called **cladograms**, to show the results, in a process called **cladistics**.

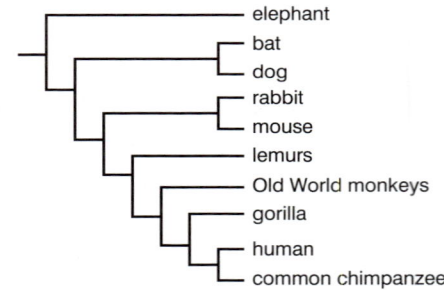

△ Fig 1.7 A cladogram to show the relatedness among several groups of mammals.

A cladogram shows the closeness in similarity of a group of organisms. The vertical scale has no meaning – it just spreads out the species so they are easy to see. The horizontal scale shows how similar the organisms are. In this cladogram, for example, humans are most similar to chimpanzees and most different from elephants. There is no attempt to show where one classification group of organisms (e.g. family) ends and another begins. All you see is the separate species and how similar or different they are.

The shape of the tree in a cladogram matches the shape of the evolutionary tree that produced these species. It is important to use characteristics that are only inherited and that are not strongly affected by the environment. For example, bats have wings and can fly, but in this cladogram their nearest relative walks on four legs. These two species were not classified by their type of movement.

Advances in theory and technology, such as working out the code in genetic material (DNA and RNA), have given scientists another way to compare how similar or different organisms are. These results can also be used to produce cladograms. Sometimes the cladograms from the genetic code are nearly identical to those produced from physical characteristics, which confirms the idea of evolutionary relationships between species. Occasionally the cladograms produce very different results. For example, there are several species of mole that live on different continents, and they all look very alike. However, analysis of the genetic code shows that on each continent the moles evolved from other larger species and so are not closely related to moles on other continents.

END OF EXTENDED

QUESTIONS

1. **EXTENDED** Describe how a cladogram is constructed, and explain what it shows.
2. **EXTENDED** Describe how cladistics differs from the 'five kingdom' method of classification.

FEATURES AND ADAPTATIONS OF VERTEBRATES

The animal kingdom can be divided into two large groups:
- **invertebrates**, which have no backbones
- **vertebrates**, which have backbones.

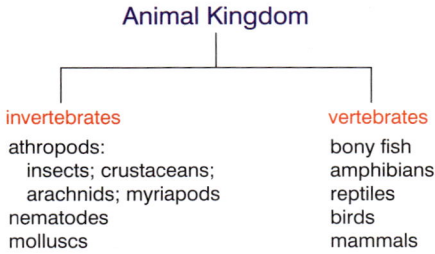

△ Fig 1.8 Some groups within the animal kingdom.

Some invertebrates are described in Features and adaptations of invertebrates. Here we will look more closely at vertebrates, which form the phylum Chordata. This phylum includes five orders:
- fish
- amphibians
- reptiles
- birds
- mammals.

Bony fish

Bony fish have:

- gills that exchange gases (oxygen and carbon dioxide) with water, covered by a flap called the operculum
- a swim bladder that controls buoyancy
- paired fins strengthened by flexible bony rays
- skin covered with overlapping scales
- lateral lines along the length of the body for detecting movement.
- external fertilisation – during reproduction, eggs and sperm are released into the water at the same time and fertilisation takes place in the water.

△ Fig. 1.9 Bony fish, such as the jackfish, have characteristics that help them swim, breathe, feed and reproduce in water.

Amphibians

Amphibians include frogs and toads, newts and salamanders. Most amphibians have two pairs of legs, but some have no legs. Amphibians lay their eggs in water and in the juvenile stage (such as the tadpole) they live in water. As adults, most amphibians live on land.

Juveniles have:

- gills that exchange gases with water
- no legs
- a tail for swimming.

As juveniles age, they change, so as adults most have:

- soft moist skin and a simple lung, both of which exchange gases with air (the gills disappear)
- legs for walking on land
- no tail
- external fertilisation.

△ Fig. 1.10 Most frogs have a short body, no tail, and webbed fingers or toes.

Reptiles

Reptiles live in a wide range of environments, from marine and fresh water through to dry deserts. They include snakes, lizards, turtles and crocodiles. Some reptiles have four legs, a few have two and snakes have none.

Reptiles have:

- lungs that exchange gases with air
- thick scaly skin that reduces loss of moisture to the air
- a body temperature that usually varies with air temperature though they may change this by their behaviour, e.g. sunbathing or sheltering in shade
- internal fertilisation where the eggs are fertilised inside the female's body during sexual reproduction
- eggs that are laid on land, and are protected by thick leathery shells.

△ Fig. 1.11 Snakes are legless reptiles that lay eggs on land.

Birds

Birds exist in almost every environment and across every continent. Most are able to fly.

Birds have:

- well-developed lungs
- feathers that cover the body and insulate the skin to reduce heat loss to air
- wings for flight
- bones that are modified to be strong but light
- an internal (core) body temperature that is kept constant, often at around 40 °C
- internal fertilisation.
- hard-shelled eggs.

△ Fig. 1.12 Most birds are well adapted to flight.

Mammals

Mammals are found in almost all environments on Earth. Most live on land, but some (e.g. whales and seals) are adapted to living completely in water. Some are very small (e.g. pygmy shrew) and some are very large (e.g. elephant). The blue whale is the largest mammal and also the largest living organism. Humans are also mammals.

Mammals have:

- well-developed lungs
- hair or fur to insulate the body, although marine mammals have little hair and are insulated by a thick layer of fat called blubber just under the skin
- a constant internal (core) body temperature at around 37 °C
- internal fertilisation.

A few mammal species (monotremes) lay eggs, though most give birth to live young. In the marsupial mammals, young are born while still very small and develop in the mother's pouch. In the placental mammals the young develop inside the mother's body, supported by the placenta. After birth all mammal young are fed on milk produced by mammary glands.

△ Fig.1.13 Even aquatic (water-living) mammals have some hair. In the walrus the hairs are adapted to sense touch.

QUESTIONS

1. Write down the key features of each of the following groups of vertebrates:

   a) bony fish

   b) amphibians

   c) reptiles

   d) birds

   e) mammals.

2. A new animal species has been discovered. It has a backbone, scaly legs, lungs and feathers. How should it be classified? Give a reason for your answer.

EXTENDED

CLASSIFYING VIRUSES, BACTERIA AND FUNGI

Viruses, bacteria and some fungi are often grouped as microorganisms, because they are too small to be seen without the use of a microscope. However, they have very different structures.

Viruses

Viruses are particles rather than cells, consisting of a protein coat surrounding genetic information in the form of either DNA or RNA. They are even smaller than bacteria: they may only be seen with an electron microscope. Viruses are parasitic, which means they can only reproduce inside the cells of another living organism that they have 'infected'. For this reason some people say that viruses are not truly 'alive'.

There are viruses that can infect every type of living thing. Many viruses are **pathogens** and cause disease in the organisms they infect. The tobacco mosaic virus is a plant virus. It prevents the formation of chloroplasts in the tobacco plant cells, which causes discolouring of the leaves.

Influenza viruses are a group of many closely related viruses that can cause 'flu' in many different animals including humans. The human immunodeficiency virus (HIV) causes the disease called AIDS (acquired immune deficiency syndrome) in humans.

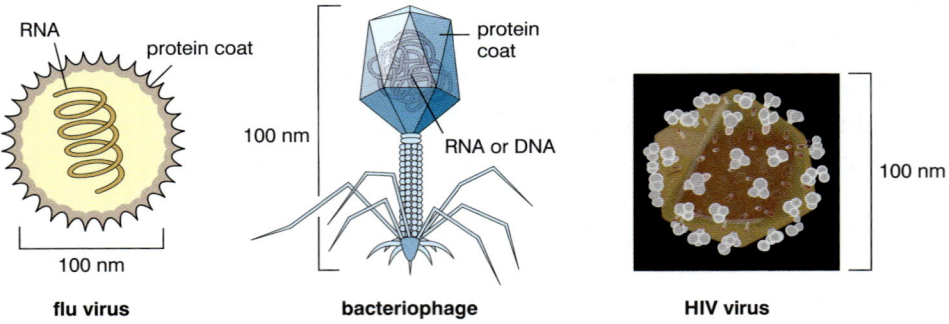

△ Fig. 1.14 1000 virus particles would fit across the width of a human hair.

1 m = 1000 mm (millimetres)

1 mm = 1000 μm (micrometres)

1 μm = 1000 nm (nanometres)

> **SCIENCE IN CONTEXT**   **HIV AND AIDS**
>
> The HIV virus is one of a group of viruses that attack and destroy cells in the body's immune system. This leaves the body open to infection by other pathogens – in the case of HIV, this causes the disease called AIDS. Many AIDS patients die not from the HIV virus, but from other infections such as tuberculosis (caused by a bacterium) because their immune systems are too weak to fight them off.
>
> The HIV virus does not survive well in the environment and is mainly transmitted from one person to another through body fluids. The most common transmission is during sexual intercourse. Transmission in blood is also possible, such as through blood transfusion, or sharing of injection needles between drug users. An infected mother can pass the HIV virus to a fetus in her uterus through the placenta, or through breast milk after birth.

Bacteria

Bacteria are single-celled, microscopic organisms that are smaller than plant and animal cells and come in many different shapes. Their cells have no nucleus, so the single circular chromosome lies free in the cytoplasm inside the cell.

Many bacteria have additional circles of genetic material, called **plasmids**. Bacterial cells are surrounded by a cell membrane and cell wall, although in different groups of bacteria the cell wall is made of different chemicals (but not cellulose, as in plants).

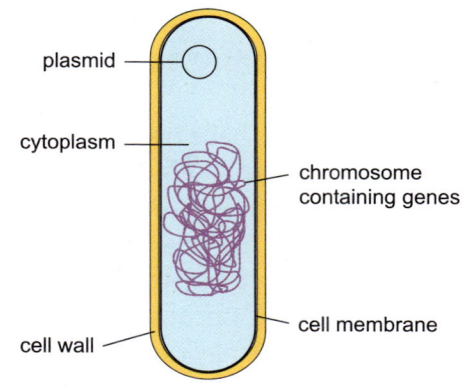

△ Fig. 1.15 Generalised structure of a bacterial cell.

Some bacteria are useful to humans: for example, *Lactobacillus bulgaricus*, a rod-shaped bacterium, is used to make yoghurt. Other bacteria are pathogens, causing diseases in plants and animals. An example of a pathogen is *Pneumococcus*, a spherical bacterium that can cause pneumonia in humans.

SCIENCE IN CONTEXT: BACTERIAL GROUPS

Bacteria are divided into two main groups according to their cell walls. Some have cell walls that absorb a stain called Gram stain. This makes the cell wall visible under the microscope, and these bacteria are called Gram-positive bacteria. Other bacteria have a different cell wall structure that doesn't take up the stain, and they are called Gram-negative bacteria. Pathogenic bacteria in the two groups are treated with different antibiotics, to help get the chemical through the different kinds of cell wall.

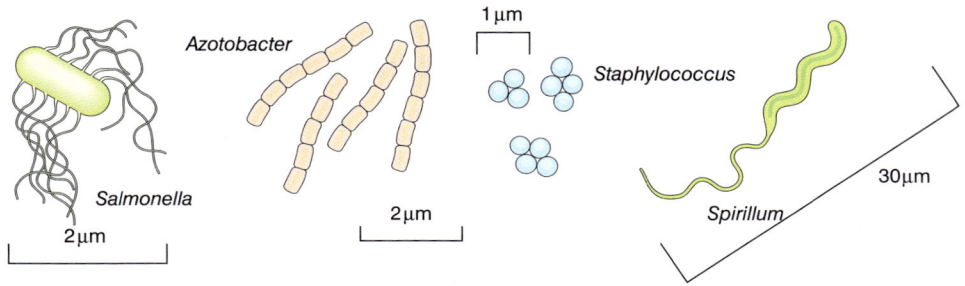

△ Fig.1.16 Different bacteria may be recognised from their shape and structure.

> **SCIENCE IN CONTEXT**
>
> ## BACTERIAL PLASMIDS
>
> Bacterial plasmids have become very useful to us in genetic engineering, where they are used as vectors (see Genetic engineering). Not all bacteria have them, but those that do transfer these small circles of genetic material to other bacteria quite easily. Plasmids may even be transferred between bacteria of different species. This is not true reproduction, because the transfer is not of the main chromosome and may not lead to production of new individuals. However, this kind of transfer may be important in the spread of antibiotic resistance between bacterial species, because some of the genes for antibiotic resistance are found in the plasmids.

Fungi

Some fungi (such as yeast) are single-celled but most have a structure consisting of fine threads known as **hyphae**. Each hypha may contain many nuclei. Several hyphae together form a **mycelium**. Many fungi may be seen without a microscope. Their cell walls are made of chitin, a fibrous carbohydrate which is different to the cellulose used in plants. Their cells do not contain chlorophyll so they cannot carry out photosynthesis. To obtain energy they secrete digestive enzymes outside the cells, onto living or dead animal or plant material, and absorb the digested nutrients. This is called **saprotrophic nutrition**. Like animals, fungi may store carbohydrate in the form of glycogen.

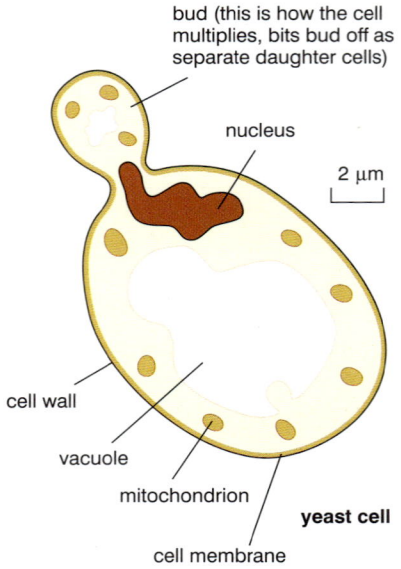

△ Fig. 1.17 The internal structure of a yeast cell that is reproducing by budding.

Examples of fungi include yeast, a single-celled fungus used by humans in baking and brewing, and *Mucor*, a fungus with the typical hyphal structure. *Mucor* is often seen as a mould growing on spoiled foods.

Some species of fungi are pathogens, which cause disease in other organisms. For example, ringworm is caused by a fungus that produces rings of itchy skin in humans. Also, many plants are damaged by rusts and moulds which are different kinds of fungi.

SCIENCE IN CONTEXT: MUSHROOMS AND TOADSTOOLS

We normally think of a mushroom and toadstool as the whole of a fungus, because this is usually all we can see. However, these are only the reproductive organs, where spores are produced. The mycelium of the fungus is usually hidden below ground or within rotting materials, where it is moist and where the hyphae can digest the surrounding tissue and absorb the nutrients that are released. The reproductive structures must be large enough so that the wind can carry the spores away to other places, and tough enough to survive the drying conditions of the air until the spores have been dispersed.

▷ Fig. 1.18 A mushroom or toadstool is only the visible part of this fungus. The rest of the structure is hidden from view.

QUESTIONS

1. Describe the basic structure of a virus.
2. Why are viruses described as particles rather than cells?
3. Compare the size of a virus with the size of a bacterium.
4. Describe the key features of a bacterial cell.
5. Which characteristics do fungi share with a) plants and b) animals?
6. Most people think of toadstools and mushrooms as the main parts of fungi. Explain why this is incorrect.

END OF EXTENDED

FEATURES AND ADAPTATIONS OF INVERTEBRATES

There are many groups of invertebrates, but some of the largest and most important groups are the insects, crustaceans, arachnids, myriapods, nematodes and molluscs. The insects, crustaceans, arachnids and myriapods belong to the phylum Arthropoda because the soft tissue of their bodies is supported by a hard external structure called an **exoskeleton**. Their limbs and bodies have joints to allow movement.

Insects

Insects are the most successful group on the planet (in terms of numbers) and are found in almost every environment.

- The insect body has three regions (head, thorax and abdomen), six jointed legs, and often two pairs of wings.
- The whole of the body is covered by a tough exoskeleton made of chitin.
- Sense organs include compound eyes, antennae that detect vibrations or, in some cases, extremely sensitive chemical detectors that can smell things over huge distances.

Some insects (such as ants and bees) live in large communities with a single individual (the queen) who produces most of the young. The individuals of these communities have complex social interactions.

∆ Fig. 1.19 A colony of bees has a complex social structure.

Crustaceans

Crustaceans are a mainly marine group including crabs, lobsters and crayfish. Woodlice are terrestrial but need to live in cool damp places to avoid drying out.

- Crustaceans have a standard body plan with head, thorax and abdomen. The head has two pairs of antennae. In some species a pair of front legs has been highly modified into pincers (chelipeds).
- Marine crustaceans grow in size by moulting their hard exoskeleton, growing rapidly and then hardening the new exoskeleton. This can occur a number of times during life.

△ Fig. 1.20 The body plan of a crab follows the standard arthropod structure.

Arachnids

The majority of arachnids are spiders, though the group also includes scorpions. Almost all arachnids live on land.

- The main body plan of spiders has two main parts, with eight legs that arise from the front part. Two structures called pedipalps at the front are used to manipulate food.
- Most spiders are carnivorous, catching flying insects in webs. Spider webs are made of a sticky protein with a strength in excess of steel wire.
- Scorpions have an elongated body. The venomous sting at the end of the tail is used for defence and to capture prey. Most scorpions are nocturnal and feed on a variety of smaller insects.

△ Fig. 1.21 A spider's body is made up of body sections called the prosoma and opithosoma.

Myriapods

Myriapod means 'many-legged ones' and this group includes centipedes and millipedes.

- Millipedes can have up to 200 pairs of legs on their bodies and range in size from microscopic to nearly 30 cm in length. They live in leaf litter and soil and generally eat plant debris.
- Centipedes have fewer legs and are predators.

△ Fig. 1.22 Myriapods can have up to 200 pairs of legs.

Annelids

Annelids are worms, such as earthworms.

- They have a body divided into many similar segments, with setae (small bristles) which allow the worm to grip on surfaces to help with movement.
- Earthworms can detect light levels and will avoid bright lights.
- The mouth can expand greatly to swallow relatively large objects such as leaves or leaf fragments.
- The waste material (worm cast) from the gut is important in maintaining soil condition.

Nematodes

Nematodes are roundworms and are tiny so we don't usually notice them. They are the most numerous multicellular organisms on earth and have adapted to nearly every ecosystem, including land, saltwater and freshwater.

- They are distinguished from annelids by the fact that they have no segments to their bodies.
- Many nematodes are parasitic, living inside other organisms and feeding off their tissues. The mouth often has adaptations to aid the nematode in its diet. Some species have a sharp tubular structure that can penetrate cells and suck out the contents.

Molluscs

Molluscs include animals that have either one shell, like snails, or two shells, such as clams, to protect the soft tissues inside and to stop them drying out. Some, such as octopuses, squid and cuttlefish, have no shells. They may be large and have soft bodies.

△ Fig. 1.23 Squid, like this Caribbean reef squid, do not have an external shell but have a hard structure inside their body.

- There is a huge range of body structure in the molluscs, but most move about using a single muscular 'foot'.
- All molluscs have a mantle, a large sac-like structure used for breathing and for excretion.
- The octopuses, squid and cuttlefish have an internal hard structure (not true bone) to support the soft tissues.

Some molluscs are filter feeders, sieving tiny pieces of food material from the water. Most are active grazers, scraping tiny algae from the surfaces of rocks. The large octopuses and squids are active hunters, with excellent eyesight, the ability to produce bursts of speed using jets of water, and tentacles with which to trap their prey.

REMEMBER

In the rest of your course, when you are describing an organism, make sure you consider how the key features of the group it belongs to help it to be well adapted to its environment.

QUESTIONS

1. Summarise the main characteristics of four groups within the phylum Arthropoda.
2. Explain why, although nematodes and annelids are both called worms because of their shape, they are not classified together.
3. Octopuses and snails are molluscs, but they look very different. Explain why they are grouped together.

FEATURES AND ADAPTATIONS OF THE FLOWERING PLANTS

The flowering plants, the angiosperms, are the most obvious group of the plant kingdom. Most of the trees, woody plants, herbaceous (soft leafy) plants and grasses that you see are flowering plants.

Flowering plants are the most successful group of plants and can survive a much wider range of environments than other groups. They produce true seeds and fruits from flowers.

They usually have a central stem bearing side branches with leaves that tend to be smaller than the large fronds of ferns and flatter and wider than the needles of pines. Roots tend to be well developed, and water-conducting tissues in angiosperms are more advanced than in any other plant group.

Reproduction in flowering plants depends on flowers, which produce two sorts of gametes: male pollen, which tends to leave the flower where it is produced; and female gametes, which remain in the ovule for the whole of their life cycle. Flowering plants also produce fruits, which aid in dispersal of the seeds (see Seed dispersal).

Cotyledons are food stores that are found in the seeds of flowering plants. In some species, after germination, they are lifted above ground and develop chloroplasts for photosynthesis until the leaves develop. **Monocotyledonous** flowering plants include grasses and have a single cotyledon. **Dicotyledonous** flowering plants (eudicotyledons) have two cotyledons and include most of the plants with visible flowers. The two groups of plants also differ in structure. Monocotyledonous plants often have long strap-like leaves with parallel veins, while dicotyledonous plants have broad leaves of many shapes with branching veins.

REMEMBER

Most dicotyledonous plants belong to the eudicotyledons. A very few species belong to ancient groups that do not share all the features of the eudicotyledons. You don't need to be able to distinguish these, but be aware that they exist.

QUESTIONS

1. Which key features distinguish the flowering plants from other plants?
2. List the features that distinguish dicotyledonous plants from monocotyledonous plants.

End of topic checklist

Key terms

cotyledon, dicotyledon, hypha, monocotyledon, mycelium, pathogen, plasmid, saprotrophic nutrition

During your study of this topic you should have learned:

○ How to define and describe the *binomial* system of naming species as a system in which the scientific name of an organism is made up of two parts showing the genus and species.

○ About the main features of the following vertebrate groups: bony fish (scales, gills for gas exchange, fins for swimming); amphibians (soft moist skin for gas exchange, gills when young and lungs as adults, lay eggs in water); reptiles (scaly skins, lungs for gas exchange, lay leathery-shelled eggs); birds (feathers, lay hard-shelled eggs); mammals (hair/fur, produce milk from mammary glands to feed their young).

○ About the main features of arthropods: insects have three main sections to their body and usually six jointed legs and four wings; crustaceans have a three-part body structure with two pairs of antennae and legs that may be modified for holding food; arachnids have segments fused into two structures and eight legs; myriapods have many segments and many pairs of legs.

○ About the main features of annelids (segmented worms, such as earthworms, that feed on fragments of leaves and other material in soil), nematodes (roundworms, without segments, that live in may different conditions) and molluscs (a muscular 'foot' for movement and mantle for breathing and excretion).

○ About the main features of flowering plants (plants that have flowers for their reproductive structures), which include dicotyledons (plants that have two food stores in their seeds) and monocotyledons (plants, such as grasses, that have only one food store in their seeds).

○ **EXTENDED** That there are other classification systems (such as cladistics, which is a way of classifying organisms by how similar or different they are, using physical characteristics or the similarity of their genetic code).

○ **EXTENDED** About the main features used in the classification of viruses, bacteria and fungi, and their adaptation to the environment.

End of topic questions

Note: The marks awarded for these questions indicate the level of detail required in the answers. In the examination, the number of marks awarded to questions like these may be different.

1. Give two reasons to explain why the classification of organisms is useful.
 (2 marks)

2. Draw up a table to compare the main features of the main groups of invertebrates.
 (12 marks)

3. Explain why a large tree and a crop such as rice or maize are both classified as plants. **(2 marks)**

4. Imagine you discovered a new animal while on an expedition to a remote island in Indonesia. Explain how you would work out how to classify it. **(2 marks)**

5. Put the following organisms into size order, starting with the smallest:
 - bacteria
 - protoctists
 - viruses
 (3 marks)

6. **EXTENDED** Explain why all viruses are parasitic (live off other living organisms).
 (2 marks)

7. **EXTENDED** Using what you know about the key features of the main vertebrate groups, draw a cladogram to show how they are related. Explain your diagram.
 (5 marks)

Simple keys

INTRODUCTION

Imagine that you have been out for a walk and seen a lovely flower that you don't recognise. Now you want to know what it is – how do you find out? You could pick up a book of wildflowers and flick through all the pictures until you find a flower that looks like the one you saw. However, plants vary, and the artist may not have drawn a picture that looks exactly like your flower, so you might have difficulty working out what it is.

What you need is an identification key. This uses clearly distinctive features to split all the flowering plants up into smaller and smaller groups until you get to the single species that you found.

△ Fig. 1.24 In order to identify this plant accurately, you would need a magnifying glass to look closely at its characteristics and an identification key.

KNOWLEDGE CHECK

✓ Know that living organisms can be grouped into species using their physical characteristics.

LEARNING OBJECTIVES

✓ Be able to use simple dichotomous keys to identify organisms.

WHAT IS A KEY?

Keys are made up of questions that allow biologists to identify unknown organisms. The questions must have simple answers (usually 'yes' or 'no') that can be answered by looking at the organism. A **dichotomous key** uses a sequence of questions each of which has two possible answers. Choosing the best answer for each question will place the unknown organism into smaller and smaller groups until the final group contains only one species. In this way an organism can be identified in the field without the need for complicated equipment.

This type of key needs careful use. As well as being easily observed, the feature chosen for each question must be constantly present. Remember that living things are highly variable – think how different one human being can be from another. It is often better to examine several examples of an organism rather than only one.

When using a key, make sure you understand exactly which feature you are meant to be observing. Always carefully consider both options given, especially where the answer is not a simple 'yes' or 'no'.

Many keys include a simple description of the organism along with its name to act as a quick check that the key has been used correctly. When using a key, if you find that the organism you are examining does not fit this description, go back over all the questions and carefully consider your answers.

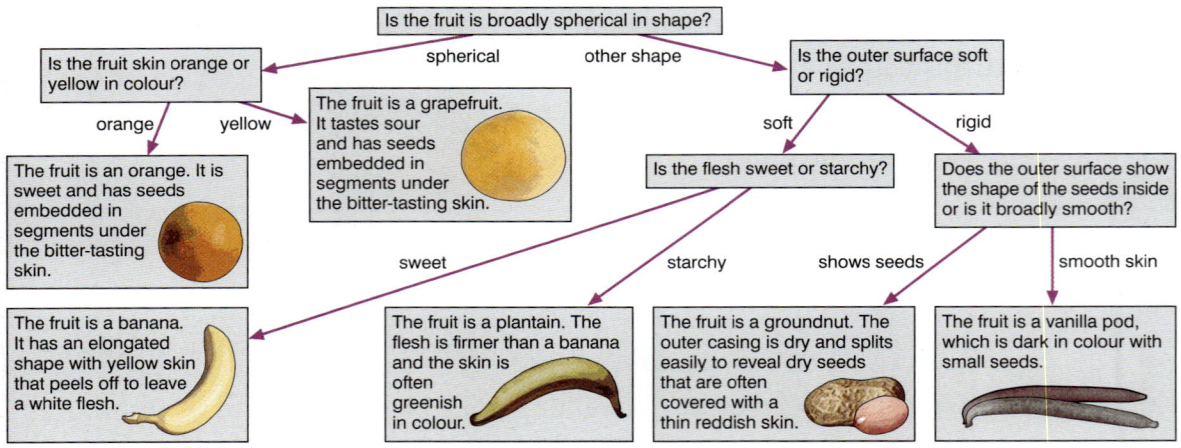

Δ Fig. 1.25. A simple dichotomous key to identify some common fruits.

QUESTIONS

1. Explain what is meant by a dichotomous key.
2. Give two reasons why a dichotomous key may not always identify an organism.

> **SCIENCE IN CONTEXT**
>
> ### IDENTIFICATION IN THE FIELD
>
> Scientists working in the field can now take photos of organisms that they cannot identify. They then email the photos to museums where they are checked against keys of organisms. If the organism is new to science, the scientist can take further pictures, or collect an individual, to take back to the lab for further study.

End of topic checklist

Key terms

dichotomous key

During your study of this topic you should have learned:

○ How to use a dichotomous key based on easily identifiable characteristics to identify organisms.

End of topic questions

Note: The marks awarded for these questions indicate the level of detail required in the answers. In the examination, the number of marks awarded to questions like these may be different.

1. Look at the key of fruits shown above. List the questions and answers that would lead to an identification of a banana. **(3 marks)**

2. a) Give two advantages of this sort of dichotomous key. **(2 marks)**

   b) Name one disadvantage of using such a key. **(1 mark)**

3. Read the following questions carefully. Choose the question that is most useful for placing an organism in the bird group. Explain your choice.

   a) Does the specimen have wings?

   b) Does the specimen have feathers?

   c) Does the specimen lay eggs? **(2 marks)**

TEACHER'S COMMENTS

a) i) It is important to know the features of different groups of organisms and be able to label these.

   A Correct – nucleus.

   B Correct – vacuole.

   C Incorrect – the outer layer of the hypha is the wall.

   D Incorrect – the membrane is the next layer within the wall, pushed up against the wall.

   E Correct – cytoplasm.

ii) Correct – the hyphae of moulds such as Mucor have a large central vacuole.

The answer is correct in that the hyphae have a wall, but this cannot be described as a 'cell wall' as the hyphae are not divided into cells. To be completely correct, the student simply had to repeat this information from part a) i).

Exam-style questions

Note: The questions, sample answers and marks in this section have been written by the authors as a guide only. The marks awarded for these questions indicate the level of detail required in the answers. In the examination, the number of marks awarded to questions like these may be different.

Sample student answer

Question 1

The diagram shows the structure of a part of an organism called *Mucor*.

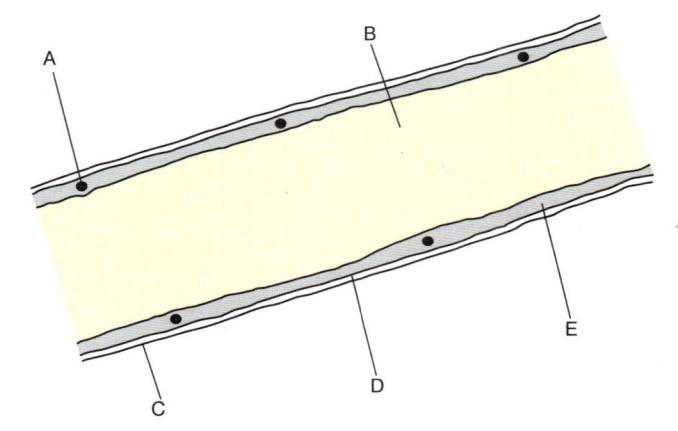

a) i) The diagram shows the structure of the organism *Mucor*. Name each of the parts A–E. Use the words in the box below. (5)

| cytoplasm | membrane | nucleus |
| starch grain | vacuole | wall |

A	*nucleus*	✓ ①
B	*vacuole*	✓ ①
C	*membrane*	✗
D	*wall*	✗
E	*cytoplasm*	✓ ①

ii) State **two** features the organism has in common with plants. (2)

Mucor has a large central vacuole. ✓ ①

Mucor has a cell wall. ✓ ①

Exam-style questions continued

   iii) Give one feature that tells you that *Mucor* is a fungus. (1)

It has many nuclei lying in the cytoplasm. ✓ ①

b) Describe how moulds such as *Mucor* feed. (4)

Mucor lives on its food, e.g. bread, and secretes enzymes into it. ✓ ①

The food is absorbed over the surface of the fungus. ✓ ①

c) Yeast is another type of fungus. State one major difference between *Mucor* and yeast. (1)

Yeast is single-celled. ✓ ①

(Total 14 marks)

9/14

> **iii)** Correct. The student could have chosen from a selection of features but was only asked to give one for the mark.
>
> **b)** The answer is correct, but the student could have added that food is digested outside the mould, and that this process is called saprotrophic nutrition.
>
> **c)** Correct – the mycelium of *Mucor* has many nuclei distributed through the cytoplasm, with no cell boundaries; yeast is made up of single cells.

Question 2

This question is about the variety of living organisms.

a) Four types of living organism are listed as A to D below:

A: bird	B: bony fish	C: mollusc	D: nematode

  i) Which **two** organisms have backbones? (2)

 ii) Which organism has a shell? (1)

iii) Which organism has jointed legs? (1)

 iv) Which organism has an exoskeleton? (1)

  v) Which organism is shaped like a tube? (1)

b) The illustration below shows an animal called the duck-billed platypus.

i) State **one** characteristic, visible in the diagram, that is typical of all mammals. (1)

   ii) State **one** characteristic of the duck-billed platypus, **not** visible in the diagram, that is typical of all mammals. (1)

   iii) State **one** characteristic of the duck-billed platypus that is not typical of other mammals. (1)

   (Total 9 marks)

Question 3

This question is about the characteristics of living things.

a) Copy and complete the sentences by writing the most appropriate word in each space.

Use only words from the box.

detect	development	energy	gravity
growth	light	location	nutrition
position	respiration	respond	sensitivity

.... is the taking in of substances needed for energy and for .... and ....

.... is a series of reactions that take place in living cells to release .... from nutrient molecules so that cells can use this to keep them alive.

.... is the ability to .... and .... to changes in external and internal conditions.

Movement causes a change in .... of the organism. In animals, this involves their entire bodies. Plants often move parts of their body in response to external stimuli, such as .... and .... (12)

b) Define the term excretion. (3)

c) Which of the characteristics shown by living things are shown by a motor car? (4)

(Total 19 marks)

Question 4

All organisms are classified.

a) Describe how scientists classify organisms into groups. (1)

b) Describe why the binomial system is used to name living organisms. (2)

Exam-style questions continued

c) The following is a list of four animal species.

| PANTHERA LEO | Homo Sapiens | *Lumbricus terrestris* | pieris brassicae |

i) Which animal species above is written correctly using the binomial system? (1)

ii) Define the term species. (2)

iii) In the correctly-named organism in the box above, which part of the name defines the species? (1)

(Total 7 marks)

EXTENDED Question 5

Not everyone agrees that viruses should be called living things. Use your knowledge of viruses and the characteristics of living things to discuss whether or not viruses should be classed as living. (4)

(Total 4 marks)

EXTENDED Question 6

When scientists discover a new species, they classify it.

a) Explain why it is important to classify all organisms. (2)

b) The diagrams show the structure of the fore limbs of some animals.

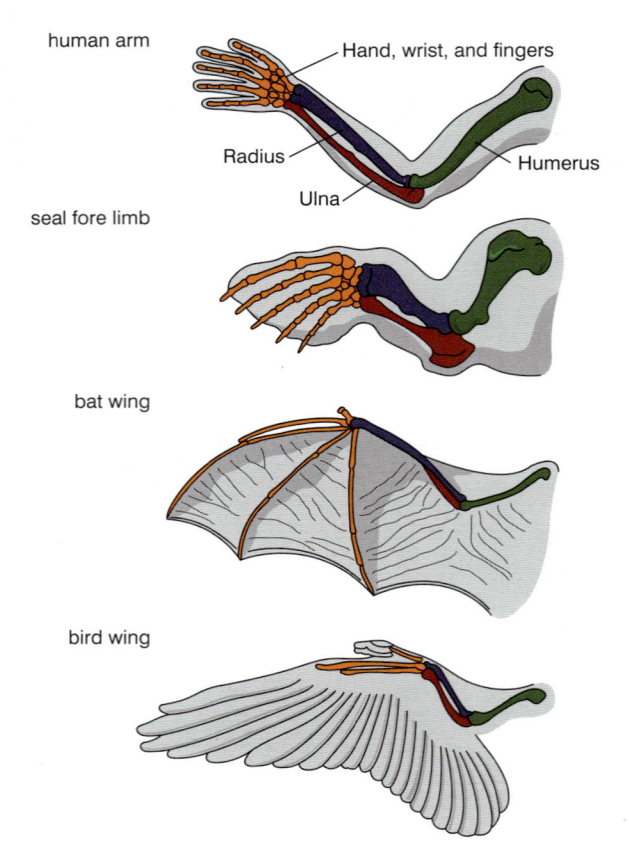

All of these organisms are vertebrates.

    i) State the major characteristic of vertebrates. (1)

    ii) Using information from the diagram, explain how scientists know that these organisms can be placed into the same group. (1)

    iii) Describe the major differences between the wing of the bat and the bird. (3)

c) What feature do scientists currently use when investigating the relationships of organisms? (1)

(Total 8 marks)

EXTENDED Question 7

The diagrams below show five different types of arthropod.

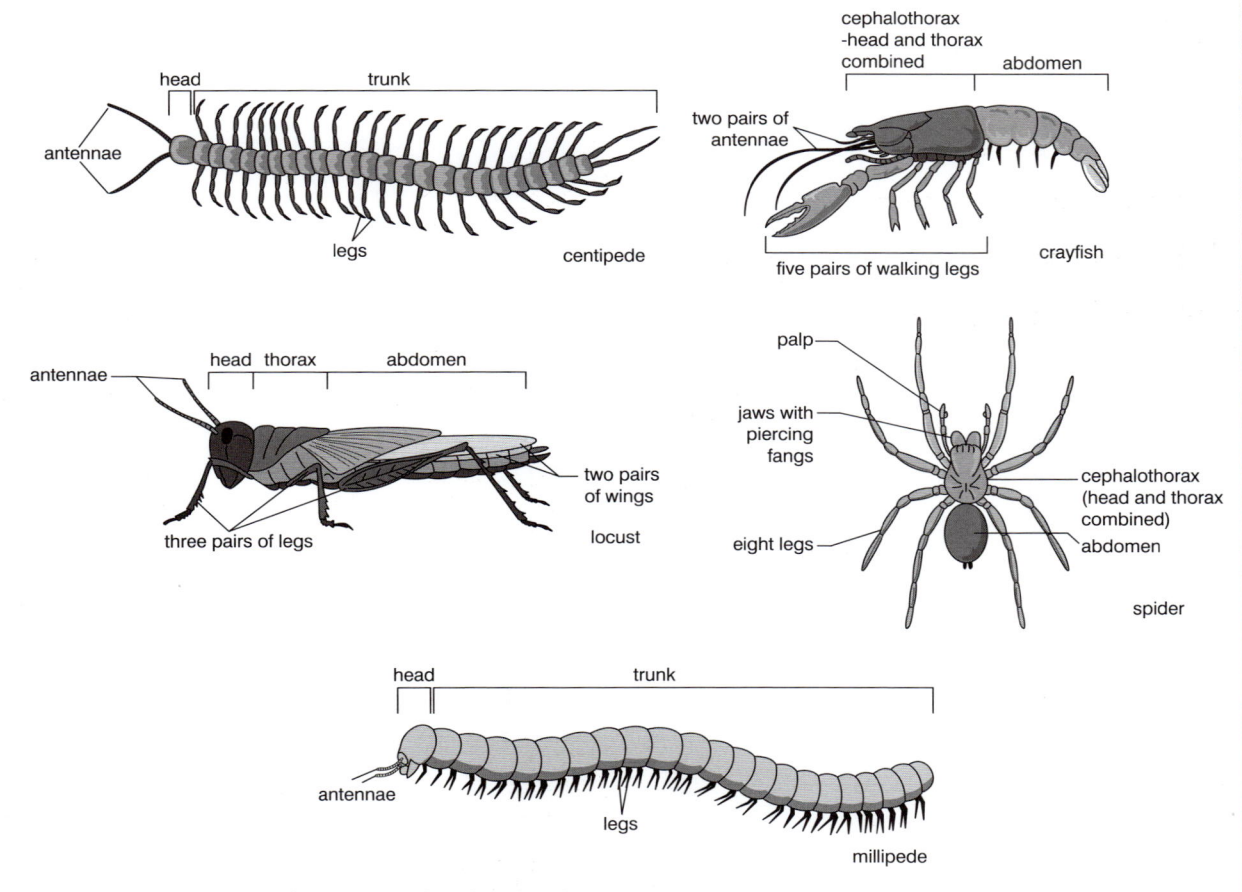

a) State **two** major features of arthropods. (2)

b) Construct a dichotomous key to identify the five organisms. (5)

(Total 7 marks)

When we talk about the 'heart' of something, in a general way, we mean the centre of something not only in position but also in the role it plays. There is good reason for this, as the human heart is not only positioned in the middle of the body, it also plays a central role in maintaining life.

The role of the heart and circulatory system is to circulate blood around the body. This delivers oxygen from the lungs, and nutrients from the digestive system, to all cells so that they can respire and carry out all the processes needed for life. The blood also removes waste products from these processes, for example delivering carbon dioxide to the lungs and urea to the kidneys for excretion and removal from the body. The heart plays a central role in staying alive and staying healthy.

STARTING POINTS

1. How is the body organised so that it can carry out the life processes effectively?
2. What are the basic molecules of life?
3. How do cell membranes control what can get into and out of the cell?
4. How do plants and humans get the food they need for growth?
5. What is cellular respiration and how do the body systems support it?
6. How are gas exchange surfaces adapted for rapid exchange of gases into and out of the body?
7. How are materials transported around the bodies of plants and humans?
8. What are the waste materials of metabolism and how are they removed from the body?
9. How do plants and humans respond to changes in the environment around them?

SECTION CONTENTS

a) Cell structure and organisation
b) Levels of organisation
c) Size of specimens
d) Movement in and out of cells
e) Enzymes
f) Nutrition
g) Transportation
h) Respiration
i) Excretion
j) Coordination and response
k) Exam-style questions

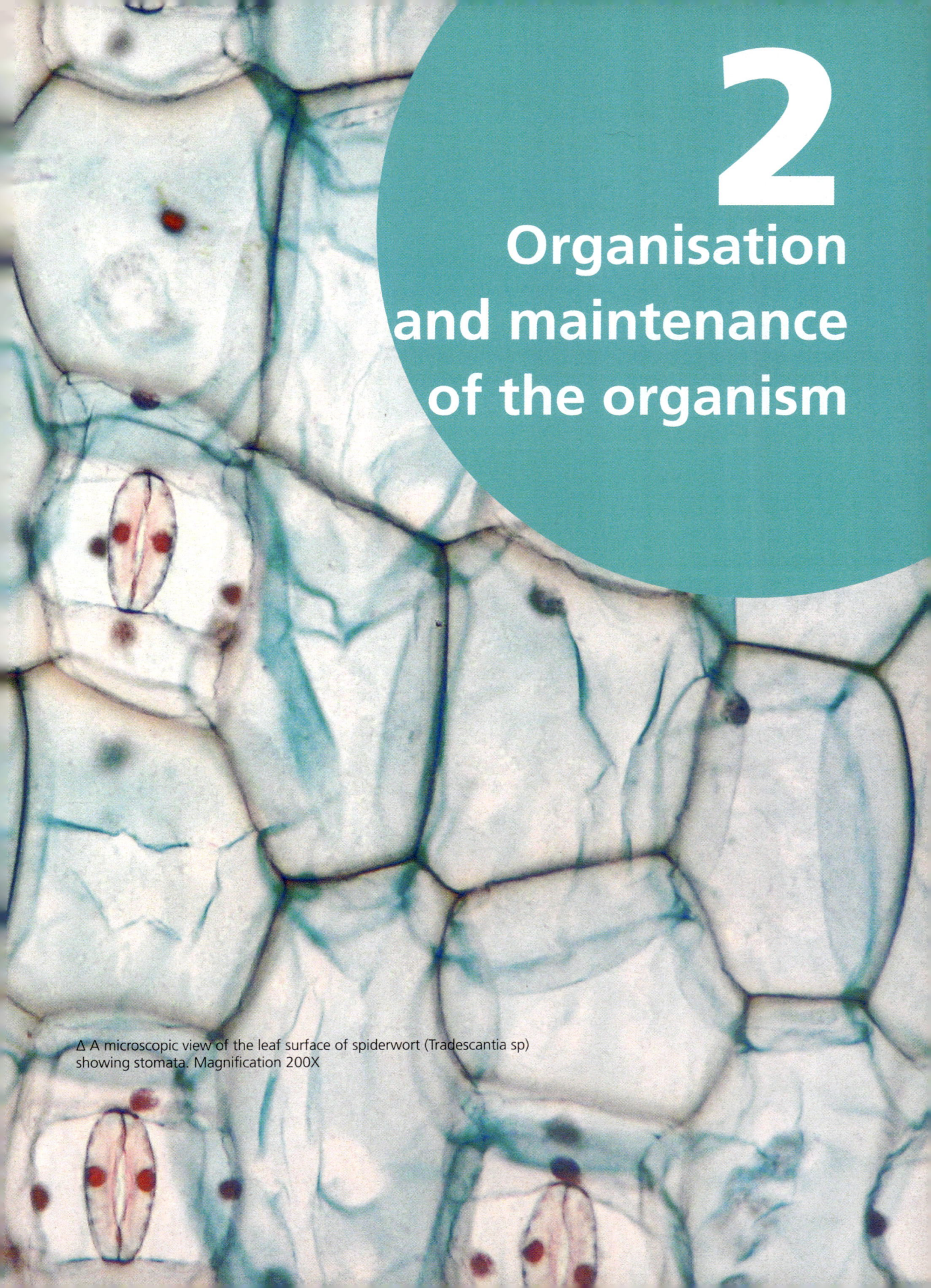

2
Organisation and maintenance of the organism

△ A microscopic view of the leaf surface of spiderwort (Tradescantia sp) showing stomata. Magnification 200X

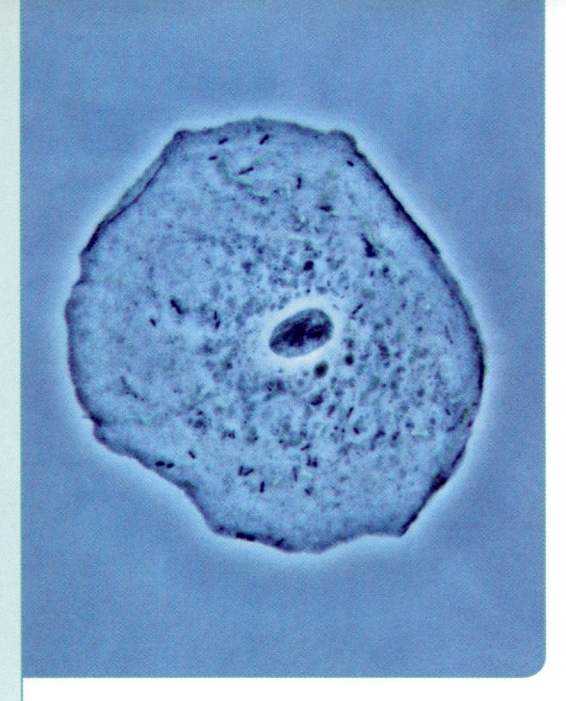

△ Fig. 2.1 All plant, protoctist and animal cells (apart from some very specialised cells), like this human cheek cell, have a cell nucleus like this.

Cell structure and organisation

INTRODUCTION

All 'complex' cells (those that contain a nucleus) in all animals, plants and protoctists on Earth have the same basic structure. Scientists say that this is because we have all evolved from a single complex cell. This cell evolved from a simple bacteria-like cell (without a nucleus) over 1600 million years ago. This is the origin of all the millions of different species of plants, animals and protoctists that live on Earth today.

KNOWLEDGE CHECK
✓ Know that most organisms are formed from many cells.
✓ Know that cells may be specialised in different ways to carry out different functions.

LEARNING OBJECTIVES
✓ Be able to state that living organisms are made of cells.
✓ Be able to identify and describe the structure of a plant cell (palisade cell) and an animal cell (liver cell), as seen under a light microscope.
✓ Be able to describe the differences in structure between typical animal and plant cells.
✓ **EXTENDED** Be able to relate the structures seen under the light microscope in the plant cell and in the animal cell to their functions.

PLANT AND ANIMAL CELLS

The diagrams below show a typical animal cell and typical plant cells. These cells all have a nucleus and cytoplasm.

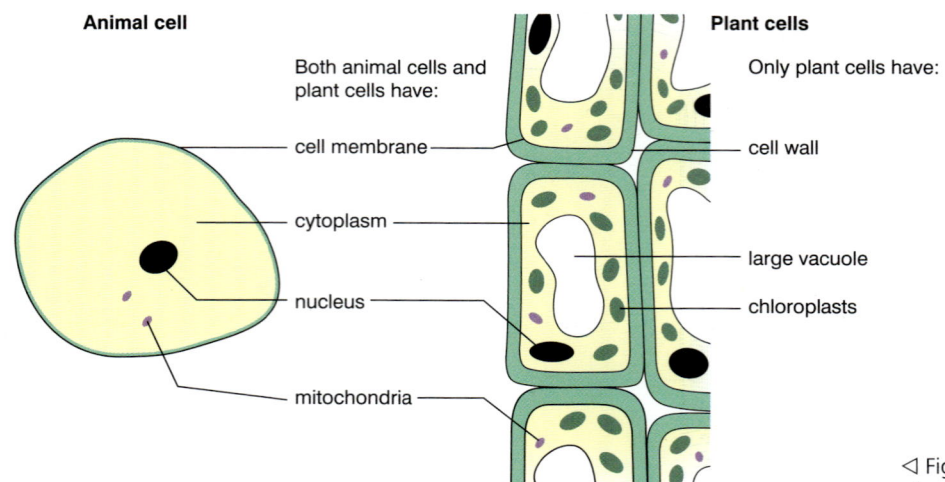

Animal cell

Both animal cells and plant cells have:
- cell membrane
- cytoplasm
- nucleus
- mitochondria

Plant cells

Only plant cells have:
- cell wall
- large vacuole
- chloroplasts

◁ Fig. 2.2 The basic structures of an animal cell and plant cells.

All living organisms are made of cells. Some, such as bacteria and some fungi, are formed from a single cell; others, such as most plants and animals, are **multicellular**, with a body made of many cells. All animal and plant cells have certain features in common:

- a **cell membrane** that surrounds the cell
- the **cytoplasm** inside the cell, in which all the other structures are found
- a large **nucleus**.

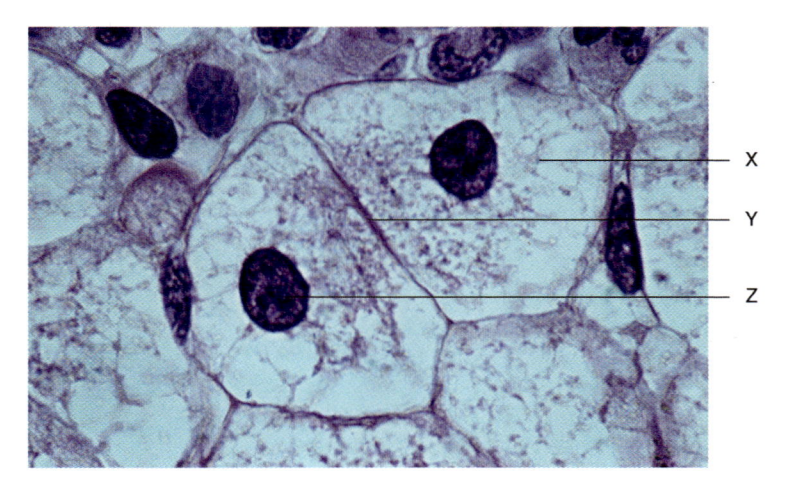

△ Fig. 2.3 A human liver cell is a typical animal cell.

Plant cells also have features that are not found in animal cells, such as:

- a **cell wall** surrounding the cell membrane
- a large central **vacuole**
- green **chloroplasts** found in some, but not all, plant cells.

Each structure in a cell has a particular role.

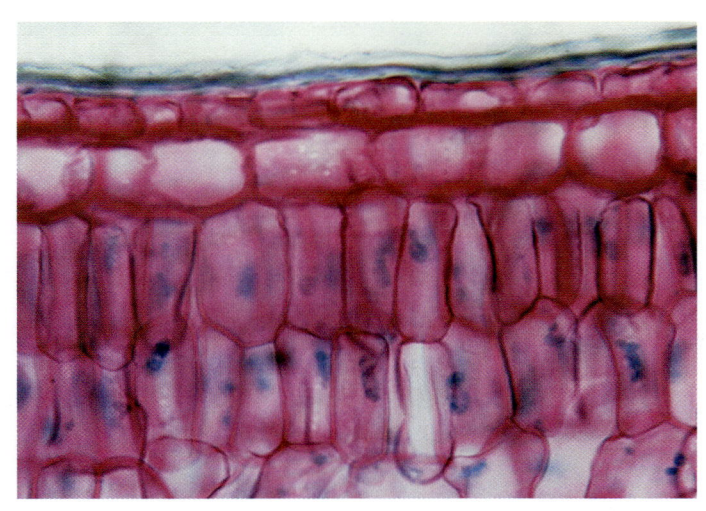

△ Fig. 2.4 A typical plant cell is a palisade cell in the upper part of a leaf. The chloroplasts are supported by the cytoplasm.

QUESTIONS

1. **a)** Using the photograph in Fig. 2.1, make a careful drawing of the cell using a sharpened pencil to make clear lines.

   b) Use Fig. 2.3 to help you label your drawing to show the three key structures of animal cells.

2. List three cell structures that are found in plant cells but not in animal cells.

EXTENDED

- The cell membrane holds the cell together and controls substances entering and leaving the cell.
- The cytoplasm supports many small organelles and is where many different chemical processes happen. It contains water, and many solutes are dissolved in it.
- The nucleus contains genetic material in the **chromosomes**. These control how a cell grows and works. The nucleus also controls cell division.
- The plant cell wall is made of cellulose, which gives the cell extra support and defines its shape.
- The plant vacuole contains cell sap. The vacuole is used for storage of some materials, and to support the shape of the cell. If there is not enough cell sap in the vacuole, the whole plant may wilt.
- Chloroplasts contain the green pigment chlorophyll, which absorbs the light energy that plants need to make food in the process known as **photosynthesis**.

QUESTIONS

1. Name the part of a plant cell that does the following:

   a) carries out photosynthesis

   b) contains cell sap

   c) stops the cell swelling if it takes in a lot of water.

END OF EXTENDED

SCIENCE IN CONTEXT — ARTIFICIAL CELLS

Scientists have discovered enough about how the structures of cells are formed and work together to let them start creating artificial cells. This has great potential for medicine, because these cells could be used, for example, to deliver drugs inside the body directly to the cells that need them. They could also be used in biotechnology, such as to make fuels that could replace fossil fuels.

Developing investigative skills

The photograph shows the view of some cells seen through a light microscope.

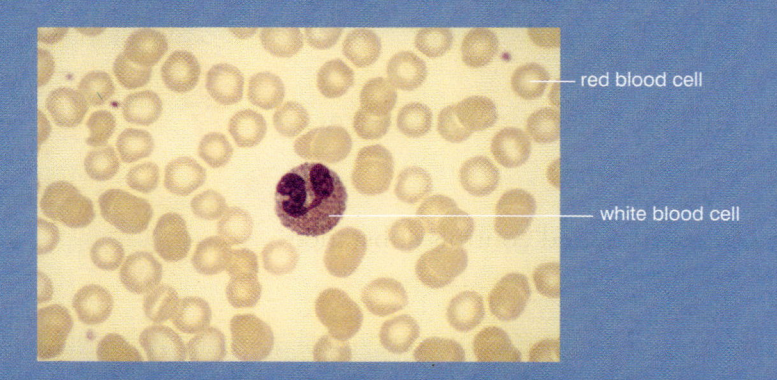

△ Fig. 2.5 Real light micrograph of red and white blood cells.

Using and organising techniques, apparatus and materials
❶ a) Describe how to set up a slide on a microscope so that the image is clearly focused.

b) Describe and explain what precautions should be taken when viewing a slide at high magnification.

c) Describe what precaution should be taken if using natural light to illuminate the slide, and explain why this is important.

Observing, measuring and recording
❷ a) Draw and label a diagram of the white blood cell shown in the light micrograph above.

b) If the ×4 eyepiece and the ×20 objective were used, calculate the magnification of the image compared with the specimen on the slide.

Handling experimental observations and data
❸ Explain how the structure of the cells shown indicates that they cannot be plant cells.

End of topic checklist

Key terms

cell membrane, cell wall, chloroplast, cytoplasm, nucleus, photosynthesis, vacuole

During your study of this topic you should have learned:

○ That living organisms are made of cells.

○ How to identify and describe the structure of a plant cell (palisade cell) and an animal cell (liver cell), as seen under a light microsope.

○ How to describe the differences in structure between typical animal and plant cells.

○ **EXTENDED** How to relate the structures seen under the light microscope in the plant cell and in the animal cell to their functions.

End of topic questions

Note: The marks awarded for these questions indicate the level of detail required in the answers. In the examination, the number of marks awarded to questions like these may be different.

1. Describe the role of the following cell structures:

    a) nucleus **(1 mark)**

    b) cell membrane **(1 mark)**

    c) cytoplasm. **(1 mark)**

2. Draw up a table to compare the structures found in plant and animal cells. **(12 marks)**

3. Here are some examples of statements written by students. Each statement contains an error. Identify the error and rewrite the statement so that it is correct.

    a) Animal cells are surrounded by a cell wall that controls what enters and leaves the cell. **(1 mark)**

    b) All plant cells contain chloroplasts. **(1 mark)**

    c) Both animal cells and plant cells contain a large central vacuole in the middle of the cell. **(1 mark)**

4. **EXTENDED** Red blood cells are unusual because they contain no nucleus. When they are damaged, they have to be replaced with new cells from the bone marrow. Explain how this is different from other cells. **(2 marks)**

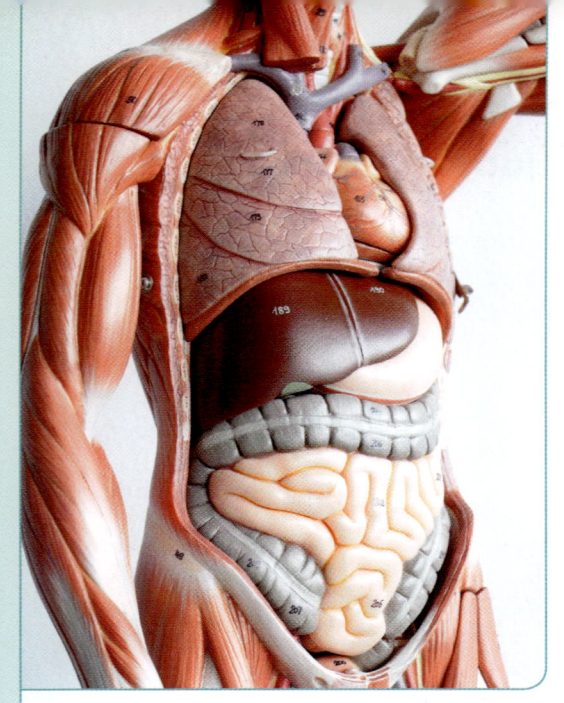

△ Fig. 2.6 The human body is made up of several systems of grouped organs, including the digestive system, the nervous system, the muscle/skeletal system and the respiratory system.

Levels of organisation

INTRODUCTION

Bringing together similar activities with the same purpose can make things much more efficient. For example, bringing teachers and students together in a school helps more students to learn more quickly than if each teacher travelled to each student's home for lessons. The same is true in the body. Having groups of similar cells in the same place as a tissue, and grouping tissues into organs, helps the body carry out all the life processes much more efficiently and so stay alive.

KNOWLEDGE CHECK

✓ Know how to define the terms tissue, organ and organ system.
✓ Know how to describe how the organisation of tissues, organs and organ systems contribute to the seven life processes.

LEARNING OBJECTIVES

✓ Be able to relate the structure of ciliated cells, root hair cells, xylem vessels, muscle cells and red blood cells to their functions.
✓ Be able to define *tissue*, *organ* and *organ system*.

ORGANISATION

Cells in multicellular organisms, such as most plants and animals, rarely work independently of all other cells.

- Similar cells are grouped in **tissues**, to perform a shared function, such as muscle cells in a human, or palisade cells in a plant leaf.
- Different tissues are grouped together to form **organs**, which carry out specific functions, for example the heart in a human, or the leaf in a plant.
- Organs are arranged in **organ systems** which carry out major body functions. Examples are the circulatory system, reproductive system, nervous system and respiratory system in humans.

For example, the heart is an organ in the human circulatory system that pumps blood around the body. The heart can contract to pump because it is formed from muscle tissue, which contains muscle cells that are specially adapted to contract. The heart also contains other tissues, such as fibrous tissue that forms the valves.

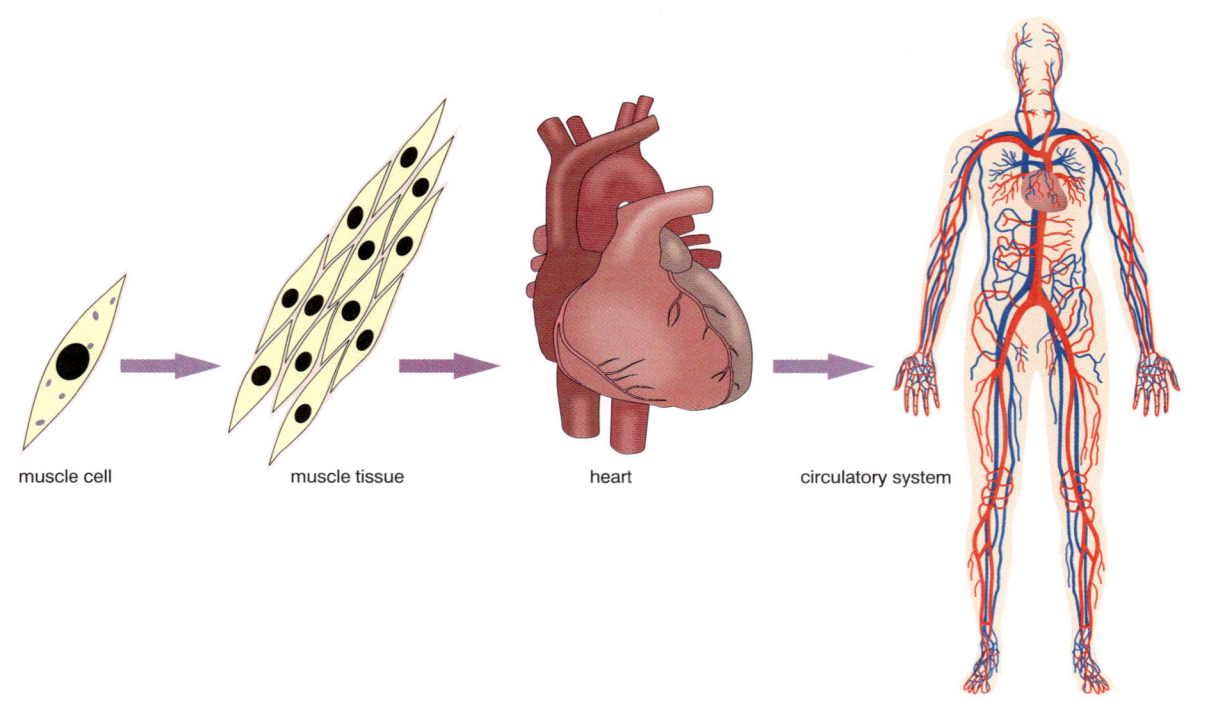

△ Fig. 2.7 The human body is organised at cell level. Muscle cells are found in muscle tissue, which may be found in the heart, which is part of the circulatory system.

REMEMBER

As you study organ systems in more detail through your course, remember to identify the organs, tissues and cell types involved in each system, so that you build a range of examples that you can use to answer questions with.

QUESTIONS

1. Give one example of each of the following in the human body:

    a) tissue   b) organ   c) system.

2. Give one example of each of the following in a plant:

    a) tissue   b) organ.

CELL SPECIALISATION FOR FUNCTION

Different types of cells carry out different jobs. Cells have special features that allow them to carry out their job. This is called **specialisation**. Good examples of specialised cells are

- ciliated cells, muscle cells and red blood cells in humans
- root hair cells and xylem cells in plants.

> **SCIENCE IN CONTEXT**
>
> ## STEM CELLS
>
> Every tissue in the human body contains a small number of unspecialised cells. These are called stem cells. The role of stem cells is to divide and produce new specialised cells within the tissue, for growth and repair. Scientists are investigating how stem cells could be given to people to mend tissue that the body cannot mend, such as the spinal cord after an accident in which it is cut. This would make it possible for a person paralysed by spinal cord damage to move their whole body again.

Ciliated cells

Cilia are tiny hair-like projections that cover the surfaces of certain types of cells. Cilia can wave from side to side. The cell coordinates this movement to produce waves across the cell surface. These waves of moving cilia can move liquid on the cell surface.

Ciliated epithelial cells that line the bronchi and bronchioles of the respiratory system move a liquid called mucus. Tiny particles of dust or bacteria that are trapped in the mucus are carried along by the cilia and pass up the tubes. They are then emptied, along with the mucus, into the oesophagus where they are swallowed and pass into the stomach. In this way the ciliated epithelium keeps the lungs clean. Smoking affects the cilia so they cannot clear away the dirty mucus in lungs as effectively as usual. This explains why many smokers cough, to get the sticky dirty mucus out of their lungs.

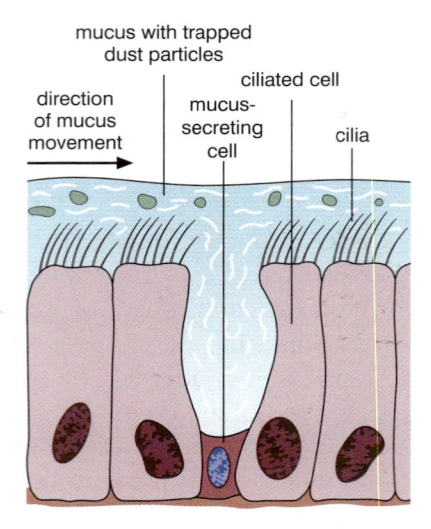

△ Fig. 2.8 Goblet cells produce mucus, and healthy cilia sweep the mucus out of the lungs.

Muscle cells

Muscle cells contain bands of two proteins called actin and myosin. These proteins react with each other and pull themselves towards each other. This produces a contraction, which is how muscles move. The muscle cells are arranged in fibres that run along the length of the muscle. The protein fibres in the cells contract when they are stimulated by a nerve impulse, which passes over the surface of the cell.

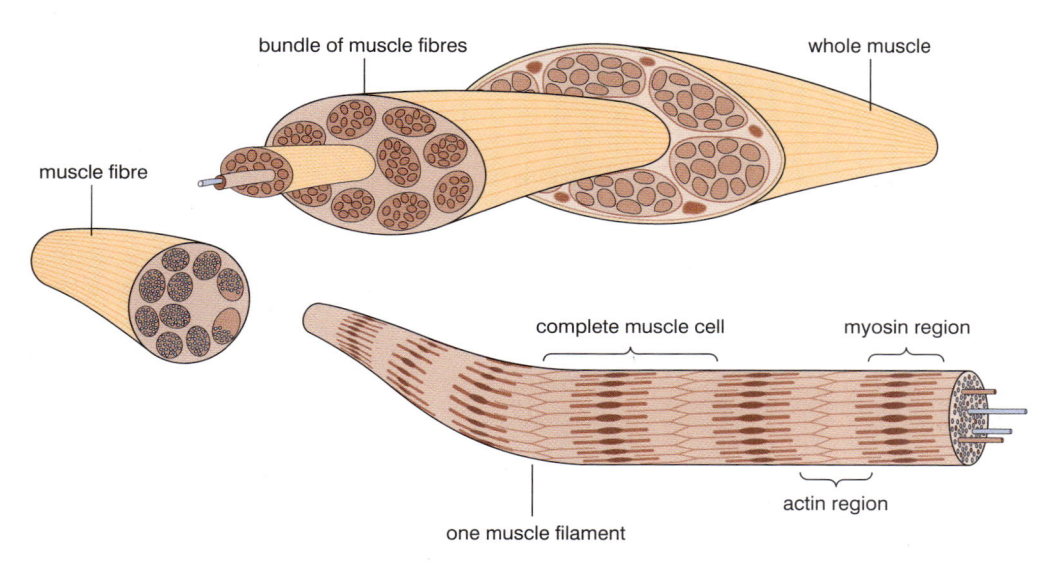

△ Fig. 2.9 A muscle filament is made of protein fibres. The cells in the filament are able to contract and shorten in length.

Red blood cells

Red blood cells in mammals are unusual because they do not have a nucleus. They are filled with a chemical called haemoglobin, which can pick up oxygen in the lungs and release it near the cells that need it deep inside the body. The biconcave (disc) shape of a red blood cell means that its innermost part is never far away from the outside. So diffusion of oxygen in and out of the cell happens very rapidly. Red blood cells are made in the red bone marrow and last only 120 days before they are destroyed in the spleen and liver. Since they have no nucleus, they cannot divide.

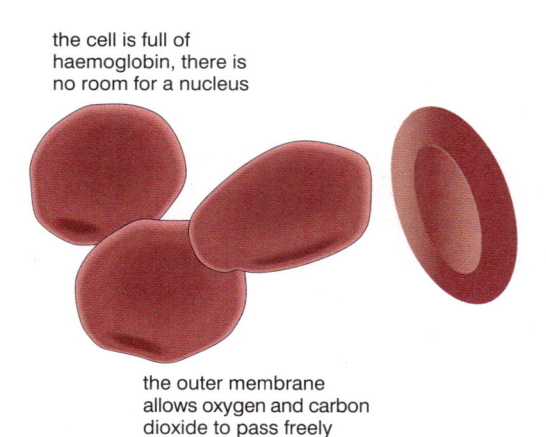

△ Fig. 2.10 Red blood cells.

Root hair cells

In many plants, water and minerals are absorbed from the soil by root hairs, which penetrate the spaces between soil particles. These hairs are very fine extensions of the surface cells of the root, just behind the growing tip. Their elongated shape increases the surface area available for absorption of water and dissolved mineral ions. As they age, root hairs develop a waterproof layer and become non-functional. New root hairs are constantly growing as the root pushes through the soil.

△ Fig. 2.11 Root hair cells greatly increase the surface area for absorption near the tips of roots.

Xylem vessels

Xylem vessels are made when xylem cells die. Before they die, the cells produce a thick cell wall that contains chemicals which make the cell wall strong and waterproof. The cells are arranged in columns in the **vascular bundles** (veins) in the plant root, stem and leaf. When the xylem cells die, the cell walls form long thin tubes through the plant, which carry water and dissolved minerals from the roots to all the other parts of the plant. The strong vessel walls also help to support the plant.

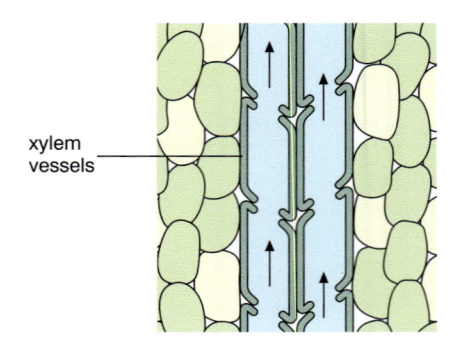

△ Fig. 2.12 Xylem vessels are long tubes of dead cells connecting the roots with all the other parts of the plant.

QUESTIONS

1. Where would you find the following cells and what do they do?
    a) ciliated cells
    b) muscle cells
    c) red blood cells
    d) root hair cells
    e) xylem vessels.

End of topic checklist

Key terms

cilia, ciliated cell, organ, organ system, red blood cell, root hair cell, tissue, xylem vessel

During your study of this topic you should have learned:

○ How to relate the structure of the following to their functions:
- ciliated cells – in the respiratory tract
- root hair cells – absorption
- xylem vessels – conduction and support
- muscle cells – contraction
- red blood cells – transport.

○ Definitions of:
- *tissue*, as a group of cells with similar structures, working together to perform specific functions
- *organ*, as a structure made up of a group of tissues, working together to perform specific functions
- *organ systems*, as a group of organs with related functions, working together to perform body functions.

End of topic questions

Note: The marks awarded for these questions indicate the level of detail required in the answers. In the examination, the number of marks awarded to questions like these may be different.

1. Put the following in order of size, starting with the largest:

    cell   organ system   organ   tissue   (3 marks)

2. Write definitions for each of these words:

    a) tissue   (1 mark)

    b) organ   (1 mark)

    c) organ system   (1 mark)

3. Draw up a table with the following headings.

System	Function	Organs in this system	Tissues in these organs	Cells in these tissues

    Complete your table as fully as you can, using up to three examples of systems in the human body.   (max 15 marks)

4. Fig. 2.13 shows a part of a leaf, which is a plant organ. The diagram is labelled to show some of the tissues. Describe the functions of the palisade and xylem tissues in this organ.   (2 marks)

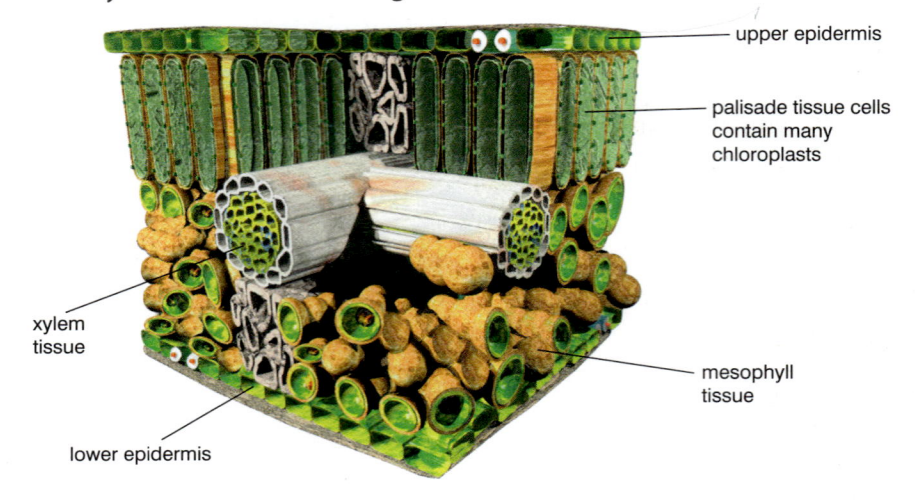

△ Fig. 2.13 Section through part of a leaf.

5. Explain as fully as you can the advantage for multicellular organisms of cell specialisation and organisation.   (3 marks)

Size of specimens

INTRODUCTION

One of the largest cells in the human body is the egg cell. This is about 0.1 mm in diameter, about the size of a full stop on a printed page and just visible to the naked eye. Other very large cells include the nerve cells in a giraffe's neck that may be several metres long, and cells of the alga *Spirogyra* which are also visible to the naked eye. One of the smallest human cells is the sperm cell, which is about 20 times smaller than an egg cell.

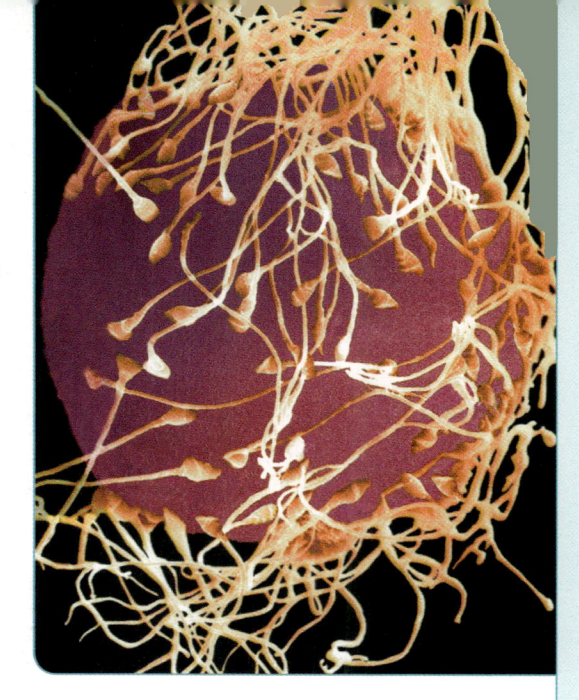

△ Fig. 2.14 Human egg cell surrounded by sperm cells during fertilisation.

KNOWLEDGE CHECK
✓ Know that a microscope can be used to magnify specimens so we can see more detail.

LEARNING OBJECTIVES
✓ Be able to calculate the magification and size of biological specimens using millimetres as units.

MAGNIFICATION

Many of the structures that you study in biology are too small to be seen by just using your eyes. You can use magnifying glasses and microscopes to examine details of plant and animal cells, and to take pictures and draw the diagrams you see in this book. But often you want to know the actual size of the specimen you are looking at. If you know the **magnification** you are using to look at a specimen, then you can work out the size of a structure.

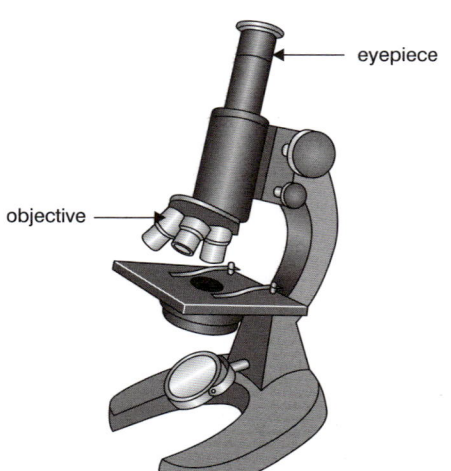

◁ Fig. 2.15 When using a microscope, the magnification of a specimen is calculated from the eyepiece and the objective used to view it.

- The magnification of an eyepiece for a light microscope may be ×4, ×5 or ×10.
- The magnification of an objective for a light microscope may be ×5, ×10, ×20 or ×40.
- The magnification of the specimen is the magnification of the eyepiece multiplied by the magnification of the objective.

WORKED EXAMPLE

If the microscope is set up with the ×5 eyepiece and ×20 objective, the magnification of a specimen viewed will be:

$$5 \times 20 = 100$$

You can work out the *size of a structure* from the image size seen under the microscope and the magnification used to view it.

actual size = observed size/magnification

The observed size is measured using a scale viewed through the microscope such as a **graticule**.

WORKED EXAMPLE

If the diameter of a cell seen under a microscope is 6 mm, and the magnification is ×400, the actual diameter of the cell is:

$6/400 = 0.015$ mm.

End of topic checklist

Key terms

graticule, magnification

During your study of this topic you should have learned:

○ How to calculate the magnification of a biological specimen seen under a microscope.

○ How to calculate the size of a biological specimen using millimetres as units.

End of topic questions

Note: The marks awarded for these questions indicate the level of detail required in the answers. In the examination, the number of marks awarded to questions like these may be different.

1. You are looking at an object that measures 0.5 mm and the image you see is 10 mm long. Your friend is looking at an object that is 0.1 mm long using the same magnification.

   What size of image does your friend see? (1 mark)

2. Imagine you are examining a specimen of blood under a microscope to look at red blood cells.

   Why might it be important to know the magnification of the lens you are using? (1 mark)

3. The image you are looking at is 1.5 mm long and you are using a magnification of 100.

   Write down the calculation you would use to work out the actual size of the object. (1 mark)

Movement in and out of cells

INTRODUCTION

If you put a red blood cell into pure water, it will eventually burst open. If you place the red blood cell into a salty solution instead, it will shrink.

Surrounding every cell is the cell membrane. Imagine the cell membrane as a leaky layer that is strong enough to hold all the contents in the cell together, but allows small particles to move through it. The cell membrane also has special 'gates' that allow certain important particles to pass through. Different cells have different kinds of 'gate' in them and so let different particles through. So cell membranes play an essential role in controlling what goes in and out of cells, and therefore they control the way that the cell functions.

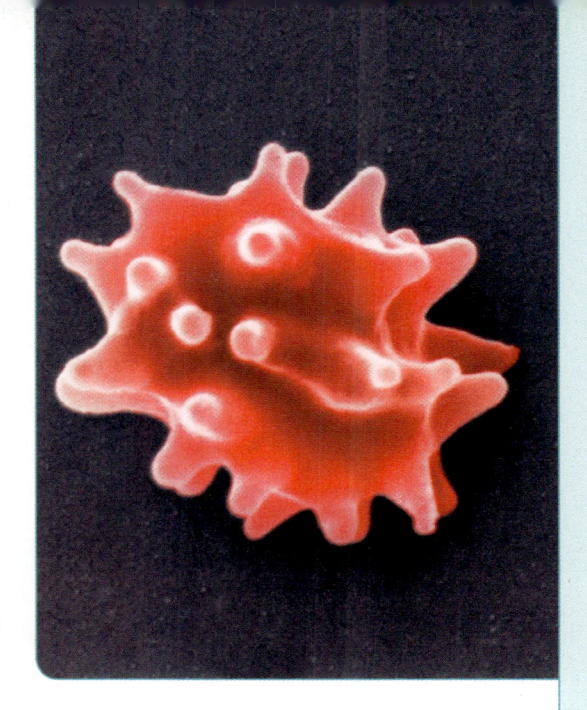

△ Fig. 2.16 A red blood cell that has been placed in a salty solution loses water and shrinks.

KNOWLEDGE CHECK
✓ Know that cells need oxygen and glucose for respiration.
✓ Know that cells need to get rid of waste substances, such as carbon dioxide from respiration.

LEARNING OBJECTIVES
✓ Be able to define diffusion as the net movement of molecules from a region of their higher concentration to a region of their lower concentration down a concentration gradient, as a result of their random movement.
✓ Be able to describe the importance of diffusion of gases and solutes and of water as a solvent.
✓ Be able to define osmosis as the diffusion of water molecules from a region of their higher concentration (dilute solution) to a region of their lower concentration (concentrated solution), through a partially permeable membrane.
✓ Be able to describe the importance of osmosis in the uptake of water by plants, and its effects on plant and animal tissues.
✓ **EXTENDED** Be able to define active transport as movement of ions in or out of a cell through the cell membrane, from a region of their lower concentration to a region of their higher concentration against a concentration gradient, using energy released from respiration.
✓ **EXTENDED** Be able to discuss the importance of active transport as an energy-consuming process by which substances are transported against a concentration gradient, e.g. ion uptake by root hairs and uptake of glucose by epithelial cells of villi.
✓ **EXTENDED** Be able to describe the importance of a water potential gradient in the uptake of water by plants.

DIFFUSION

Substances such as water, oxygen, carbon dioxide and food are made of particles (**molecules** or ions).

The particles in liquids and gases are constantly moving around. This means that they eventually spread out evenly. For example, if you dissolve sugar in a cup of water, even if you do not stir it, the sugar molecules eventually spread throughout the liquid. This is because all the molecules are moving around, colliding with and bouncing off other particles.

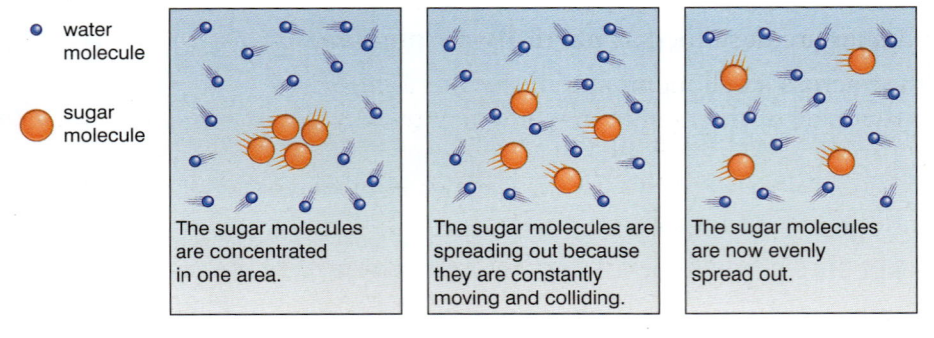

- water molecule
- sugar molecule

The sugar molecules are concentrated in one area.

The sugar molecules are spreading out because they are constantly moving and colliding.

The sugar molecules are now evenly spread out.

△ Fig. 2.17 Diffusion of sugar molecules in a solution.

The sugar molecules have spread out from an area of high concentration, when they were added to the water, to an area of low concentration. Eventually, although all the particles are still moving, the overall picture is that the sugar molecules are evenly spread out and there is no longer a **concentration gradient**.

Net movement is when there are more particles moving in one direction than another. **Diffusion** is defined as the net movement of molecules from a region of their higher concentration to a region of their lower concentration.

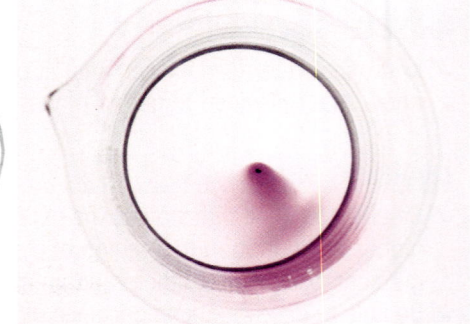

Diffusion can only occur when there is a difference in concentration between two areas. Particles are said to move *down* their concentration gradient. This happens because of the random movement of particles and needs no energy from the cell. It is a **passive** process.

△ Fig. 2.18 Potassium manganate(VII) crystal dissolving in still water.

Diffusion is important for living organisms because it allows gases such as oxygen and carbon dioxide to be exchanged between the organism and the environment. It is also the process by which **solutes** dissolved in water (such as mineral ions and food molecules) enter a living organism and waste substances (such as urea) are excreted from the organism into the environment.

Water is an essential **solvent** for living organisms because so many substances dissolve in it. This means they can be transported around organisms, in the xylem and phloem of plants, and in the blood of animals. The cytoplasm also has a high concentration of water, making it possible for many metabolic reactions to take place in the cell cytoplasm.

Diffusion in cells

Cells are surrounded by membranes. These membranes are leaky, so they let tiny particles pass through them. Large particles can't get through, so cell membranes are described as **partially permeable**.

Tiny particles can *diffuse* across a cell membrane if there is a difference in concentration on either side of the membrane (a concentration gradient). For example, in the red blood vessels in the lungs there is a *low* oxygen concentration inside the red blood cells (because they have given up their oxygen to cells in other parts of the body) and a *high* oxygen concentration in the alveoli of the lungs. Therefore oxygen diffuses from the alveoli into the red blood cells.

Other examples of diffusion include:

- carbon dioxide entering leaf cells
- digested food substances from the small intestine entering the blood.

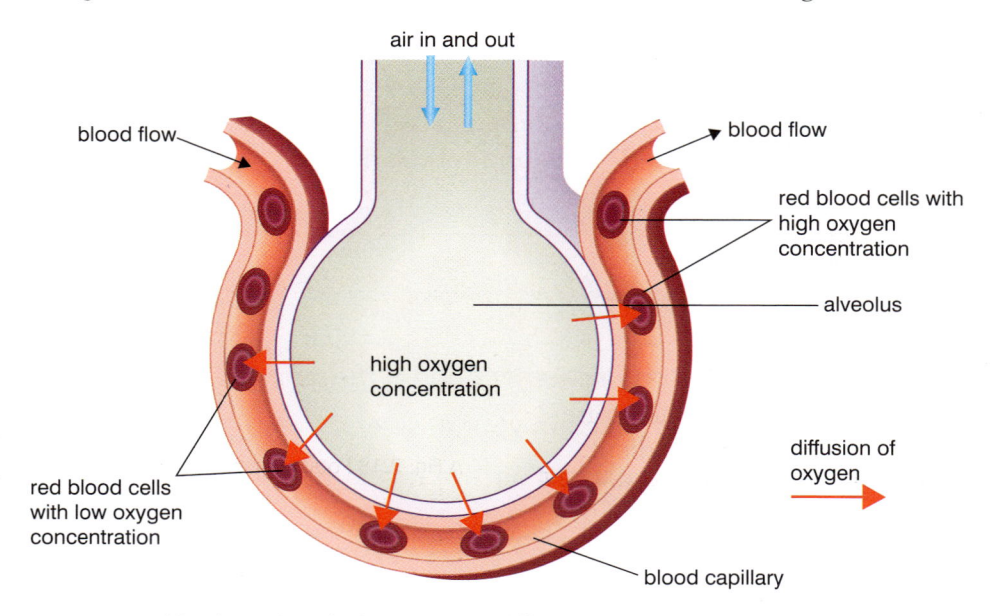

△ Fig. 2.19 In blood vessels in the lungs, oxygen diffuses down its concentration gradient from the air in the lungs into red blood cells.

SCIENCE IN CONTEXT

KIDNEY FAILURE AND HAEMODIALYSIS

The kidneys are organs that depend on filtration and diffusion to produce urine and keep the concentration of many substances in the blood at a fairly constant level. People who suffer from kidney failure are unable to do this, and are very quickly at risk from the build-up of waste products, such as urea, in the body as a result of cell processes. In high concentrations these waste products can damage body cells and lead to death.

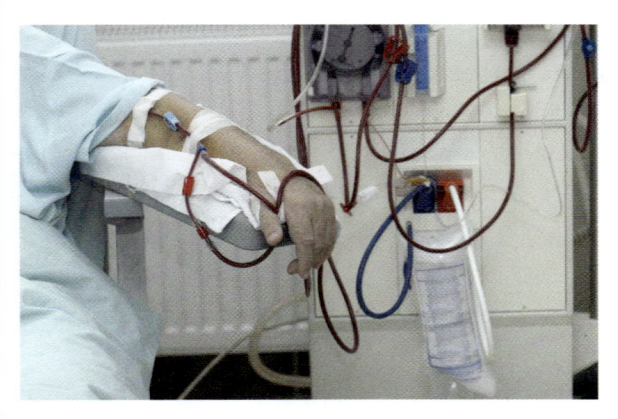

◁ Fig. 2.20 A patient undergoing kidney dialysis.

Haemodialysis is an artificial way of cleaning the blood. In this treatment, substances diffuse out of the blood into dialysis fluid in a machine called a dialyser. The concentration of substances in the dialysis fluid must be correct, so that all the waste products are removed and other substances are returned to the body at the right concentration.

QUESTIONS

1. In your own words, define the terms *net movement* and *diffusion*.
2. Is diffusion a passive or active process? Explain your answer.
3. Explain why some particles but not others can diffuse through cell membranes.

OSMOSIS

Water molecules are small enough to diffuse through the holes in cell membranes. However, because water molecules are so important to cells, and may be diffusing in a different direction to other molecules, this kind of diffusion has a special name: **osmosis**. Like diffusion, osmosis is a passive process and is a result of the random movement of particles.

Water molecules diffuse from a place where there is a high concentration of water molecules (such as a dilute sucrose sugar solution) to where there is a low concentration of water molecules (such as a concentrated sucrose sugar solution). Osmosis is defined as the net movement of water molecules from a region of their higher concentration to a region of their lower concentration across a partially permeable membrane.

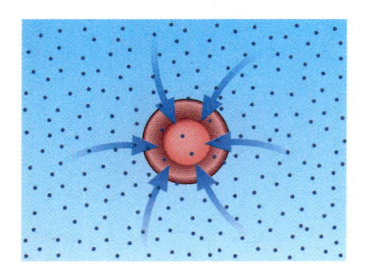

△ Fig. 2.21 A red blood cell in pure water takes in water by osmosis. The blue arrows show the direction of osmosis.

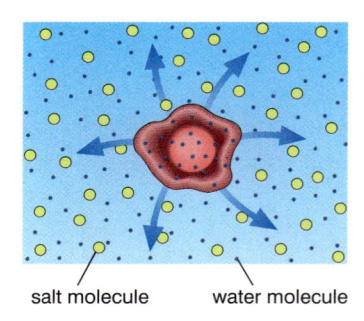

△ Fig. 2.22 A red blood cell in a concentrated salt solution will lose water as a result of osmosis. The blue arrows show the direction of osmosis.

Concentrations in solutions

Many people confuse the concentration of a solution with the concentration of the water. Remember, in osmosis it is the *water molecules* that you are considering, so you must think of the concentration of water molecules in the solution instead of the concentration of solutes dissolved in it.

- A low concentration of dissolved solutes means a high concentration of water molecules.
- A high concentration of dissolved solutes means a low concentration of water molecules.

So the water molecules are moving from a *high concentration (of water molecules)* to a *low concentration (of water molecules)*, even though this is often described as water moving from a low-concentration *solution* to a high-concentration *solution*.

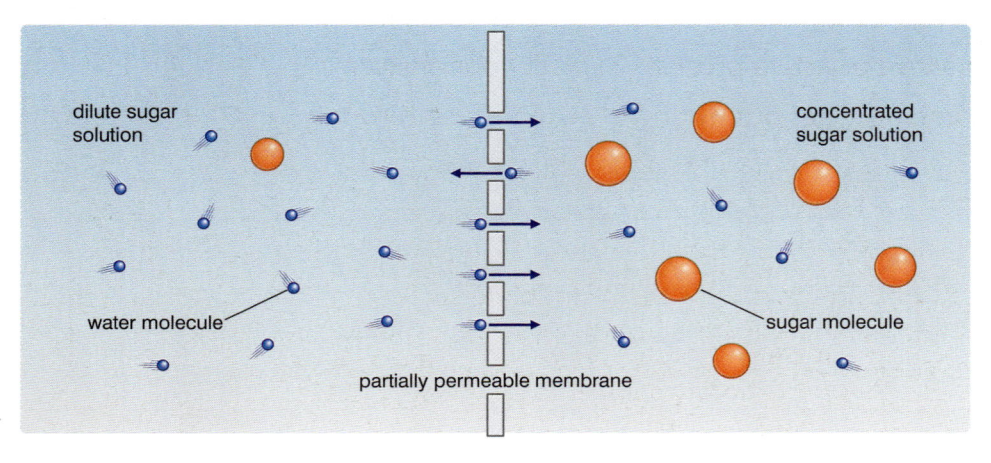

△ Fig. 2.23 In this system, the sugar molecules are too large to diffuse through the partially permeable membrane. However, the water molecules can pass through the membrane. So there is diffusion of water molecules from the side with their higher concentration to the side with their lower concentration. The diffusion of water molecules is known as osmosis.

Osmosis in animal and plant cells

If an animal cell is placed in a solution that has a higher concentration of solute (and so a lower concentration of water molecules) than the cytoplasm inside the cell, water will leave the cell by osmosis and the cell will shrink (Fig. 2.21). In living cells this causes problems because many reactions inside the cell take place in solution.

Even bigger problems occur if an animal cell is placed in a solution that has a lower concentration of solute (and so a higher concentration of water molecules) than the cytoplasm inside the cell. Osmosis will cause water to enter the cell and increase its volume (Fig 2.20). If too much water enters, the pressure inside the cell may become too much for the cell membrane to hold, and the cell may burst.

Plant cells are surrounded by cell walls outside the cell membrane. Cell walls are completely permeable, which means that water and solute molecules pass easily through them. In a solution of high concentration of solute (low concentration of water), water will leave a plant cell by osmosis and the cell will become **flaccid**. This can be seen in plant cells as the cytoplasm reduces in volume inside the cell, and so reduces the pressure on the cell wall. However, the cell doesn't shrink, unlike in animal cells, because the plant cell wall controls the structure of the cell. Instead, the whole plant wilts.

In a solution of low concentration of solute (high concentration of water), water will enter the plant cell by osmosis. However, the plant cell does not eventually burst. The strong cell wall provides strength when the cytoplasm is full of water and **turgid**, preventing the cell from expanding any further and bursting. Turgid cells also provide strength, making the plant stand upright with its leaves held out to catch the sunlight.

△ Fig. 2.24 When the cells of the plant are not full of water, the cell walls are not strong enough to support the plant, and the plant collapses (wilts). When the cells are fully turgid, the plant stands upright.

Developing investigative skills

Dandelions are weed plants in northern Europe that have a hollow fleshy stem. Strips of dandelion stem about 5 cm long and 3 mm wide were placed in sodium chloride solutions of different concentration. (Note: care should be taken with sharp instruments used for cutting; dandelion parts are very mildly poisonous and should not be eaten and hands should be washed after handling.) After 10 minutes, the strips looked as shown in the diagram. (Note that the outer layer of a dandelion stalk is 'waterproofed' with a waxy layer to protect them from water loss to the environment.)

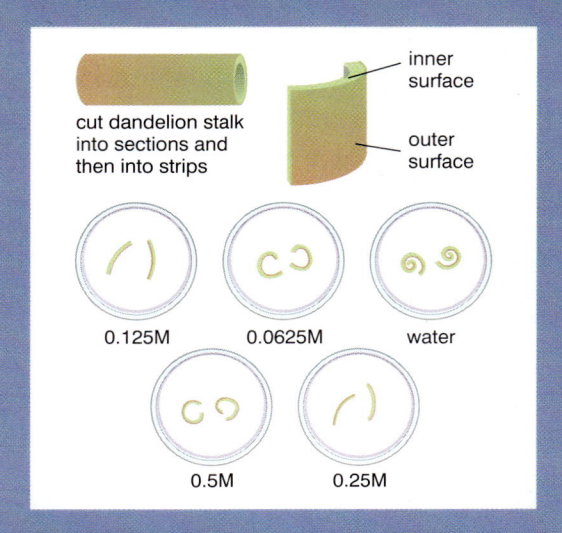

◁ Fig. 2.25 Investigating osmosis.

Using and organising techniques, apparatus and materials

❶ Write a plan for an experiment to carry out this investigation. Your plan should include:
   a) instructions on how to prepare the stem samples
   b) instructions on how to keep the stem samples until the experiment starts.

Observing, measuring and recording

❷ Using the diagram, describe the results of this investigation.

Handling experimental observations and data

❸ Explain as fully as you can the results of this investigation.
❹ Use the results to suggest the normal concentration of cell cytoplasm. Explain your answer.

SCIENCE IN CONTEXT: STOMATA

Stomata (single: stoma) are the pores in the surface of a leaf (usually the undersurface) that allow air to move into and out of the leaf. This provides the oxygen for respiring cells and carbon dioxide for photosynthesising cells, and allows water vapour that has evaporated from cell surfaces inside the leaf to diffuse out into the atmosphere.

Each stoma is surrounded by two guard cells. These control the opening and closing of the stoma. Usually stomata are open during the day, and close at night. The stoma opens and closes as the guard cells change shape. During the day the guard cells gain water from surrounding cells as a result of osmosis. This makes the cells turgid and, because the inner edge of the guard cell does not stretch, the cells curve and create a space between them which is the stoma. During the night, the guard cells lose water by osmosis. The cells lose their turgidity and collapse a little, closing the stoma between them.

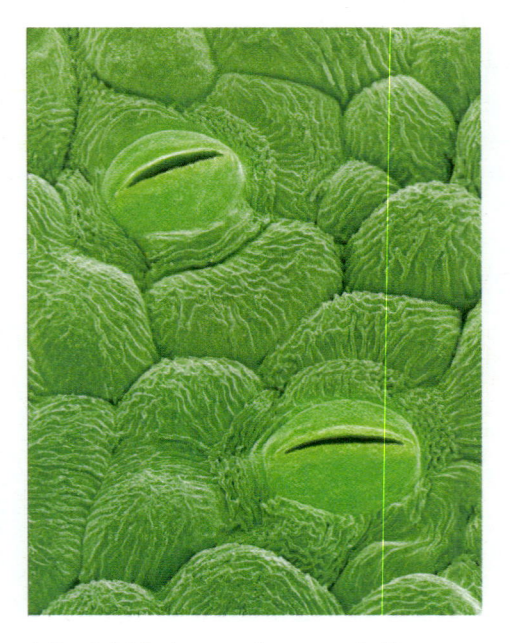

△ Fig. 2.26 Each stoma is surrounded by two guard cells.

REMEMBER

For the highest marks, you will need to explain diffusion and osmosis in terms of particles and their concentration gradients. Be clear that, even when diffusion and osmosis stop because there is no concentration gradient, the particles in the solution continue to move. However, there is no net movement any longer.

Osmosis and plant uptake of water

Plant roots are surrounded by soil water, and the cytoplasm inside root hair cells has a lower concentration of water molecules than the soil water. So water molecules will cross the cell membrane of root hair cells and enter the plant as a result of osmosis.

The water entering the root hair cells increases their concentration of water molecules, so that it is higher than the concentration of water molecules in surrounding cells. As a result, water molecules cross the cell membranes into the other cells. This raises their concentration of water molecules, and makes it higher than the concentration in cells even further into the root tissue, so water molecules cross those cell membranes. Osmosis continues from the surface root hair cells, deeper and deeper into the root tissue until the water molecules reach the xylem vessels. Once they have entered the xylem vessels, the water molecules are transported away from the root towards the leaves.

The process of **transpiration** from the leaves (see Losing water: transpiration) helps to draw the water up the xylem tubes from the roots, so maintaining the concentration gradient between the root cells and xylem vessels.

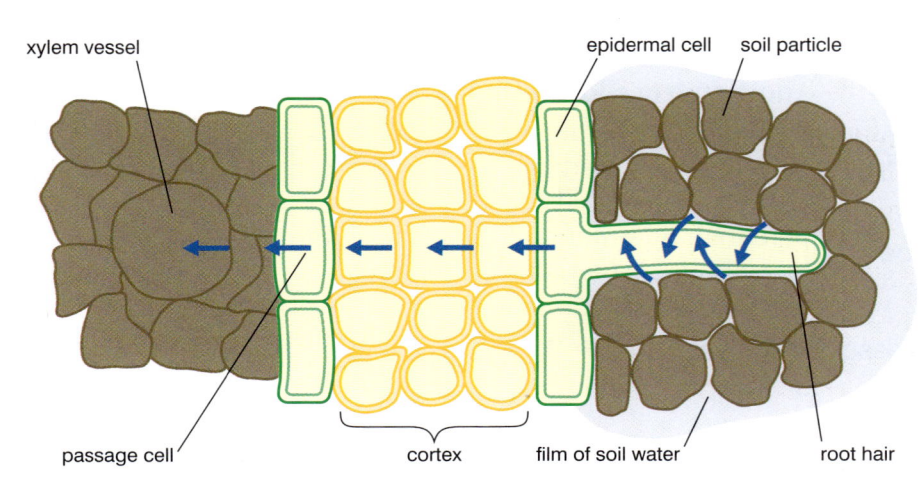

△ Fig. 2.27 Water enters a root through a root hair cell, and crosses from cell to cell inside the root, by osmosis until it reaches a xylem vessel.

REMEMBER

Look for how to apply your understanding of diffusion and osmosis to the processes that you learn about in the rest of the course. Make sure you refer to them appropriately in exam answers. For example, osmosis, which causes a change in shape of cells, is important in the opening and closing of stomata in plants.

QUESTIONS

1. In your own words, define the term *osmosis*.
2. Explain how osmosis is a) similar to and b) different from diffusion.
3. Draw a labelled diagram to show what happens to the water molecules when a red blood cell is placed in a solution that has a higher concentration of solute than the cytoplasm of the cell.
4. Explain the role of the plant cell wall in supporting a plant with turgid cells.

EXTENDED

Water potential gradient

A cell's ability to draw water into itself is called its **water potential**. Pure water has a water potential of zero. As solutes are added to the water, its water potential falls and it becomes more negative. So a concentrated sugar solution has a *lower* water potential (more negative) than pure water.

When two regions of different water potential are separated by a partially permeable membrane, water moves from the region of higher water potential to lower water potential. Water molecules move *down* the water potential gradient.

The movement of water molecules into and through the root of a plant can be explained in terms of water potential gradient, because the soil water has the highest water potential and the cells nearest the xylem vessels have the lowest water potential. So water molecules move into the root hair cell, and across the cells of the root, down the water potential gradient. The water potential gradient is maintained as the xylem vessels remove water molecules from the root due to transpiration from the leaves.

QUESTIONS

1. **EXTENDED** Describe the uptake of water from the soil by plant roots in terms of water potential gradient.

ACTIVE TRANSPORT

Sometimes cells need to absorb particles *against a concentration gradient*, from a region of low concentration into a region of high concentration. For example, root hair cells take in nitrate ions from the soil even though the concentration of these ions is higher in the plant cells than in the soil. Also glucose, which is in higher concentration in digested food in the gut than in body cells, needs to be absorbed against its concentration gradient. The nitrate ions in plants and glucose molecules in animals are essential for healthy growth. The cells use energy released from respiration to absorb these substances, so this is an *active* process and is called **active transport**.

Active transport occurs when special **carrier proteins** in the surface of a cell pick up particles from one side of the membrane and transport them to the other side. You can see this happening in Fig. 2.28.

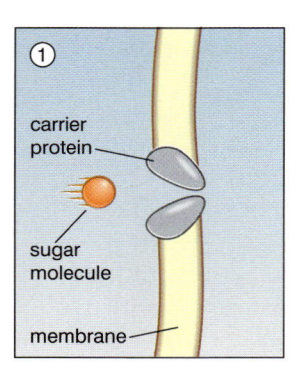

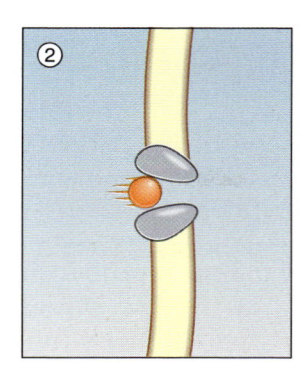

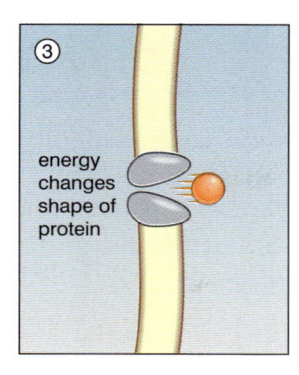

△ Fig. 2.28 Active transport of a sugar molecule through a carrier protein in a membrane.

> **REMEMBER**
>
> The energy for active transport comes from cell respiration. There is a simple test to show whether a substance is being absorbed by an active process or a passive process. This test is done by treating the cells with a metabolic poison that stops the cells respiring. For example, treating root hair cells with cyanide stops the uptake of nitrate ions but doesn't affect osmosis.

QUESTIONS

1. **EXTENDED** Explain what is meant by *active transport*.
2. **EXTENDED** Give one example of active transport in plant cells and one in animal cells, and explain why this is important for the organism.

END OF EXTENDED

End of topic checklist

Key terms

active transport, carrier protein, concentration gradient, diffusion, flaccid, net movement, osmosis, partially permeable, passive, turgid, water potential

During your study of this topic you should have learned:

○ How to define *diffusion* as the net movement of molecules from a region of their higher concentration to a region of their lower concentration down a concentration gradient, as a result of their random movement.

○ How to describe the importance of diffusion of gases and solutes and of water as a solvent.

○ **EXTENDED** How to define *active transport* as movement of ions in or out of a cell through the cell membrane, from a region of their lower concentration to a region of their higher concentration against a concentration gradient, using energy released from respiration.

○ **EXTENDED** How to discuss the importance of active transport as an energy-consuming process by which substances are transported against a concentration gradient, e.g. ion uptake by root hairs and uptake of glucose by epithelial cells of villi.

○ How to define *osmosis* as the diffusion of water molecules from a region of their higher concentration (dilute solution) to a region of their lower concentration (concentrated solution), through a partially permeable membrane.

○ How to describe the importance of osmosis in the uptake of water by plants, and its effects on plant and animal tissues.

○ **EXTENDED** How to describe the importance of a water potential gradient in the uptake of water by plants.

End of topic questions

Note: The marks awarded for these questions indicate the level of detail required in the answers. In the examination, the number of marks awarded to questions like these may be different.

1. An old-fashioned way of killing slugs in the garden is to sprinkle salt on them. This kills the slugs by drying them out. Explain why this works. **(2 marks)**

2. Copy and complete this to compare diffusion, osmosis and active transport.

	Diffusion	**Osmosis**	**Active transport**
Active or passive?			
Which molecules move?			
Requires special carrier proteins?			

(9 marks)

3. Which of the following are examples of diffusion, osmosis or neither?

    a) Carbon dioxide entering a leaf when it is photosynthesising. **(1 mark)**

    b) Food entering your stomach when you swallow. **(1 mark)**

    c) A dried out piece of celery swelling up when placed in a bowl of water. **(1 mark)**

4. There are many membranes within a cell, separating off organelles that produce substances such as hormones and enzymes, or where cell processes such as photosynthesis and respiration occur. Explain fully the importance of these membranes and why it is an advantage to the cell to have them. **(3 marks)**

5. a) **EXTENDED** If you measured the rate of respiration of plant root hair cells, would you expect it to be:

    i) the same as

    ii) more than

    iii) less than the rate of respiration of other plant cells? **(1 mark)**

    b) Explain your answer to part **a)**. **(1 mark)**

Enzymes

INTRODUCTION

Many of our staple foods, such as rice, potato, pasta or bread, contain large quantities of starch. Take a mouthful of one of these, plain, without anything else, and at first there is not much taste. If you continue chewing on it for a few minutes, to mix it with saliva and reduce it to a slush, you will find it starts to taste sweeter. This is because saliva has enzymes that start to break down the starch into smaller, sweet sugar molecules. Enzymes are essential in digestion, to break down the large molecules in our food into molecules small enough for diffusion through the cells of the gut wall and into our bodies.

△ Fig. 2.29 The starch in this pasta will be broken down by digestion.

KNOWLEDGE CHECK
✓ Know that food is digested in the gut into smaller molecules.

LEARNING OBJECTIVES
✓ Be able to define the term *catalyst* as a substance that speeds up a chemical reaction and is not changed by the reaction.
✓ Be able to define *enzymes* as proteins that function as biological catalysts.
✓ Be able to investigate and describe the effect of changes in temperature and pH on enzyme activity.
✓ **EXTENDED** Be able to explain enzyme action in terms of the 'lock and key' model.
✓ **EXTENDED** Be able to explain the effect of changes in temperature and pH on enzyme activity.
✓ **EXTENDED** Be able to describe the role of enzymes in the germination of seeds, and their uses in biological washing products and in the food industry (including pectinase and fruit juice).
✓ **EXTENDED** Be able to outline the use of microorganisms and fermenters to manufacture the antibiotic penicillin and enzymes for use in biological washing powders.
✓ **EXTENDED** Be able to describe the role of the fungus *Penicillium* in the production of the antibiotic penicillin.

ENZYMES AS CATALYSTS

A **catalyst** is a substance that changes the speed of a chemical reaction, but is left unchanged at the end of the reaction. For example, catalysts are often used in industrial processes to make chemicals such as ammonia.

Living cells also use catalysts to speed up the reactions that happen inside them. We call these **metabolic reactions** because they are the reactions of the metabolism (all the processes that keep a living organism alive).

Catalysts that control metabolic reactions are **enzymes,** and because they work in living cells they are called **biological catalysts**. Enzymes help cells carry out all the life processes quickly. Without enzymes, these metabolic reactions would happen too slowly for life to carry on.

Enzymes are proteins that function as biological catalysts. Enzymes are also **specific**, which means each enzyme only works with one substance (or a similar group of substances), called its **substrate**.

- Amylase is a type of carbohydrase enzyme produced in the mouth which starts the digestion of starch in food into simple sugars.
- Proteases are digestive enzymes that break down proteins into smaller units.
- Lipases are digestive enzymes that break down lipids in foods.

QUESTIONS

1. Define the term *catalyst*.
2. Explain what is meant by a biological catalyst.
3. Explain why cells need enzymes.

ENZYMES AND TEMPERATURE

Enzymes work best at a particular temperature, called their **optimum temperature**. For many enzymes in the human body, particularly those that work in the organs in the core (centre) of the body such as the heart, liver, kidneys and lungs, the optimum temperature is about 37 °C.

At lower temperatures the substrate and enzyme molecules have less energy and move around more slowly. So they collide and interact less frequently, which means the reaction happens more slowly. At temperatures that are too high, the structure of an enzyme changes and it will not work. This is a permanent change, and when it happens the enzyme is said to be **denatured**.

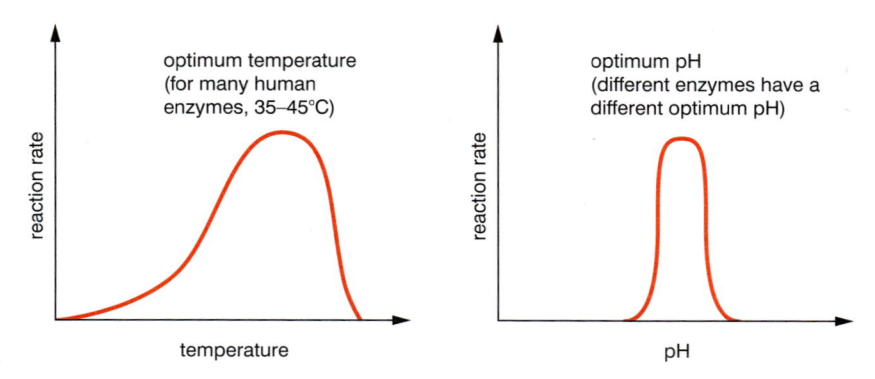

△ Fig. 2.30 Many enzymes work best at optimum temperature and pH.

> ### REMEMBER
> Remember the relationship between enzyme activity, temperature and pH, particularly when discussing excretion (see Excretion) and homeostasis (see Homeostasis), as this helps to explain why maintaining particular conditions in the body is so important for health.

Investigating the effect of temperature on enzymes

The effect of temperature on an enzyme can be tested by measuring the rate of action of the enzyme at different temperatures. One method is shown in the box below. Alternatively, you could use the following method to investigate the optimum temperature of amylase.

Starch is broken down to glucose by the enzyme amylase. Starch reacts with iodine solution by turning it blue-black. Glucose does not react with iodine solution, leaving it brown or orange. If you mix starch solution with amylase solution, placing different tubes of the mixture in water baths of different temperature and taking a sample for testing with iodine solution every minute or so, you can see at which temperature the amylase works fastest. The first sample to stop reacting with iodine solution is the one that comes from the tube kept at the optimum temperature for amylase.

Developing investigative skills

Developed black-and-white negative film consists of a celluloid backing covered with a layer of gelatin. Where the film has been exposed the gelatin layer contains tiny particles of silver which make that area black. Gelatin is a protein and is easily digested by proteases.

Strips of exposed film were soaked in protease solution at different temperatures. (Note: enzymes can cause severe allergies in some people; care should be taken when using hot water baths.) When the gelatin had been digested, the silver grains fell away from the celluloid backing, leaving transparent film. The table shows the results.

Tube	Temperature/°C	Time to clear
1	10	6 min 34 sec
2	20	3 min 15 sec
3	30	2 min 43 sec
4	40	3 min 55 sec
5	50	8 min 33 sec

Using and organising techniques, apparatus and materials
❶ Describe how you would set up this investigation to get results like those shown in the table.

Observing, measuring and recording
❷ Draw a graph using the data in the table.
❸ Describe the shape of the graph.
❹ Explain the shape of the graph.

Handling experimental observations and data
❺ How could you modify this experiment to get a more accurate estimate of the optimum temperature for this enzyme?

ENZYMES AND PH

Enzymes also often work best at a particular pH, called their **optimum pH**. Extremes of (very high or very low) pH can slow down the rate of action of the enzyme and even denature them.

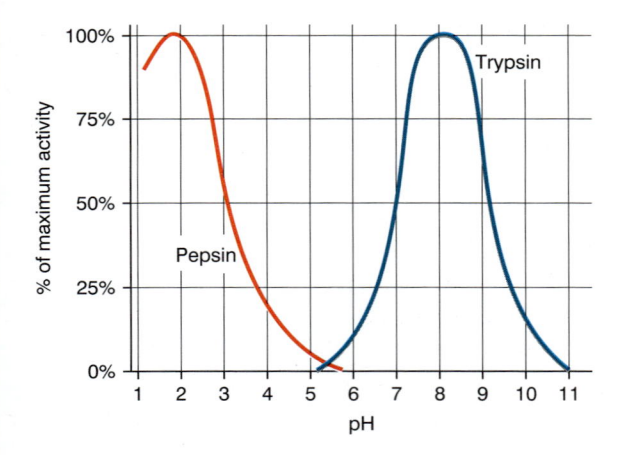

◁ Fig. 2.31 Pepsin (a protease enzyme found in the stomach) and trypsin (a protease enzyme produced by the pancreas and released into the small intestine) have different optimum pHs.

Different enzymes have different optimum pHs, depending on where they are normally found in the body. Pepsin digests proteins in the stomach, which is a highly acidic environment. Trypsin digests proteins in the small intestine, where conditions are more alkaline.

Investigating the effect of pH on enzymes

You can investigate the effect of pH on amylase enzyme using a similar method to the one above for temperature.

Set up one tube for each pH to be investigated and add buffer solution which will keep the contents at a particular pH. Add starch solution to each tube, and then amylase solution. Take a sample from each tube every minute or so and test for starch using iodine solution. The sample that is the first to stop turning iodine solution blue-black comes from the tube where digestion of starch to glucose was fastest, and therefore from the tube kept at the optimum pH for that enzyme.

QUESTIONS

1. Describe the effect of temperature on the rate of an enzyme-controlled reaction.
2. Compare the optimum pHs for pepsin and trypsin, shown in the graph, and explain the differences.

EXTENDED

THE 'LOCK AND KEY' MODEL OF ENZYME ACTION

Enzymes are proteins. Like all proteins, they have a three-dimensional shape produced by the way the molecule folds up. Enzymes are unusual proteins in that they have a space in the molecule with a particular 3D shape, called the **active site**. This site matches the shape of the reacting molecule, called the **substrate**.

The **'lock and key' model** of enzyme action explains the way that enzymes work in terms of how the substrate fits into the active site and is then changed. The important feature of this model is that the substrate (the 'key') fits tightly into the active site of the enzyme (the 'lock'), forming an enzyme–substrate complex. This makes it easier for the bonds inside the substrate to be rearranged to form the product. Once the products are formed, they no longer fit the active site, so they are released, leaving the active site free to bind with another substrate molecule.

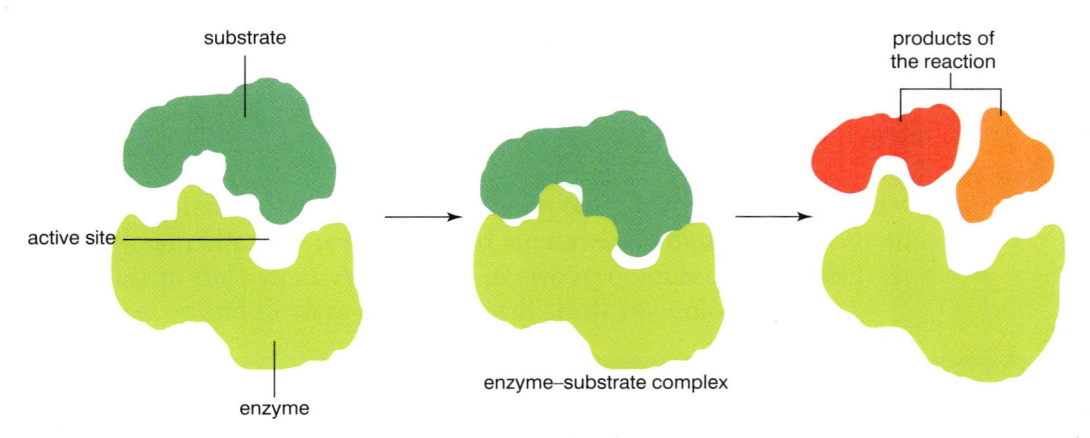

△ Fig. 2.32 The 'lock and key' model of enzyme action, during the breakdown of one substrate molecule into two product molecules.

This model works for both types of reaction:
- where one substrate molecule is broken down into two or more product molecules, e.g. in digestion of food molecules
- where two or more substrate molecules are joined together to make a product molecule, such as in the synthesis of hormones.

The 'lock and key' model helps to explain the fact that enzymes are specific, because only a substrate with the correct shape can fit into the active site and so be affected by the enzyme.

QUESTIONS

1. **EXTENDED** Explain what is meant by the 'lock and key' model of enzyme action.

EXPLAINING ENZYME ACTIVITY

The effect of temperature

If we understand the effect of temperature on particles (atoms and molecules), we can explain how changing temperature affects the rate of a reaction. Heat energy causes particles to move. If they move freely, they will bump into surrounding particles. If they are held within larger molecules by bonds, they vibrate.

An enzyme molecule and substrate molecule can only form an enzyme–substrate complex when they bump into each other and the substrate fits into the active site.

- At a low temperature the enzyme and substrate molecules move slowly, so they may take a long time to bump into each other and start the reaction.
- As the temperature increases, the molecules gain more energy and move faster, so the chance of them bumping into each other and reacting increases. The rate of reaction increases up to the optimum temperature.
- Beyond the optimum temperature, the atoms in the enzyme molecule are vibrating so much that they start to change the shape of the active site. This means the substrate doesn't fit so well, so the chances of an enzyme–substrate molecule forming decreases. The rate of reaction decreases.
- If the temperature increases too far, the bonds between atoms in the enzyme molecule start to break, changing the shape of the active site permanently and denaturing the enzyme.

The effect of pH

Proteins are made of amino acids joined together in a chain. The amino acids then interact with surrounding amino acids, which causes the chain to fold up into the 3D shape of the enzyme.

Some of the interactions between amino acids in the enzyme molecule depend on the pH of the surrounding solvent. So the shape of the enzyme depends on the surrounding pH. If the pH changes too much from the optimum pH, the shape of the enzyme, and particularly its active site, will change. Then the substrate will not fit so well into the active site and the rate of reaction will decrease.

QUESTIONS

1. **EXTENDED** Use the 'lock and key' model to explain why enzymes are specific.

2. **EXTENDED** Explain the effect of temperature on enzyme activity:

    a) at temperatures below the optimum

    b) at temperatures above the optimum.

3. **EXTENDED** Use the 'lock and key' model of enzyme action to explain the effect of pH on pepsin (see Fig 2.30).

ROLE OF ENZYMES IN GERMINATION

Until **germinating** seeds develop green leaves that can carry out photosynthesis, they rely on food stored inside the seed to provide energy from respiration for growth. Also, until they have a properly developed root, they cannot absorb mineral nutrients from the soil water for use in making proteins and other molecules needed in new cells. Different seeds store their food as different chemicals, but they are all complex compounds.

Species	Complex chemicals in the seed's food store
Rice, wheat, barley	Starch and protein
Sunflowers, cashews, sesame seeds	Oils
Peanuts, walnuts, Brazil nuts	Proteins and oils
Soya beans, lentils, peas	Protein

△ Table 2.1 Contents of food stores in the seeds of different species.

In order to use the food stores for respiration and growth, these compounds need to be broken down into the smaller units that make them: that is, simple sugars (such as glucose), amino acids and fatty acids. Enzymes are used to break down the complex compounds. The enzymes only become active when water has entered the seed.

QUESTIONS

1. **EXTENDED** Describe the role of enzymes in seed germination.
2. **EXTENDED** Explain why these enzymes are not needed once the seedling has developed leaves and a root.

USEFUL ENZYMES

In the food industry

We use many enzymes to produce our food, for example:

- amylases that break down starch – in baking to make softer bread, and in brewing beer
- proteases that break down proteins – in baby foods to make them easier to digest
- chymosin in rennet breaks down milk protein – used to curdle milk when making cheese.

Pectinase is an enzyme that breaks down some of the chemicals in plant cell walls. It is added to fruit pulp before it is pressed to extract the juice. As the cell walls are broken down, more juice can be extracted more easily, making it a more effective process. It also produces a clearer juice.

△ Fig. 2.33 When juice is extracted from bananas, the enzyme pectinase helps to break down the cell walls, producing more juice from the fruit (in the jar on the left). The juice is also clearer than that produced without pectinase (in the jar on the right).

Biological washing products

Many of the stains on our clothes come from food, or biochemicals such as those in blood, sweat or grass sap. These are difficult to remove. Getting clothes clean again used to mean lots of scrubbing with soap and washing at high temperatures.

△ Fig. 2.34 Biological washing powders contain enzymes that break down biochemicals in stains so that they dissolve in water.

These complex biochemicals can all be broken down using enzymes. Adding enzymes to washing products not only makes it easier to clean clothes, it also makes it possible to get them cleaner at lower temperatures. This saves time and energy, because the water for washing doesn't need to be heated for so long, and the wash cycle of the washing machine can be shorter.

> **SCIENCE IN CONTEXT**
>
> ### ENZYMES FOR WASHING CLOTHES
>
> The first enzyme used commercially in washing products was introduced in the 1960s. It was a protease that broke down protein-based stains such as blood, and it was extracted from a bacterium. This enzyme could withstand a temperature of up to 60 °C and a high pH, which suited the conditions that most people used to wash clothes in those days.
>
> Since then a much wider range of enzymes has been added to washing products, to digest fats, starches and other molecules. Enzymes that break down cellulose in cottons are also added to make natural fibres, such as cottons, smoother after washing.
>
> Different organisms contain enzymes that have different optimum temperatures and pHs. So scientists are researching many organisms, to find those that can be used to produce the best enzymes for washing products most easily.

QUESTIONS

1. **EXTENDED** Describe and explain the effect of adding pectinase to fruit when it is pulped to produce juice.
2. **EXTENDED** Give two advantages of adding enzymes to washing products.

Microorganisms and fermenters

Microorganisms can be used to make chemicals on a large scale. They are grown in ideal conditions inside a large vessel called a **fermenter**. Examples of chemicals made this way include:

- the fungus *Penicillium,* which produces the antibiotic penicillin (see below)
- bacteria, such as *Bacillus subtilis*, which produce enzymes used in biological washing products.

Growing microorganisms like this makes it easy to control the conditions in which they grow and also to extract large quantities of the products that we need.

Conditions inside the fermenter are controlled to help the microorganisms grow as quickly as possible.

- Nutrients that the fungal cells need for growth are added to replace those that are used up. These include an energy source, such as glucose, for respiration. They also include other nutrients needed for the synthesis of new cell materials, such as a nitrogen source (e.g. ammonia) for making proteins and nucleic acids.
- The reactions of respiration release heat energy. If the temperature of the mixture gets too high, this could affect the rate of enzyme action and slow down the rate of growth. A cooling jacket is used to remove excess heat, and the temperature inside the fermenter is monitored continually.
- The pH is continually monitored because if it varies too far from the optimum pH for the enzymes, the rate of growth will slow. If needed, the pH of the solution is adjusted using buffer chemicals.
- *Penicillium* and other species of bacteria that are cultured in fermenters are aerobic. So they need a continuous supply of oxygen, in the form of air bubbled through the mixture in the fermenter.
- The mixture in the fermenter is continually **agitated** by stirring, to make sure the cells don't all settle to the bottom and so that all the materials in the fermenter are well mixed.
- Before a new batch of culture is added to the fermenter, it is sterilised by passing steam through it. Also all solutions added to the fermenter are sterilised. These **aseptic precautions** make sure that no other microorganisms are added to the fermenter, which could affect the growth of the microorganism in the culture.

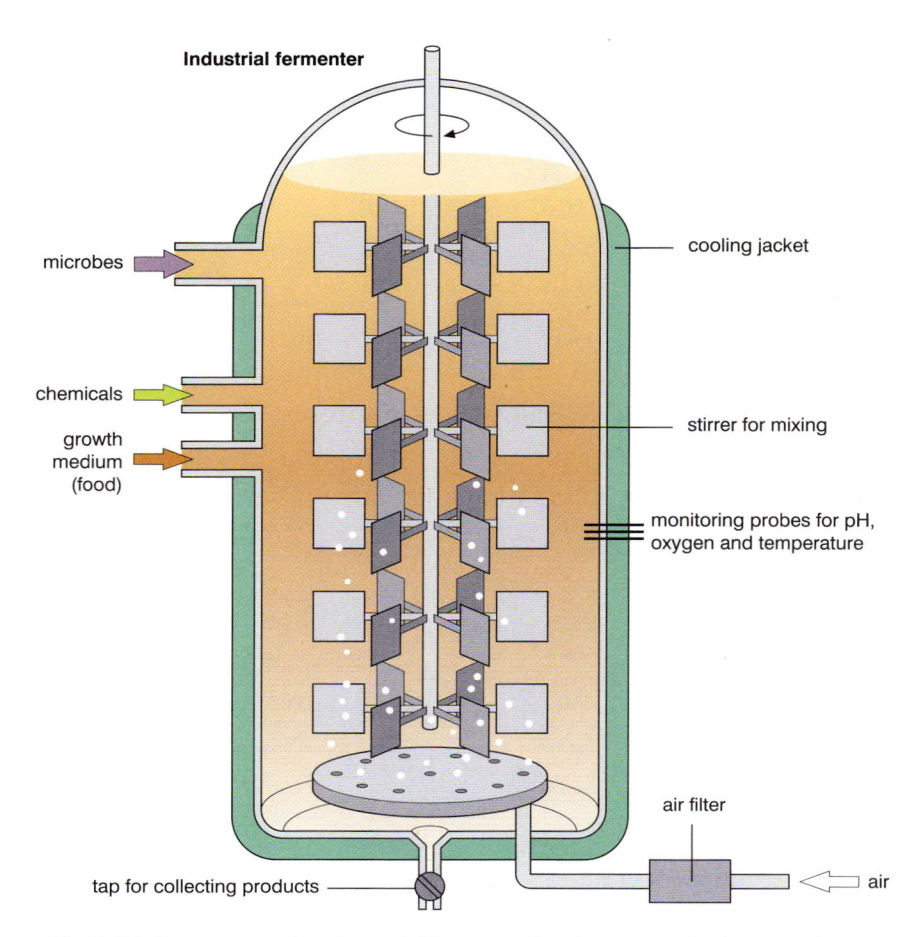

△ Fig. 2.35 The structure of an industrial fermenter for the growth of microorganisms.

Once the process has been going long enough for the mixture in the fermenter to contain enough product, some of it is drained through a tap at the bottom of the fermenter. The mixture is then processed to extract the product. More culture and nutrients can be added to the fermenter to replace what has been drained off, so that the process can continue without stopping.

> **REMEMBER**
>
> Make sure you can explain what each feature of an industrial fermenter does and why it is needed.

The source of penicillin

Penicillin is an **antibiotic**, which means it kills living organisms or stops them growing. Penicillin kills bacteria without damaging human cells, so it can be used inside the human body. It is a common and important medicine, used to treat many kinds of bacterial infection.

Penicillin is produced by a fungus (mould) called *Penicillium* and is secreted into the environment surrounding the fungal cells. The fungus is grown in ideal conditions inside a fermenter, and the penicillin is extracted from the liquid surrounding the fungus.

QUESTIONS

1. **EXTENDED** Explain what is meant by an *industrial fermenter* and what it is used for.

2. **EXTENDED** Give two examples of products that are made in biofermenters.

3. **EXTENDED** Describe and explain the advantages of using a fermenter to produce a product.

4. **EXTENDED** State three conditions that need to be monitored in a fermenter and explain how they are controlled.

5. **EXTENDED** Explain why aseptic precautions are used to prepare a fermenter for use.

6. **EXTENDED** Explain how penicillin is produced from the fungus *Penicillium* on a large-scale.

END OF EXTENDED

SCIENCE IN CONTEXT

ALEXANDER FLEMING AND PENICILLIN

Since medieval times, bread mould had been known in folk medicine to be useful for treating some infections. However, the discovery that penicillin was the chemical in bread mould that affects the growth of some bacteria was only made by Alexander Fleming in 1928. He noticed that bacteria growing on an agar plate had been killed in the area surrounding a growth of bread mould. He identified the mould as the fungus *Penicillium notatum*.

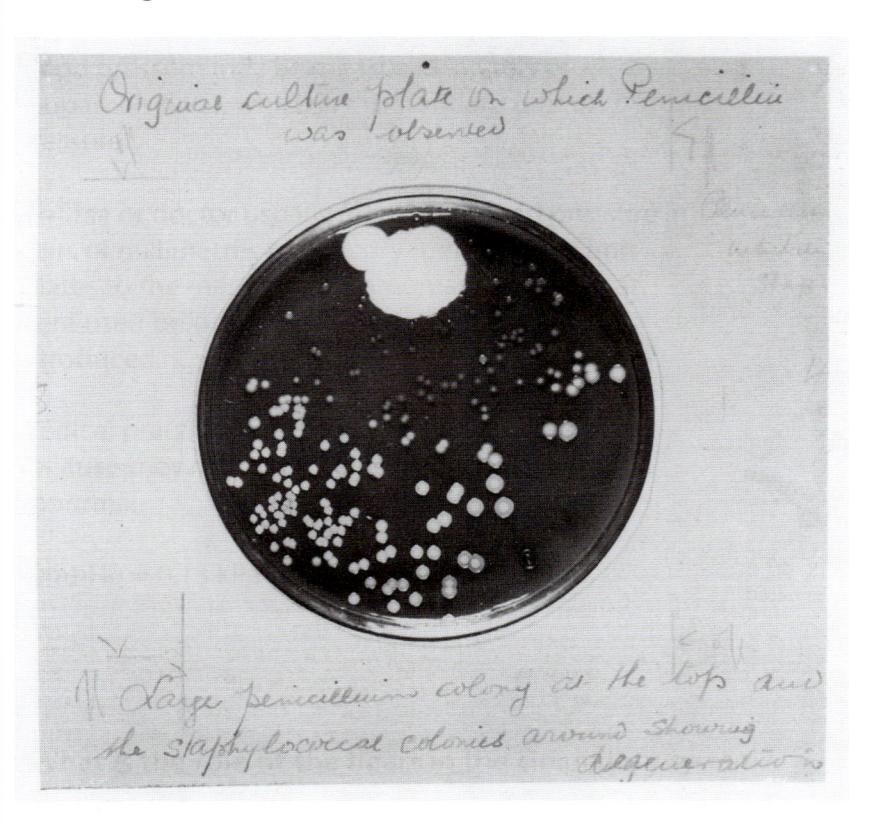

△ Fig. 2.36 Fleming's original agar plate, with the large mould area at the top of the plate and small bacterial colonies on the rest of the plate. It is clear that the bacterial colonies nearest the mould are dying.

It took until 1940 to purify penicillin from an extract of the fungus, and another few years to find a way to produce large quantities of it by growing the mould in a suitable mix of nutrients. During the last two years of the Second World War, it was made in large enough quantities for use in hospitals, and is considered to have saved the lives of many injured soldiers. Since then it has prevented millions of deaths from bacterial infections.

Scientists continue to research penicillin and develop new synthetic chemicals that have a similar structure but are more effective. This is becoming increasingly important as bacteria develop resistance to antibiotics that have been in use for some time.

End of topic checklist

Key terms

active site, agitate, antibiotic, aseptic precautions, biological catalyst, catalyst, denature, enzyme, fermenter, germination, 'lock and key' model, optimum pH, optimum temperature, specific, substrate

During your study of this topic you should have learned:

○ How to define the term *catalyst* as a substance that speeds up a chemical reaction and is not changed by the reaction.

○ How to define *enzymes* as proteins that function as biological catalysts.

○ How to investigate and describe the effect of changes in temperature and pH on enzyme activity.

○ **EXTENDED** How to explain enzyme action in terms of the 'lock and key' model.

○ **EXTENDED** How to explain the effect of changes in temperature and pH on enzyme activity.

○ **EXTENDED** How to describe the role of enzymes in the germination of seeds, and their uses in biological washing products and in the food industry (including pectinase and fruit juice).

○ **EXTENDED** About the use of microorganisms and fermenters to manufacture the antibiotic penicillin and enzymes for use in biological washing powders.

○ **EXTENDED** How to describe the role of the fungus *Penicillium* in the production of the antibiotic penicillin.

End of topic questions

Note: The marks awarded for these questions indicate the level of detail required in the answers. In the examination, the number of marks awarded to questions like these may be different.

1. There are over 75 000 different enzymes in the human body. Explain why we need so many. **(2 marks)**

2. Describe how you would investigate the optimum temperature for a particular enzyme. **(4 marks)**

3. Scientists researching new enzymes for biological washing products extract enzymes from bacterium found growing in a hot spring (temperature around 80 °C) and algae that grow in the Arctic Ocean (temperature around 4 °C).

    a) Suggest the optimum temperature for enzymes extracted from:

    i) the bacterium

    ii) the algae.

    Explain your answers. **(3 marks)**

    b) Sketch a graph to show the effect of temperature on the rate of reaction for an enzyme from humans. Label the value of the optimum temperature on your graph. **(2 marks)**

4. The body has many mechanisms for keeping internal conditions within limits. One of the internal conditions that is controlled is the concentration of carbon dioxide in the blood. Carbon dioxide gas is acidic and highly soluble.

    a) Which process in cells produces carbon dioxide? **(1 mark)**

    b) How is this gas removed from the body? **(1 mark)**

    c) What would you expect to happen to the amount of carbon dioxide in the body during exercise? Explain your answer. **(1 mark)**

    d) What effect would this have on conditions inside cells if the carbon dioxide was not removed? **(1 mark)**

    e) What problem would this cause for enzymes and cell processes they control? **(2 marks)**

5. Explain the shape of the graph you drew for Question 3b. **(5 marks)**

6. **EXTENDED** Which organism that the scientists were researching in Question 3 might be the most useful for producing enzymes for biological washing products? Explain your answer as fully as you can. **(3 marks)**

7. **EXTENDED** Describe fully how the scientists might produce these enzymes on a large scale. **(2 marks)**

8. **EXTENDED** Explain why, before germination can begin, a seed needs to take in water. **(2 marks)**

Nutrition

INTRODUCTION

At a certain time in the summer in Canada the rivers are full of salmon fish returning to breed. Their sudden appearance attracts large numbers of bears, which at other times of the year feed on nuts, grasses, insects and other small animals. The salmon are full of protein and fat and are an essential source of nutrition for the bears.

△ Fig. 2.37 Canadian brown bears fishing for salmon.

KNOWLEDGE CHECK

✓ Know that plants make their own food in their leaves using photosynthesis.
✓ Know that plant structures, such as the leaf and root cells, are adapted for their functions in nutrition.
✓ Know that animals eat other organisms to get the food they need for their life processes.
✓ Know that the organs, tissues and cells of the digestive system are adapted to digest and absorb nutrients from food.
✓ Know that food may be chemically or mechanically digested before absorption.
✓ Know that different groups of people need different diets.

LEARNING OBJECTIVES

✓ Be able to define *nutrition* as the taking in of nutrients for growth and tissue repair.
✓ Be able to list the elements in carbohydrates, fats and proteins and describe the synthesis of large molecules from smaller basic units.
✓ Be able to describe tests for starch, reducing sugars, protein, fats.
✓ Be able to list the main sources of, and describe the importance of, carbohydrates, fats, proteins, vitamins C and D, calcium and iron, fibre, water.
✓ Be able to describe the deficiency symptoms for vitamins C and D and the mineral salts calcium and iron.
✓ Be able to define photosynthesis as the fundamental process by which plants manufacture carbohydrates from raw materials using energy from light.
✓ Be able to state the word equation for the production of simple sugars and oxygen.
✓ Be able to investigate the necessity for chlorophyll, light and carbon dioxide for photosynthesis, using appropriate controls.
✓ Be able to describe the intake of carbon dioxide and water by plants.
✓ Be able to explain that chlorophyll traps light energy and converts it into chemical energy for the formation of carbohydrates and their subsequent storage.
✓ Be able to identify the structures in a leaf and describe their roles in photosynthesis, gas exchange, transport and support.

- ✓ Be able to describe the importance of nitrate and magnesium ions in plants and the uses, and dangers of overuse, of nitrogen fertilisers.
- ✓ Be able to state what is meant by the term *balanced diet* and describe a balanced diet related to age, sex and activity of an individual.
- ✓ Be able to describe the effects of malnutrition in relation to starvation, coronary heart disease, constipation and obesity.
- ✓ Be able to discuss ways in which food production has increased due to modern technology.
- ✓ Be able to define the terms *ingestion* and *egestion*.
- ✓ Be able to identify the main regions of human alimentary canal and associated organs and their functions in relation to *ingestion*, *digestion*, *absorption*, *assimilation* and *egestion*.
- ✓ Be able to define *digestion* as the breakdown of large insoluble food molecules into small, water-soluble molecules using mechanical and chemical processes.
- ✓ Be able to identify the types of human teeth and describe their structure and functions.
- ✓ Be able to state the causes of dental decay and describe the proper care of teeth.
- ✓ Be able to describe the process of chewing.
- ✓ Be able to describe the role of longitudinal and circular muscles in peristalsis.
- ✓ Be able to outline the role of enzymes and bile in digestion.
- ✓ Be able to state the significance of chemical digestion in the alimentary canal in producing small, soluble molecules that can be absorbed.
- ✓ Be able to state where, in the alimentary canal, amylase, protease and lipase enzymes are secreted.
- ✓ Be able to state the functions of a typical amylase, a protease and lipase, listing the substrate and end-products.
- ✓ Be able to define *absorption* as movement of digested food molecules through the wall of the intestine into the blood or lymph and identify the small intestine as the region for the absorption of digested food.
- ✓ Be able to describe the significance of villi in increasing the internal surface area of the small intestine.
- ✓ Be able to define *assimilation* as movement of digested food molecules into the cells of the body where they are used, becoming part of the cells.
- ✓ Be able to describe the role of the liver in the metabolism of glucose.
- ✓ Be able to describe the role of fat as an energy storage substance.
- ✓ **EXTENDED** Be able to describe the use of microorganisms in the food industry.
- ✓ **EXTENDED** Be able to describe the uses, benefits and health hazards associated with food additives.
- ✓ **EXTENDED** Be able to state the balanced equation for photosynthesis in symbols.
- ✓ **EXTENDED** Be able to investigate and state the effect of varying light intensity, carbon dioxide concentration and temperature on the rate of photosynthesis.
- ✓ **EXTENDED** Be able to define the term *limiting factor* as something present in the environment in such short supply that it restricts life processes.
- ✓ **EXTENDED** Be able to explain the concept of limiting factors in photosynthesis.
- ✓ **EXTENDED** Be able to explain the use of carbon dioxide enrichment, optimum light and optimum temperatures in glasshouse systems.
- ✓ **EXTENDED** Be able to explain the effects of nitrate ion and magnesium ion deficiency on plant growth.

- ✓ **EXTENDED** Be able to discuss the problems of world food supplies and the problems that contribute to famine.
- ✓ **EXTENDED** Be able to describe how fluoride reduces tooth decay and explain arguments for and against adding fluoride to public water supplies.
- ✓ **EXTENDED** Be able to describe the structure of a villus, including the role of capillaries and lacteals.
- ✓ **EXTENDED** Be able to state the role of the hepatic portal vein in the transport of absorbed food to the liver.
- ✓ **EXTENDED** Be able to identify the role of the small intestine and colon in absorption of water.
- ✓ **EXTENDED** Be able to define deamination as removal of the nitrogen-containing part of amino acids to form urea, followed by release of energy from the remainder of the amino acid.
- ✓ **EXTENDED** Be able to state that the liver is the site of breakdown of alcohol and other toxins.

NUTRIENTS AND NUTRITION

Nutrition is one of the seven characteristics of living organisms. Nutrition is the taking in of nutrients (organic substances and mineral nutrients) that contain the raw materials or energy needed by the organism. Nutrition also includes the **absorption** and **assimilation** of these nutrients into new body tissue during growth and tissue repair.

Plant nutrition does not include taking in organic substances because plants make their own carbohydrates using photosynthesis. However, plants need to take in mineral ions through their roots to make other substances needed for healthy growth and tissue repair.

In animal nutrition, organic substances and mineral ions taken in as food are digested, absorbed and assimilated into the body to make the substances needed for healthy growth and tissue repair.

CARBOHYDRATES, PROTEINS AND LIPIDS

Most of the molecules found in living organisms fall into three main groups: carbohydrates, proteins and lipids, which are commonly called fats and oils. All of these molecules contain carbon, hydrogen and oxygen. In addition, all proteins contain nitrogen and some also contain sulfur.

Carbohydrate molecules are made up of small basic units called **simple sugars**. These are formed from carbon, hydrogen and oxygen atoms, sometimes arranged in a ring-shaped molecule. One example of a simple sugar is glucose.

Simple sugar molecules can link together to form larger molecules. They can join in pairs, such as sucrose (the 'sugar' we use in our food). They can also form much larger molecules called polysaccharides, such as starch and glycogen, which are both long chains of glucose molecules.

Protein molecules are made up of long chains of **amino acids** linked together. There are 20 different kinds of amino acids in plant and animal cells. These can join in long chains, in any order, to make all the different proteins within the plant or animal body. Examples include the structural proteins in muscle, as well as enzymes that help control cell reactions.

A **lipid** is what we commonly call a fat or oil. Fats are solid at room temperature, whereas oils are liquid, but they have a similar structure. Both are made from basic units called **fatty acids** and **glycerol**. There are three fatty acids in each lipid, and the acids vary in different lipids. Lipids are important in forming cell membranes and many other molecules such as fats in storage cells.

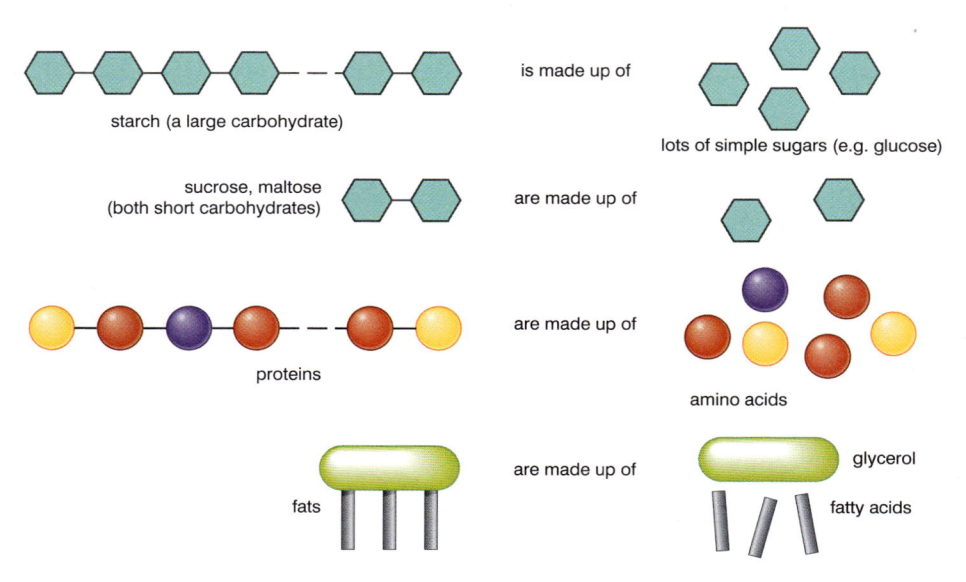

△ Fig. 2.38 Large biological molecules are formed from small sub-units.

QUESTIONS

1. What are the basic units of:
    a) lipids
    b) carbohydrates
    c) proteins?
2. Using the diagram of food molecules above, give two differences between a protein and a carbohydrate.

TESTS FOR FOOD MOLECULES

Simple tests can indicate whether or not a food contains particular food molecules, such as starch, glucose, proteins or lipids.

Test for starch

Starch is the storage molecule of plants and is found in many foods made from plant tissue. When iodine/potassium iodide solution is mixed with a solution of food containing starch, or dropped onto food containing starch, the solution changes from brown to dark blue. This happens when even small amounts of starch are present and can be used as a simple test for the presence of starch. This change is easiest to see if the test is examined against a white background, such as on a white spotting tile.

◁ Fig. 2.39 The blue-black colour shows there is starch in the biscuit.

Test for glucose

Glucose is a 'reducing sugar' that is important in respiration and photosynthesis. So it is commonly found in plant and animal tissues, and therefore in our food. Its presence can be detected using Benedict's solution.

The pale blue Benedict's solution is added to a prepared sample and heated to 95 °C. If it changes colour or forms a precipitate (solid particles in the liquid), this indicates the presence of reducing sugars. A green colour means there is only a small amount of glucose in the solution. A medium amount of glucose produces a yellow colour. A significant amount of glucose produces a precipitate that is an orange-red colour.

△ Fig. 2.40 Benedict's reagent with a range of concentration of sugars (very low in the tube on the left, getting more concentrated towards the right).

The test using Benedict's reagent produces an orange-red precipitate for any 'reducing sugar', such as the simple sugars fructose and galactose, and the disaccharides (made from two basic units; *di-* means two or double) lactose and maltose. So it is not exclusively a test for glucose. But as glucose is usually the most common sugar, this is the test most commonly used for it. Sugars such as sucrose (table sugar), which are formed from two simple sugar molecules, are not reducing sugars and so will not react with Benedict's reagent.

Test for protein

The **biuret test** is used to check for the presence of protein. A small sample of the food under test is placed in a test tube. An equal volume of biuret solution is carefully poured down the side of the tube. If the sample contains protein, a blue ring forms at the surface. If the sample is then shaken, the blue ring disappears and the solution turns a light purple.

Test for fat

This test depends on the fact that fats do not dissolve in water but do dissolve in ethanol. The test sample is mixed with ethanol. If fat is present, it will dissolve in the ethanol to form a solution. This liquid is poured into a test tube of water, leaving behind any solid that has not dissolved. If there is any fat dissolved in the ethanol, it will form a cloudy white precipitate when mixed with the water.

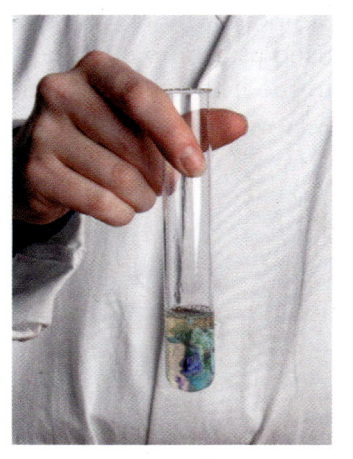

△ Fig. 2.41 A positive biuret test for protein.

QUESTIONS

1. Describe what you would see if you tested samples of the following with **i)** Benedict's solution, **ii)** iodine solution:

   a) glucose syrup

   b) cake made with wheat flour, table sugar (sucrose), fat and eggs.

   Explain your answers.

2. Explain how you would test the seed from a walnut tree to see if it contained stores of

   a) fat

   b) protein.

PLANT NUTRITION

Photosynthesis

Plant tissue contains the same types of chemical molecules (carbohydrates, proteins and lipids) as animal tissue. However, while animals eat other organisms to get the nutrients to make these molecules, plants make these molecules from basic building blocks, beginning with the process of photosynthesis.

In **photosynthesis**, plants combine the raw materials carbon dioxide (from the air) with water (absorbed from the soil) to form glucose, which is both a simple sugar and also a carbohydrate. This process transfers light energy (usually from sunlight) into chemical energy in bonds in the glucose. Photosynthesis is fundamental to most living organisms on Earth, because most organisms other than plants get their energy from the chemical energy in the food that they eat.

Oxygen is also produced in photosynthesis and, although some is used inside the plant for respiration (releasing energy from food), most is not needed and is given out as a **waste product**.

The sunlight is absorbed by the green pigment **chlorophyll** in plants.

The process of photosynthesis can be summarised in a word equation:

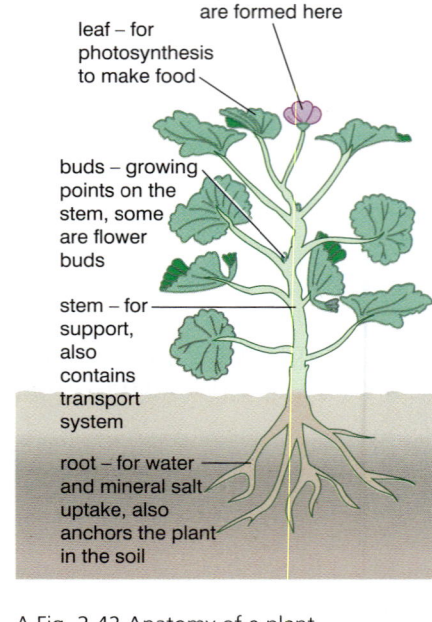

$$\text{carbon dioxide} + \text{water} \xrightarrow[\text{light energy}]{\text{chlorophyll}} \text{glucose} + \text{oxygen}$$

△ Fig. 2.42 Anatomy of a plant.

- flower – needed for reproduction, seeds are formed here
- leaf – for photosynthesis to make food
- buds – growing points on the stem, some are flower buds
- stem – for support, also contains transport system
- root – for water and mineral salt uptake, also anchors the plant in the soil

EXTENDED

It can also be summarised as a balanced symbol equation:

$$6CO_2 + 6H_2O \xrightarrow[\text{light energy}]{\text{chlorophyll}} C_6H_{12}O_6 + 6O_2$$

END OF EXTENDED

REMEMBER

For higher marks you will need to know and be able to balance the chemical equation for photosynthesis.

Glucose is soluble in water, so it cannot be stored in cells. It also affects osmosis. Much of the glucose formed by photosynthesis is converted into other substances, including **starch**. Starch molecules are large carbohydrates made of lots of glucose molecules joined together. Starch is insoluble and so can be stored without affecting water movement into and out of the cells by **osmosis**. Some plants, such as potato and rice plants, store large amounts of starch in particular parts of the plant (tubers or seeds). We use these parts as sources of starch in our food.

Some glucose is converted to **sucrose** (a type of sugar formed from two glucose molecules joined together). This is still soluble, but not as reactive as glucose, so it can easily be carried in solution around the plant in the phloem.

The energy needed to join simple sugars to make larger carbohydrates comes from respiration.

QUESTIONS

1. Write the balanced symbol equation for photosynthesis.
2. Annotate your equation to show where each of the reactants come from, and each of the products go to.
3. Explain why the transfer of energy from sunlight to chemical energy in plant cells is essential for life on Earth.

INVESTIGATING PHOTOSYNTHESIS

You can use the iodine test to show that photosynthesising parts of a plant produces starch. Before carrying out this test, though, you must leave the plant in a dark place for 24 hours. This will make sure that the plant uses up its stores of starch (this is known as destarching). This means that any starch identified by the test is the result of photosynthesis during the investigation.

- The production of starch after photosynthesis can be shown simply by placing a destarched plant in light for an hour. Remove one leaf and place it in boiling water for about one minute to soften it. Then place the leaf in boiling ethanol heated in a beaker of boiling water, not over a Bunsen burner – ethanol fumes are flammable! This removes the chlorophyll in the leaf. When the leaf has lost its green, wash it in cold water before placing it in a dish and adding a few drops of iodine solution. The leaf should turn blue/black, indicating the presence of starch.

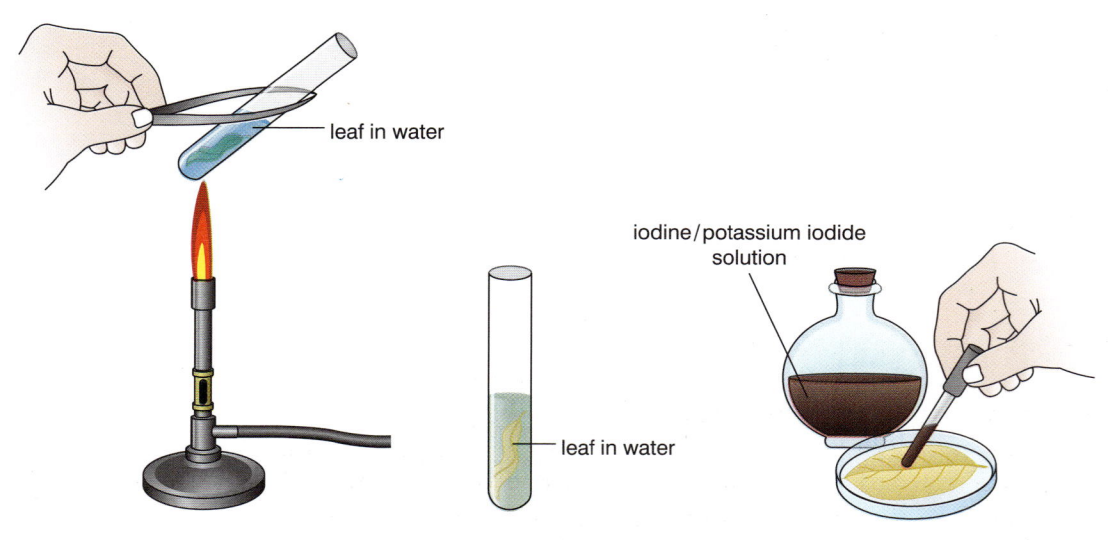

Δ Fig. 2.43 Preparing and testing a leaf for starch.

- You can adjust the investigation above to show the need for light by covering part of the leaf before bringing the destarched plant into the light. Only the part of the leaf that received light should test positive for the presence of starch, showing that photosynthesis is linked to the production of starch.
- This investigation can also be adjusted to show the need for chlorophyll by using variegated leaves. Variegated leaves are partly green (where the cells contain chlorophyll) and partly white (where there is no chlorophyll). A variegated leaf after this investigation will show the presence of starch where there was chlorophyll and not in the parts of the leaf that had no chlorophyll.
- A simple test to show the need for carbon dioxide can be carried out by setting up two bell jars on glass sheets. Sodium or potassium hydroxide reacts with the carbon dioxide, removing it from the air. So a dish of one of the hydroxides is placed in one bell jar. Carbon dioxide is added to the other bell jar by burning a candle in it, which also removes some of the oxygen. Similar de-starched plants are placed in each bell jar, and the base of the jar sealed to the glass sheet, such as with petroleum jelly. After a few hours in light, a leaf from each plant is tested for starch, which should show that the plant with the least carbon dioxide produces little starch.

△ Fig. 2.44 Light was excluded from all of the leaf except an L-shaped window. After exposure to light, only the L-shape tests positive for starch.

△ Fig. 2.45 Only the green parts of a variegated leaf can photosynthesise, as shown by the leaf on the right which has been tested for starch.

QUESTIONS

1. Describe a test that would show the need for chlorophyll in photosynthesis.
2. What precautions should be taken when boiling ethanol to remove chlorophyll in a leaf? Explain your answer.
3. How could you show that plants need carbon dioxide for photosynthesis?

EXTENDED

Factors affecting the rate of photosynthesis

The rate at which a process can occur depends on how quickly the required materials can be supplied. Photosynthesis needs light energy. As night approaches and daylight fades, or on a very cloudy day, the rate at which chlorophyll absorbs energy decreases and photosynthesis slows down. Light has become a **limiting factor** for the process.

If there is plenty of light, such as on a sunny day, then light will not be a limiting factor. Instead the rate of photosynthesis may be controlled by the rate of diffusion of carbon dioxide from the air into the photosynthesising cells in the leaf.

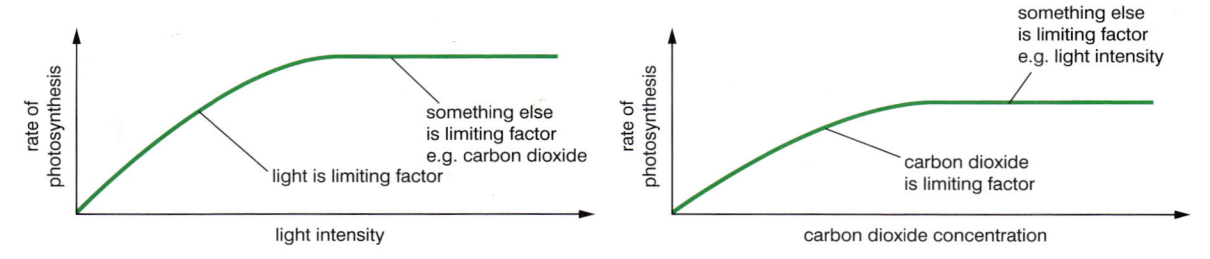

△ Fig. 2.46 The rate of photosynthesis is affected by (left) light intensity and (right) carbon dioxide concentration, up to a point when something else becomes a limiting factor.

We can test whether this is true by adding more carbon dioxide. Farmers sometimes do this in glasshouses to increase the growth rate of crops. If the rate of photosynthesis increases, then carbon dioxide was the limiting factor.

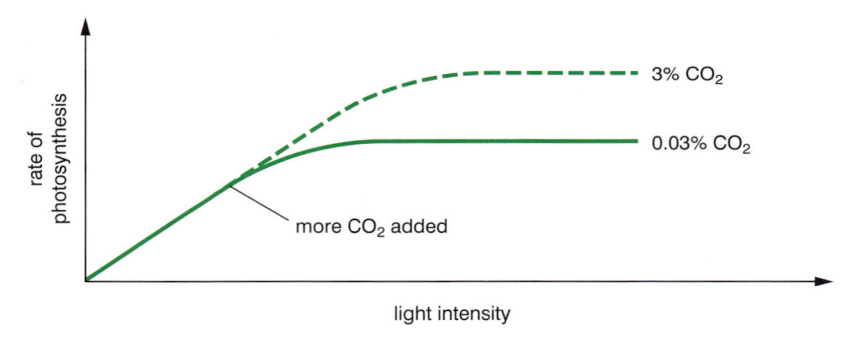

△ Fig. 2.47 The effect on rate of photosynthesis of adding more carbon dioxide to a leaf with increasing light intensity.

Photosynthesis involves several chemical reactions. Like all chemical reactions, the rate of reaction is affected by temperature, which affects the energy of the reacting particles and how quickly they bump into each other. So, on a cool day, or in the early morning when the air hasn't yet heated up, temperature may limit the rate of photosynthesis. If the temperature rise is too high, however, the enzymes that control the rate of reactions start to become denatured and so the reactions go more slowly.

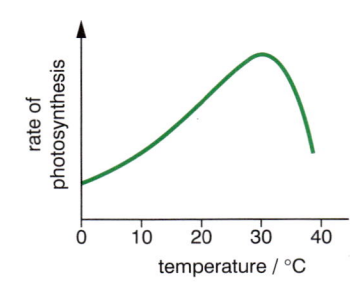

◁ Fig. 2.48 The effect of temperature on the rate of photosynthesis.

> **SCIENCE IN CONTEXT**
>
> ## GROWING PLANTS IN GREENHOUSES
>
> Farmers and plant growers want their crops to grow well, but in open fields it is not usually possible to control the amount of carbon dioxide or light the plants receive, or the temperature at which they are growing. However, if the plants are grown in sheltered conditions, such as in glasshouses, then it can be possible to change conditions, such as by:
>
> △ Fig. 2.49 Plants in a greenhouse.
>
> - using artificial lighting so that the plants can continue growing at a maximum when conditions are cloudy or even at night
> - enriching the atmosphere around the plants with carbon dioxide by burning a fuel such as propane gas, coal or oil
> - using a heating system to increase the temperature to an optimum for photosynthesis.
>
> Remember that enzymes have an optimum temperature at which they work, so glasshouses and polytunnels may need to be ventilated to release hot air if the temperature rises too high, otherwise the rate of photosynthesis will decrease.

Investigating the rate of photosynthesis

Measuring starch production is an indirect measurement of photosynthesis, because starch is made from the glucose produced in photosynthesis. You can investigate photosynthesis more directly by measuring the amount of oxygen produced by a plant. The oxygen is usually collected over water, and the simplest investigations may be done using aquatic plants (plants that grow in water), such as pondweed, using the apparatus shown in the Developing Investigative Skills box investigating the effect of light on photosynthesis.

- To prove that photosynthesis produces oxygen, simply use the glowing splint test on the gas collected. The splint should re-ignite showing that the gas is oxygen.
- The investigation can be adjusted to test for the effect of light intensity on the rate of photosynthesis as described in the investigation that follows.
- The investigation can be adjusted to test for the effect of carbon dioxide concentration by adding different amounts of sodium hydrogencarbonate to the water and measuring the rate at which bubbles of oxygen are produced.
- The investigation can be adjusted to test for the effect of temperature by placing the beaker of pondweed in water baths of different temperature and measuring the rate at which bubbles of oxygen are produced.

In each of these investigations, all other factors that may affect the rate of photosynthesis must be controlled and kept constant as far as possible.

Developing investigative skills

You can investigate the effect of light on photosynthesis by shining a light on a water plant and measuring how quickly bubbles are given off, as shown in the diagram.

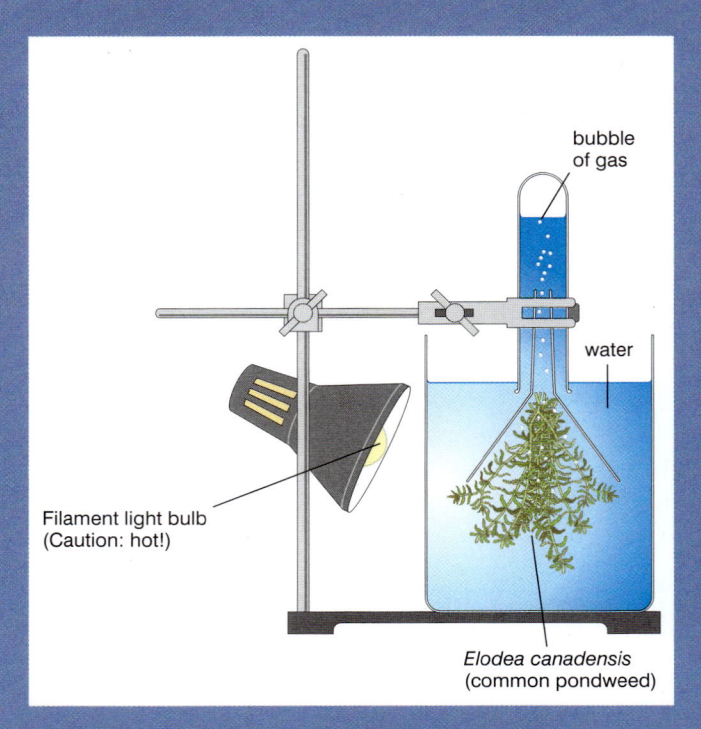

△ Fig. 2.50 Apparatus needed for the investigation into the effect of light on photosynthesis.

The results below were gathered using this apparatus.

	Distance to lamp/cm				
	5	10	15	20	25
Gas bubbles given off in 5 minutes	67	57	40	20	4

Using and organising techniques, apparatus and materials

1 a) Explain why the rate of producing bubbles can be used as a measure of the rate of photosynthesis.

  b) Explain how you would identify the gas produced by the plant.

Observing, measuring and recording

2 a) Use the data in the table to draw a suitable graph.

  b) Describe and explain the shape of the graph.

Planning and evaluating investigations

3 Light is not the only factor that can affect the rate of photosynthesis.

  a) Which other factor might have had an effect on these measurements.

  b) Suggest how the method could be changed to avoid this problem.

QUESTIONS

1. **EXTENDED** Explain what is meant by a *limiting factor*.
2. **EXTENDED** Describe how each of the following factors affects the rate of photosynthesis.
   a) light intensity
   b) carbon dioxide concentration
   c) temperature.
3. **EXTENDED** Explain why the factors have the effect you described in question 2.

END OF EXTENDED

LEAF STRUCTURE

Photosynthesis takes place mainly in the leaves, although it can occur in any cells that contain green chlorophyll. Leaves are adapted to make them very efficient as sites for photosynthesis, gas exchange, transport and support.

The diagram shows the external adaptations of the leaf that help to maximise the rate of photosynthesis, gas exchange with the air, transport of nutrients and water into and out of the leaf, and support.

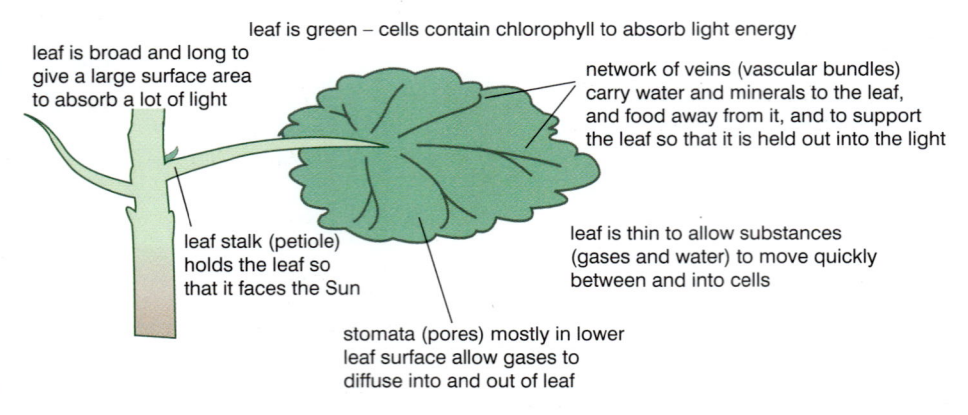

△ Fig. 2.51 Adaptations of the leaf of a plant for photosynthesis.

The diagram shows the arrangement of cells and tissues inside a leaf.
- The waxy **cuticle** that covers the leaf, particularly the upper surface, prevents the loss of water from epidermal (surface) cells and helps stop the plant from drying out.
- The transparent **epidermis** allows as much light as possible to reach the photosynthesising cells within the leaf.
- The **palisade cells**, where most photosynthesis takes place, are tightly packed together, in the uppermost half of the leaf so that as many as possible can receive sunlight.

- **Chloroplasts** containing chlorophyll are concentrated in the palisade cells in the uppermost half of the leaf to absorb as much sunlight as possible. Chloroplasts can move in the cytoplasm so they can get into a position where they receive the most light.
- The **spongy mesophyll cells** and air spaces in the lower part of the leaf provide a large internal surface area to volume ratio so that exchange of carbon dioxide and oxygen between the cells and the air in the leaf is as rapid as possible.
- Many pores or **stomata** (singular: stoma) allow gases to diffuse into and out of the leaf quickly.
- The **vascular bundles** form the veins in the stem and leaf. The thick cell walls of the xylem tissue in the bundles helps to support the stem and leaf.
- **Phloem** tissue transports sucrose, formed from glucose in photosynthesising cells, away from the leaf. **Xylem** tissue transports water and minerals to the leaf from the roots.

△ Fig. 2.52 The leaves of trees are often arranged so that they do not overlap each other, which makes it possible for the tree to capture as much light energy as possible.

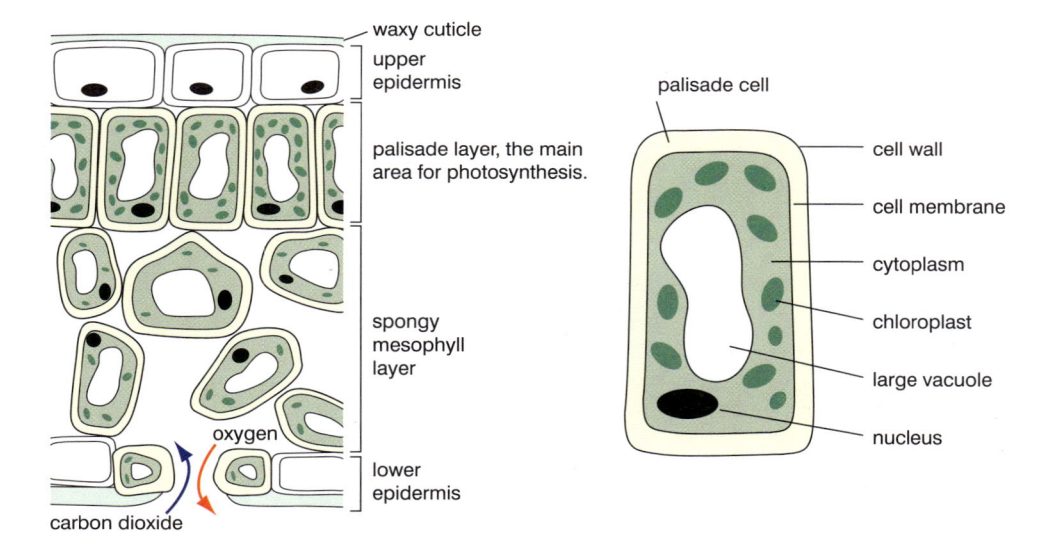

△ Fig. 2.53 Cells in the section of a leaf.

QUESTIONS

1. List as many adaptations of a plant leaf for photosynthesis as you can.
2. Explain why a large surface area inside the leaf is essential for photosynthesis.
3. Explain why a transparent epidermis is an adaptation for photosynthesis.

MINERAL REQUIREMENTS OF PLANTS

Photosynthesis produces carbohydrates, but plants also contain many other types of chemical. Carbohydrates contain only the elements carbon, hydrogen and oxygen, but the amino acids that make up proteins also contain nitrogen. So plants need a source of nitrogen in the form of nitrate ions. Other chemicals in plants contain other elements: for example, chlorophyll molecules contain magnesium and nitrogen. Without a source of magnesium ions and nitrate ions, a plant cannot produce chlorophyll and so cannot photosynthesise.

These additional elements are dissolved in water in the soil as **mineral ions**. The plant absorbs the mineral ions through their roots.

EXTENDED

The mineral salts in the soil water surrounding a root hair cell are at a lower concentration than inside the root hair cell. This means they cannot cross the root hair cell membrane into the plant by diffusion. So plants must use **active transport** to transfer mineral ions into root hair cells from soil water. Once the mineral ions have entered the root hair cell, they can then diffuse from cell to cell to the xylem, which transports them to other parts of the plant where they are needed.

Mineral deficiencies

Plants that are not absorbing enough mineral ions show symptoms of deficiency. For example:

- a plant with a nitrogen deficiency has stunted growth
- a plant with magnesium deficiency has leaves that are yellow between the veins, particularly in older leaves as the magnesium is transported in the plant to the new leaves.

△ Fig. 2.54 A plant with nitrogen deficiency.

△ Fig. 2.55 A plant with magnesium deficiency.

END OF EXTENDED

Nitrogen fertilisers

Farmers add nitrogen-rich fertilisers to fields to increase the growth of crop plants and the amount of food they produce (the **yield**). Nitrogen compounds in the fertiliser dissolve easily in soil water and are then absorbed by the plants. However, if too much fertiliser is added, the extra will dissolve in soil water and seep through the ground into nearby waterways, such as ponds, lakes, streams or rivers. In addition, if it rains heavily soon after the fertiliser is used, much of it will dissolve in the rain water and be washed away quickly into nearby waterways.

In the waterways, the fertiliser causes **eutrophication**, which is the enriching of the water with nutrients. This affects aquatic (water) plants in the same way as the crops, encouraging them to grow faster. Eutrophication leads to a number of problems:

- Rapid growth of weeds can clog the waterways.
- Some algae, whose growth is stimulated by fertilisers, produce toxic compounds that can poison animals (including people).
- The death of aquatic plants and animals due to low oxygen concentration in the water (see Overuse of fertilisers).
- Nitrogen in drinking water supplies may cause problems for young babies if they drink it.

EXTENDED

Students set up an investigation into the effect of nitrates on plant growth. They chose two plants that were as similar as possible. Over two months, the plants received the same amount of heat, light, water and carbon dioxide. However, one plant was given a liquid nitrogen feed in the water and the other only received distilled water. The images show the results at the end of two months.

1. Describe the differences between the two plants as fully as you can in terms of:

   a) the cells in the leaves

   b) the growth of the plants.

2. Explain the effect of nitrogen deficiency on the cells in the leaves.

3. Using your knowledge of plant nutrition, explain as fully as you can why these two plants grew differently.

4. Millions of tonnes of nitrogen-containing fertiliser are added to crop fields each year. Explain as fully as you can what would happen if this was not done.

△ Fig. 2.56 Two plants raised with equal amounts of heat, light, water and carbon dioxide. The bigger one also got liquid nitrogen feed.

END OF EXTENDED

QUESTIONS

1. Explain why plants need a supply of mineral ions.
2. Explain what plants use the following mineral ions for:
   a) nitrogen ions and b) magnesium ions.
3. **EXTENDED** Describe and explain the deficiency symptoms in a plant for the following mineral ions:
   a) nitrogen
   b) magnesium.
4. a) **EXTENDED** Explain why farmers add nitrogen-rich fertilisers to crop fields.
   b) Describe two problems caused by fertilisers entering waterways.

NUTRITION IN HUMANS

Essential nutrients

To keep healthy, humans need a diet that includes all the nutrients used by our cells and tissues, such as:

- **proteins** – these are digested (broken down) to amino acids. The amino acids are used to form other proteins needed by cells, including enzymes. Protein sources include eggs, milk and milk products (cheese, yoghurt, etc.) meat, fish, legumes (peas and beans), nuts and seeds.
- **carbohydrates** – these are digested to simple sugars for use in respiration. This releases energy in our cells and enables all the life processes to take place. Good sources of carbohydrate include rice, bread, potatoes, pasta and yams.
- **lipids** – these are deposited as fat in many parts of the body, including just below the skin. Some fat helps to maintain body temperature. Fat is also a store of energy to supply molecules for respiration if the diet does not contain enough energy for daily needs. Fat is present in meat and also in oils, milk products (butter, cheese, yoghurt), nuts, avocadoes and oily fish.
- **vitamins** and **minerals** – these substances are needed in tiny amounts to keep the body working properly. Most vitamins and minerals cannot be produced by the body and cooking food destroys some vitamins. For example, vitamin C is best supplied by eating raw fruit and vegetables.
- **fibre** – which is made up of the cell wall of plants. Good sources are leafy vegetables, such as cabbage, and unrefined grains such as brown rice and wholegrain wheat. It adds bulk to food so that it can be easily moved along the digestive system by peristalsis. This is important in preventing constipation. Fibre is thought to help prevent bowel cancer.
- **water** – which is the major constituent of the body of living organisms and is necessary for all life processes. Water is continually being lost through excretion and sweating, and must be replaced regularly through food and drink in order to maintain health. Most foods contain some water, but most fruit contain a lot of water.

Essential vitamins and minerals	Job	Good food source	Deficiency disease
Vitamin C	For healthy skin, teeth and gums, and keeps lining of blood vessels healthy	Citrus fruit, green vegetables, potatoes	Scurvy (bleeding gums and poor healing of wounds)
Vitamin D	To help absorption of calcium for strong bones and teeth	Fish, eggs, liver, cheese and milk	Rickets (softening of the bones)
Calcium	Needed for strong teeth and bones, and involved in the clotting of blood	Milk and eggs	Rickets (softening of the bones)
Iron	Needed to make haemoglobin in red blood cells	Red meats, liver and kidneys, leafy green vegetables such as spinach	Anaemia (reduction in number of red blood cells; person soon becomes tired and short of breath)

△ Table 2.2 Vitamins and minerals, their roles and sources and deficiency diseases.

EXTENDED

Vitamin D is not only taken in from the diet; it is also made naturally in skin that is exposed to sunlight. The paler skin of northern peoples is an adaptation to gathering more light in order to produce more vitamin D.

In the body, vitamin D is used to produce a hormone that controls the uptake of calcium from food. Lack of vitamin D results in softened bones – in children this produces rickets where the long bones of the legs, which take most of a person's weight, bend outwards in a characteristic bowed leg effect. Lack of vitamin D causes softened bones in adults too, and may have other effects, although these are not fully understood. On the other hand, too much vitamin D in the diet can affect the heart, kidneys and other organs.

1. Explain why rickets is less common in countries nearer the Equator than it used to be in higher latitude countries such as the UK.
2. Rickets used to be common in higher latitude countries among children in poorer families. Suggest a reason for this relationship.
3. Children were given regular doses of vitamin D, in the form of fish liver oil, during the winter. Explain why this helped to prevent rickets.
4. Today, more people in these countries have diets that are less dependent on carbohydrates. Explain how this has reduced the risk of rickets.

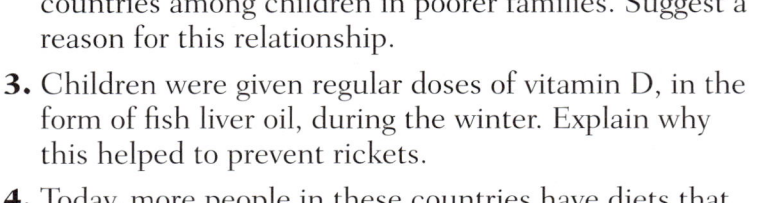

△ Fig. 2.57 Rickets – vitamin D deficiency.

5. Doctors in higher latitude countries are now finding vitamin D deficiency among women who wear clothing that covers all of their skin, and in people who eat a vegetarian diet. Describe and explain as fully as you can the reason and the treatment for this.

END OF EXTENDED

SCIENCE IN CONTEXT

KWASHIORKOR

Kwashiorkor is a condition found in young children in areas where the diet contains very little protein. It typically occurs in children that had been breast-fed but are then weaned after the birth of another baby. Breast milk contains proteins, but after weaning the child may get a carbohydrate-rich diet with little protein that is often lacking in some vitamins and minerals. Typical symptoms include swelling of the feet and abdomen, wasting muscles, thinning hair and loss of teeth. Liver damage may occur, and treatment requires careful adjustment of the diet so as not to damage the liver even more.

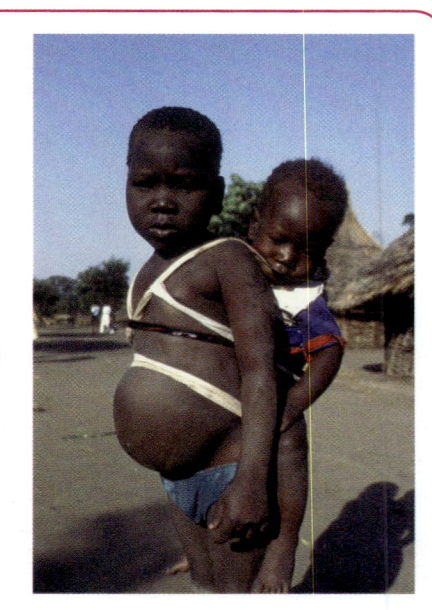

△ Fig. 2.58 A child suffering from kwashiorkor, a protein deficiency.

QUESTIONS

1. Which three groups of food molecules do we need most of in a healthy diet?
2. Give examples of foods that are good sources of each group of nutrients.
3. Which other substances are needed in our diet?
4. Explain the role of each of these substances in our diet.

The right balance

A **balanced diet** contains all of the nutrients we need in the right proportions to stay healthy. Since most foods contain more than one kind of nutrient, trying to work out what a balanced diet looks like can be difficult. Governments use images of food on a plate like these to guide people on what proportions of food to eat.

△ Fig. 2.59 A healthy meal contains a good balance of the foods your body needs and nothing in too large an amount.

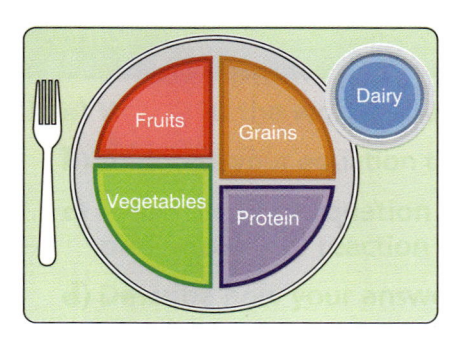

△ Fig. 2.60 Guidance from the USDA (United States Department of Agriculture) on the proportions of different nutrients in a balanced diet.

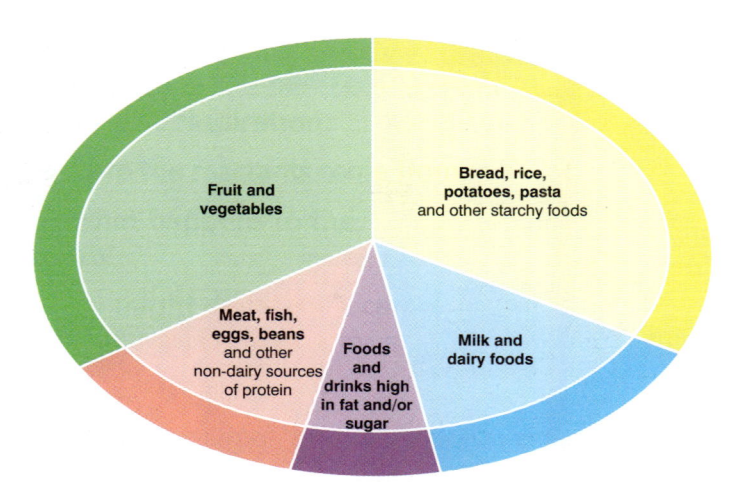

△ Fig. 2.61 Guidance from the UK Government on the proportions of different nutrients in a balanced diet.

Different groups of people have different needs for nutrients at different times in their lives, so this balance can change. For example, children need a higher proportion of protein than adults because they are still growing rapidly. Also, some groups of people have a greater need for a specific nutrient. During pregnancy, for example, women need more iron than usual, to supply what the growing baby needs for making blood cells.

Even with the right proportions of nutrients in our foods, our diets may still be unhealthy. This is because many of our foods, particularly carbohydrates but also fats and proteins, contribute to the energy our bodies need. If we eat food that supplies more energy than we use, the extra will be deposited as energy stores of fat. This can lead to **obesity** (being very overweight), which is related to many health problems such as heart disease and diabetes. Controlling the portion size at each meal, not eating snacks between meals and increasing levels of exercise all help to reduce the risk of becoming overweight.

Energy requirements depend on body size, stage of development and level of exercise, as shown in Table 2.3.

	Energy used in a day/kJ	
	Male	**Female**
6-year-old child	7500	7500
12–15-year-old teenager	12 500	9700
Adult manual worker	15 000	12 500
Adult office worker	11 000	9800
Pregnant woman		10 000

△ Table 2.3 Daily energy requirements for different people.

Developing investigative skills

Combustion (burning) of foods releases heat energy. The word equation for combustion is:

food + oxygen → carbon dioxide + water (+ heat energy)

This reaction is similar to respiration inside cells, so we can use a combustion experiment to model the energy that is released from foods during respiration.

A crisp/potato chip and leaf of a plant were tested in an investigation to see which released the most energy by combustion. Here are the results.

	Crisp	Leaf
Mass of sample/grams	22	12
Temperature of water after burning/ °C	27	16
Temperature of water before burning/°C	15	15
Temperature rise/°C	12	1
Energy released by the sample/joules	1260	105

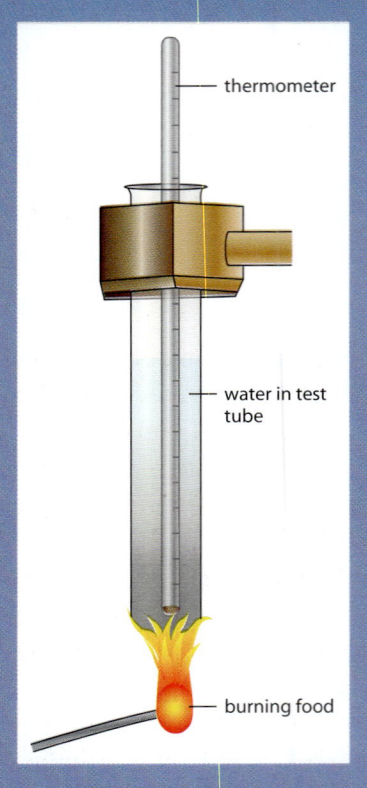

△ Fig. 2.62 Apparatus for burning food.

Using and organising techniques, apparatus and materials

1 a) Look at the diagram and describe what happens during the experiment.

b) Identify any areas of safety that should have been considered and suggest how risks could be controlled.

Observing, measuring and recording

2 a) Use the results to calculate the energy released per gram of each sample.

b) Explain why you need to do this.

3 Which part of the plant released the most energy per gram?

4 Suggest why some animals that eat the leaves of plants for most of the year change to eating seeds when they are available.

Handling experimental observations and data

5 The apparatus shown in Fig. 2.62 does not give accurate results for the amount of energy in the burning material. Explain why, and suggest a method that would increase the accuracy of the results.

Malnutrition

The term **malnutrition** literally means 'bad nutrition'. It applies to any diet that will lead to health problems. A diet that is too high in energy content, and leads to obesity, is one form of malnutrition, because obesity increases the risk of several diseases.

Malnutrition can occur if one or more nutrients is in too high a proportion in the diet. For example, a high proportion of saturated fats in the diet can lead to deposits of cholesterol forming on the inside of arteries, increasing blood pressure and also increasing the risk of coronary heart disease (see Coronary heart disease).

Malnutrition also occurs if there is too low a proportion of any of the substances needed for health. For example, a lack of a vitamin or a mineral can cause deficiency diseases, as shown in Table 2.2. Too little fibre in the diet can lead to **constipation**, where food moves too slowly through the alimentary canal, increasing the risk of diseases such as diverticulitis and bowel cancer.

Starvation occurs when there is too little energy provided by the diet. In this state, the body will start to break down its energy stores. Initially this uses the fat stores but, when those have run out, the body will start to break down muscle tissue to produce substances that can be used in respiration. This can damage the muscle tissue of the heart, and also the immune system, increasing the risk of many diseases.

△ Fig. 2.63 Starvation is most common in places where crops have failed due to drought or people are displaced as a result of war. However, it can also happen in people who choose to starve themselves in crash diets or as a result of conditions such as anorexia.

QUESTIONS

1. Explain why different groups of people need different amounts of nutrients. Give examples in your answer.
2. Explain why a healthy diet needs to consider energy as well as nutrients.
3. Explain why the following are considered as results of *malnutrition*:

    a) obesity

    b) starvation

    c) constipation.

THE HUMAN DIGESTIVE SYSTEM

Eating food involves several different processes:

- **ingestion** – taking food and drink into the body (through the mouth in humans)
- **digestion** – breaking down of large food molecules into smaller water-soluble molecules
- **absorption** of digested food molecules from the intestine into the blood and lymph
- **assimilation** – moving absorbed food molecules into cells where they can be used to produce other molecules or in respiration
- **egestion** – removal of substances that were ingested but not absorbed, such as undigested material or digested material that is not absorbed, which form **faeces**.

All these different processes take place in different parts of the **alimentary canal**.

The alimentary canal is a continuous tube through the body, from the mouth where food is ingested, through the oesophagus, stomach, small intestine and large intestine, to the anus where faeces are egested. This tube is about 8 m long in an adult. You could say that materials in the alimentary canal aren't truly in the body. Not until food molecules are absorbed do they cross cell membranes into body tissue. Then they can be assimilated and waste products *excreted* through other organs.

The digestive system includes the alimentary canal and the other organs that contribute to digestion, such as the liver, pancreas and gall bladder. Table 2.4 describes the functions of each of the organs in the digestive system.

Part of digestive system	What happens there
Mouth	Teeth and tongue break down food into smaller pieces.
Salivary glands	Produce liquid saliva which moistens food so it is easily swallowed and contain the enzyme amylase to begin breakdown of starch.
Oesophagus	Each lump of swallowed and chewed food, called a bolus, is moved from the mouth to the stomach by waves of muscle contraction called peristalsis.
Stomach	Acid and protease enzymes are secreted to start protein digestion. Movements of the muscular wall churn up food into a liquid.
Liver	Cells in the liver make bile. Amino acids not used for making proteins are broken down to form urea which passes to the kidneys for excretion. Excess glucose is removed from the blood and stored as glycogen in liver cells.
Gall bladder	Stores bile from the liver. The bile is passed along the bile duct into the small intestine where it neutralises the stomach acid in the chyme.
Pancreas	Secretes digestive enzymes in an alkaline fluid into the duodenum.
Small intestine (duodenum and ileum)	Secretions from the gall bladder and pancreas enter the first part of the small intestine (**duodenum**) to complete the process of digestion. Digested food molecules and water are absorbed in the **ileum**.
Large intestine (colon)	Water and some minerals are absorbed from the remaining material.
Rectum	The remaining, unabsorbed, material (faeces), plus dead cells from the lining of the alimentary canal and bacteria, is compacted and stored.
Anus	Faeces is egested through a sphincter.

△ Table 2.4 The functions of the human digestive system.

The distance from mouth to anus is about eight metres.

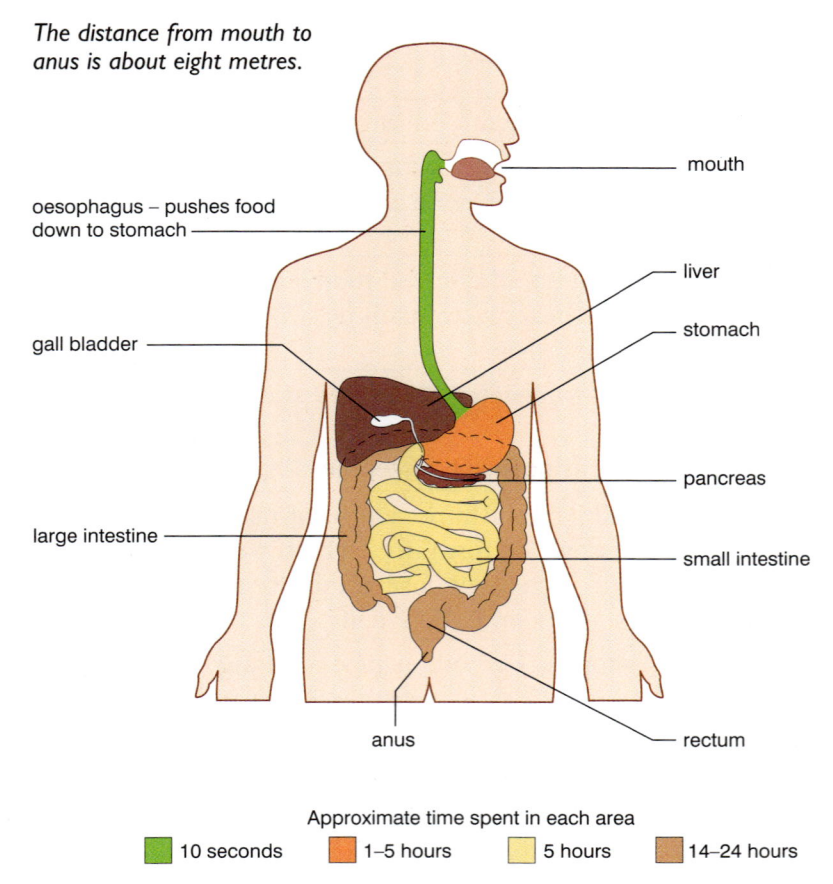

- mouth
- oesophagus – pushes food down to stomach
- liver
- gall bladder
- stomach
- pancreas
- large intestine
- small intestine
- anus
- rectum

Approximate time spent in each area

- 10 seconds
- 1–5 hours
- 5 hours
- 14–24 hours

△ Fig. 2.64 The human digestive system.

Food moves along the alimentary canal because of the contractions of the muscles in the walls of the alimentary canal. This is called **peristalsis**. Fibre in the food keeps the bolus bulky and soft making peristalsis easier.

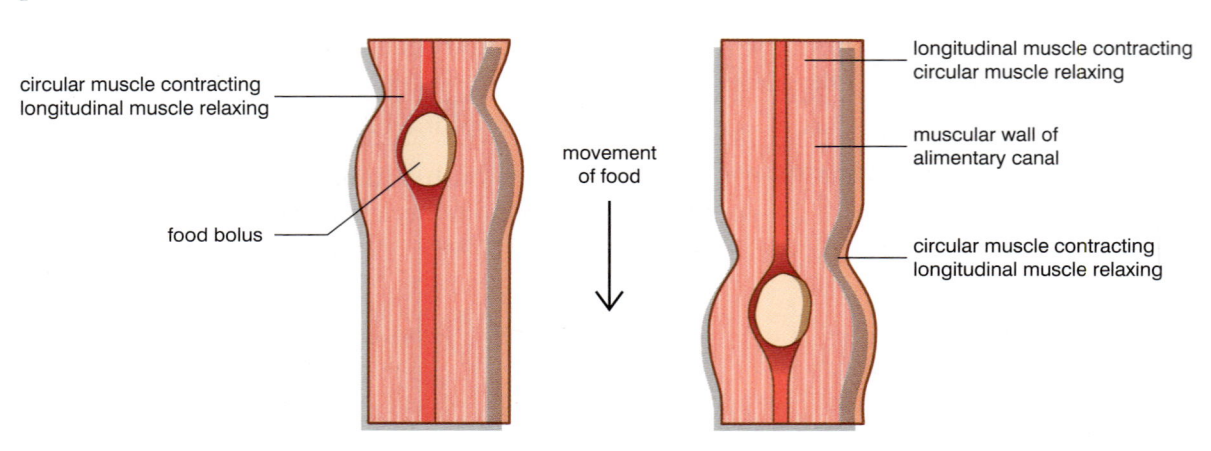

- circular muscle contracting longitudinal muscle relaxing
- food bolus
- movement of food
- longitudinal muscle contracting circular muscle relaxing
- muscular wall of alimentary canal
- circular muscle contracting longitudinal muscle relaxing

△ Fig. 2.65 Peristalsis moves food along the digestive system.

QUESTIONS

1. Sketch the diagram of the digestive system shown in Fig. 2.64. Label the organs, and add notes to each organ to explain its function in the system.
2. Explain the difference between egestion and excretion.
3. Explain how the muscles of the alimentary canal wall move food.

Digestion

In order for food to be useful to us, it must enter the blood so that it can travel to every part of the body.

Many of the foods we eat are made up of large, **insoluble** molecules that cannot cross the wall of the alimentary canal and the cell membranes of cells lining in the blood vessels. This means they have to be broken down into small, **soluble** molecules that can easily cross cell membranes and enter the blood. Breaking down the molecules is called **digestion**.

There are two types of digestion.

1. **Mechanical and physical digestion** occurs mainly in the mouth, where food is broken down physically into smaller pieces by the biting and chewing action of the teeth. It also happens in the small intestine where **bile** helps to emulsify fats, which means break them into small droplets (see Bile).
2. **Chemical digestion** is the breakdown of large food molecules into smaller ones using chemicals such as enzymes.

Some molecules, such as glucose, vitamins, minerals and water, are already small enough to pass through the alimentary canal wall and do not need to be digested.

Mechanical and physical digestion
Human teeth

Teeth cause the mechanical digestion of food in the mouth. Humans are omnivores that eat a varied diet of plant and animal material, so we have a range of tooth type to help us bite off and chew these different materials:

- **incisors** at the front of the mouth are chisel-shaped, for biting off food (particularly good with plant material)
- four pointed **canines** pierce and hold food, particularly meat, so that it can be chewed
- **premolars** help with the cutting off of tough foods, such as meat, and grinding of plant material on a small grinding surface
- **molars** at the back of the mouth have large grinding surfaces for chewing.

In each jaw, we have four incisors, two canines, four premolars and between four and six molars.

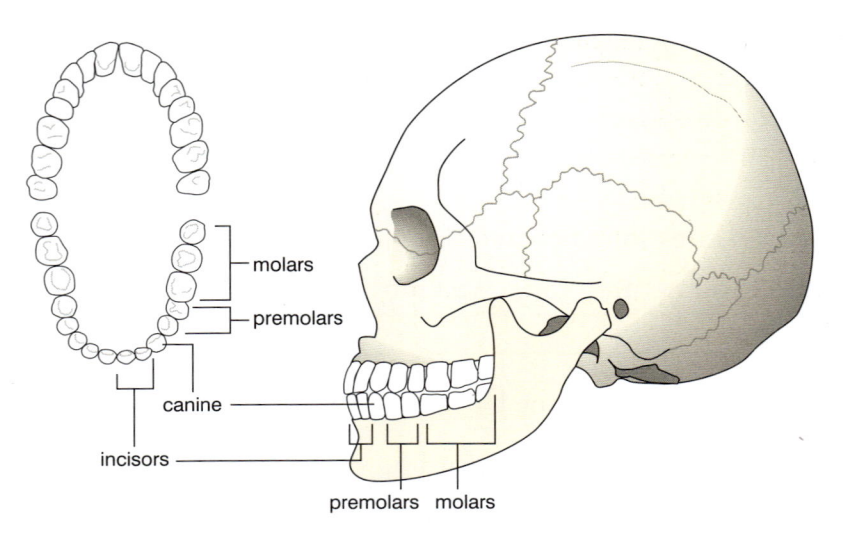

△ Fig. 2.66 Plan and side view of teeth in a human skull.

Like other mammals, humans have two sets of teeth: the 'milk' teeth of childhood are replaced by permanent (adult) teeth from the age of about 6. The molars at the far back of the mouth, sometimes called the 'wisdom teeth' may not grow until adulthood, or even never.

Dental decay and tooth care

Although tooth enamel is very hard, it is vulnerable to attack by acids. Acids are naturally present in fruits and other foods. Bacteria living in the spaces between teeth, in crevices on the tooth surface, and at the edges of the gums also make acids. Food particles get lodged in these crevices and bacteria grow on them, forming plaque. Plaque makes it even easier for bacteria to grow and produce acids right against the surface of the teeth. These acids corrode the tooth enamel, and expose the softer dentine underneath. This can cause pain (toothache) because the nerves in the pulp cavity are affected by acid, heat or cold more easily. The enamel on teeth does not extend far below the gum edge, so when plaque forms there the dentine can quickly be attacked. The links between the tooth and its socket can be weakened and the tooth might fall out.

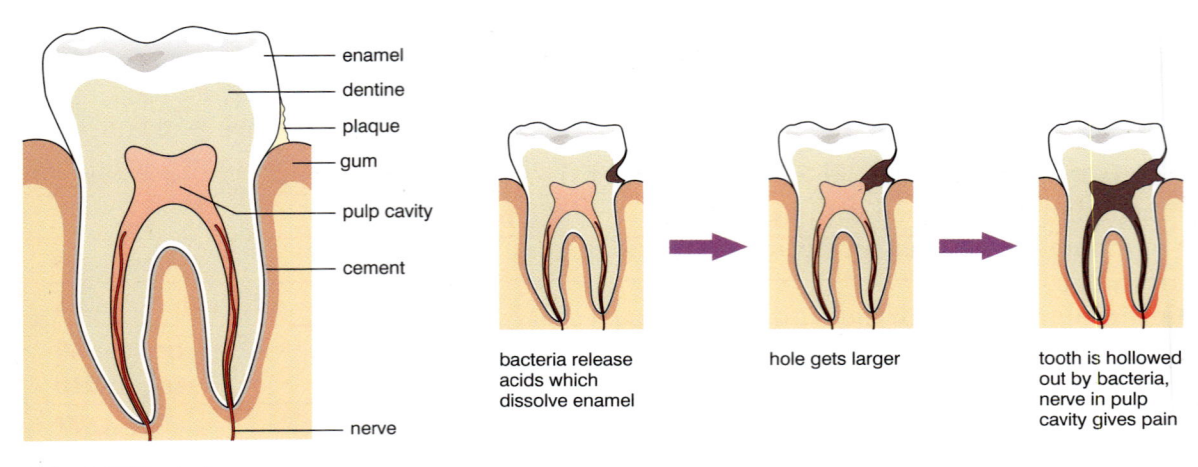

△ Fig. 2.67 Tooth decay.

Brushing your teeth regularly helps to remove build-up of plaque and also any bits of food stuck between the teeth or at the edge of the gum. This reduces the risk of bacteria producing acid that damages the enamel. Using toothpaste helps because:

- it is alkaline and neutralises acids near the teeth
- it contains antibacterial substances such as mint
- it usually contains a mild abrasive which helps to remove plaque
- it may contain **fluoride**, which helps to strengthen enamel and reduce acidic damage.

EXTENDED

Adding fluoride to water supplies

Fluoride is a soluble substance naturally found in rivers and other water sources, where it has dissolved out of the surrounding soil and rock. Depending on the type of rock or soil, water sources in different areas may contain very different amounts of fluoride.

Fluoride has been shown to strengthen tooth enamel and so reduce tooth decay. For this reason, water authorities may add fluoride to public water supplies, particularly if the natural concentration of fluoride is very low. However, in large amounts taken over a long period, fluoride can cause brown mottling of teeth. It has also been suggested that it may cause health problems such as risk of bone fractures in older women, although not all research supports this suggestion.

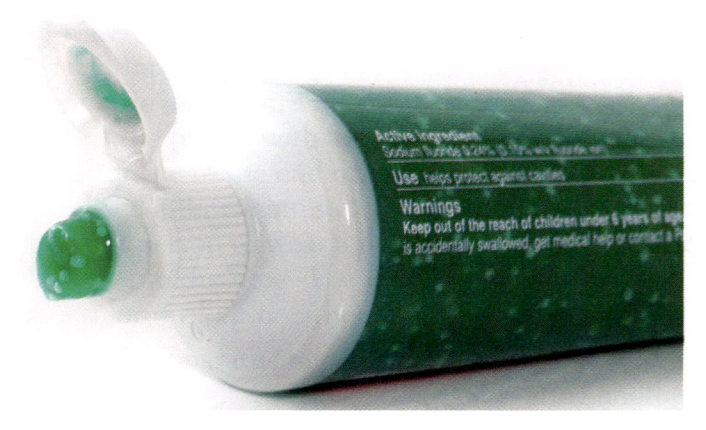

△ Fig. 2.68 Some people think that fluoride should be added to water that naturally has a low concentration, to improve tooth care for everyone. Other people think they should be allowed to choose whether or not they have extra fluoride, such as by using a fluoride toothpaste.

END OF EXTENDED

Bile

Bile is a substance produced by cells in the liver. It is stored in the gall bladder until it is needed and then passes along the bile duct into the small intestine.

Bile is important in the physical digestion of fats. Fats do not mix well with aqueous (water-based) mixtures such as the digesting food, and so remain as large droplets. This produces a small surface area for lipase enzymes to work on, which slows down the rate of digestion. Bile **emulsifies** fats, breaking them up into much smaller droplets, so that the rate of digestion is much faster.

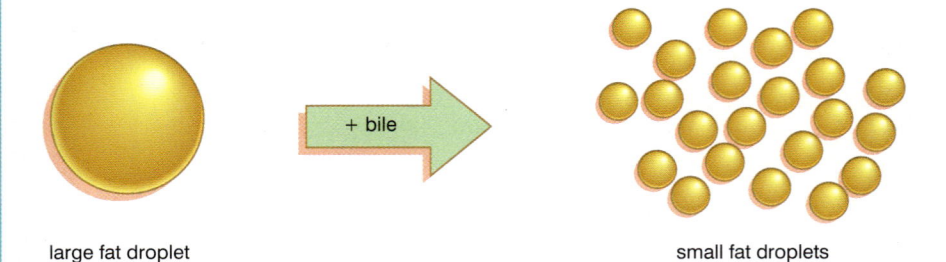

large fat droplet   small fat droplets

△ Fig. 2.69 Bile lowers the surface tension of large droplets of fat so that they break up. This part of the digestive process is called emulsification.

QUESTIONS

1. Explain the difference between chemical and mechanical/physical digestion.
2. Describe the different tooth types in a human mouth and explain how their structure is related to their function.
3. Explain why regular brushing of teeth helps prevent tooth decay.
4. **EXTENDED** Give one advantage and one disadvantage of adding fluoride to public water supplies.

Chemical digestion

Chemical digestion in the alimentary canal is the result of **enzymes**. **Digestive enzymes** are a group of enzymes that are produced in the cells lining parts of the digestive system and are **secreted** (produced) into the alimentary canal to mix with the food.

The digestive enzymes include:

- carbohydrases that break down carbohydrates, one example of which is **amylase**
- **proteases**
- **lipases**.

(Note: the *-ase* at the end means it is an enzyme, and the first part usually names the substrate that the enzyme works on.)

Each of the food groups (carbohydrates, proteins and fats) contains many different molecules. As each enzyme is specific to its substrate, this means that in each group of digestive enzymes there are many different enzymes.

Different enzymes are made in different parts of the digestive system as shown in Table 2.5.

Enzyme	Where produced	Substrate	Final products*
Amylase	Salivary glands (mouth) pancreas	Starch	Glucose
Protease (many types)	Stomach wall pancreas	Proteins	Amino acids
Lipase (many types)	Pancreas	Fats and oils (lipids)	Fatty acids and glycerol

△ Table 2.5 Digestive enzymes.

*These are the soluble substances produced at the end of digestion. The substances in food and drink may go through many stages of digestion by different enzymes as they pass through the alimentary canal.

REMEMBER

Remember that different enzymes work better in different conditions. The enzymes that digest food in the stomach work best in acid conditions. Special cells in the lining of the stomach secrete hydrochloric acid into the stomach to create the right conditions for the enzymes. The acid is also helpful in killing microorganisms taken in with the food.

In the duodenum, the enzymes from the pancreas work best at slightly alkaline conditions. So the acidic food mix that enters the duodenum from the stomach has to be neutralised. Bile is highly alkaline, so when it is added to the contents of the duodenum, it neutralises the acid from the stomach and makes the digesting food slightly alkaline.

QUESTIONS

1. Explain why enzymes are needed in the digestive system.
2. a) Which enzyme has starch as its substrate?
   b) Which products are formed by the digestion of starch by this enzyme?
3. Describe the roles of a) stomach acid and b) bile in digestion.

> **SCIENCE IN CONTEXT**
>
> **LACTOSE**
>
> Lactose is the disaccharide sugar in milk (from *lactis*, meaning milk), which is broken down in the alimentary canal by the enzyme lactase to the simple sugars glucose and galactose.
>
> Human babies, like all young mammals, produce lactase to help them digest the lactose in breast milk. In most mammals, lactose production decreases as the young mature, because the adult diet does not include milk. This also happens in adults from many human cultures where adults generally do not drink milk, such as in Southeast Asia. However, there are human cultures in Europe, India and parts of East Africa, where mammals such as sheep, goats or cattle are kept to supply meat and milk for food. In these human groups the adults continue to produce lactase and are able to digest the lactose in milk. Adults who cannot do this are *lactose intolerant*. Bacteria in the alimentary canal break down the lactose, producing gas which causes great discomfort.

Absorption of food

After food has been digested in the duodenum, small food molecules can diffuse across the wall of the ileum and be absorbed into the body. The intestine is over 6 metres long in an adult human, but to increase the rate of transport of food molecules across the ileum wall, its surface area is increased by millions of finger-like projections called **villi** (singular: villus). The total surface area for absorption in the small intestine of an adult human is about 250 m^2, which is about the size of a tennis court.

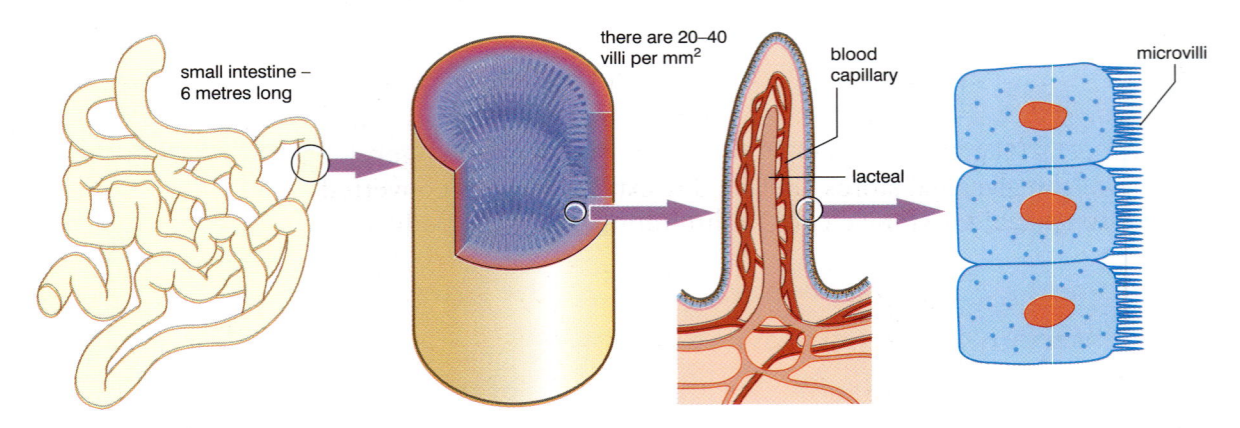

△ Fig. 2.70 The structure of the small intestine (ileum).

EXTENDED

Villi have other adaptations that help to increase the rate of diffusion.
- They are covered in a thin layer of cells, so that digested food molecules do not have to travel far to be absorbed into the body and into the blood in the capillaries in the villi.

- The surface of the cells covering the villi is extended into microscopic finger-like projections called **microvilli**. This greatly increases the surface area of each villus for absorption.
- They are well supplied with blood capillaries, taking absorbed food molecules from the small intestine to the rest of the body and supplying fresh blood. This keeps the concentration gradient between the digested food in the intestine and the cells in the body as high as possible.
- Villi also contain **lacteals** (part of the lymphatic system), which carry fat droplets separate from the rest of the food molecules because fat does not dissolve well in blood.

Note that in addition to simple diffusion, glucose is also actively transported across the cell membranes of the epithelial (lining) cells of the villi. This increases the rate of absorption of glucose into the body.

Food molecules that have been absorbed into the blood are carried directly to the liver by the hepatic portal vein where many types of molecule are assimilated.

About 80% of water in the contents of the ileum is also absorbed into the blood, which includes water in the food and drink plus water in all the secretions that have been added as they have moved from the mouth to the ileum. This amounts to between 5 and 10 dm^3 of water absorbed every day. Another 0.3–0.5 dm^3 of water is absorbed as the remaining gut contents pass through the **colon** (large intestine). The relatively dry faeces that pass into the rectum are egested through the anus.

END OF EXTENDED

ASSIMILATION

Assimilation is the various processes by which digested and absorbed food is used in the body.

Most of the nutrients absorbed from the small intestine into the blood pass straight to the liver. If glucose concentration of the blood is high, the liver cells take up surplus glucose and convert it to **glycogen** for storage. If the glycogen stores are full, any extra glucose is converted into fat in special fat-storage cells in different parts of the body.

Active cells need amino acids to build proteins, but when more are absorbed than are needed, the excess (surplus) amino acids cannot be stored. Some kinds of amino acids can be converted into other kinds that are needed. The rest are broken down in the liver for excretion.

EXTENDED

Deamination and detoxification

Deamination is the removal of the nitrogen-containing part of an amino acid which takes place in liver cells. Other parts of the amino acid are can be used to release energy in respiration. However, the nitrogen-containing part has to be excreted, because its breakdown product is alkaline and would change the pH of body fluids. It is converted into a less harmful form called **urea**. This chemical is carried in the blood to the kidneys and excreted in urine.

The liver also breaks down other harmful substances that are absorbed into the body from the alimentary canal, including alcohol which is toxic to cells in high concentration, and toxins within the food or produced by bacteria in the food which could poison the body.

END OF EXTENDED

QUESTIONS

1. Define absorption and assimilation.
2. Describe how villi maximise the surface area of the small intestine wall.
3. Explain why a large surface area is needed in the small intestine.
4. Describe the role of the liver in the metabolism of glucose and amino acids.
5. **EXTENDED** Describe the role of capillaries and the lacteal in a villus.
6. **EXTENDED** What is the role of the hepatic portal vein in the body?
7. **EXTENDED** Why is deamination of excess amino acids important?
8. **EXTENDED** Where in the alimentary canal does absorption of water take place?

FOOD SUPPLY

Increasing food production

Our food comes mainly from crop plants and animals such as sheep, fish and chickens. Over thousands of years, humans have developed many technologies to increase food production. Modern technologies include:

- use of agricultural machinery, some of which are designed for specific tasks such as sowing seed, spreading fertiliser or harvesting the crop – this makes it possible to carry out these tasks much more quickly and over larger areas
- fertilisers to give plants the nutrients they need for rapid and healthy growth, and so produce a greater harvest (yield) – we now make over 500 million tonnes of fertilisers by chemical processes each year to spread on crop fields
- killing pests. Pests are animals that damage crop plants by eating them, which reduces their growth and yield. Eliminating the pests using chemicals called **pesticides** increases the amount of food produced from a crop
- using **herbicides** (plant-killing chemicals) to kill weeds. Plant growth is also reduced if the plant is competing with many neighbouring plants (weeds) for water and nutrients from the

soil – herbicides that specifically kill the weeds will allow the crop plants to grow better and produce a bigger yield
- **artificial selection** (see Selection). This can be used to produce crop plants and animals with better characteristics that result in more food for us.

Using all these technologies has helped world food production to grow continually, as shown in Fig. 2.71.

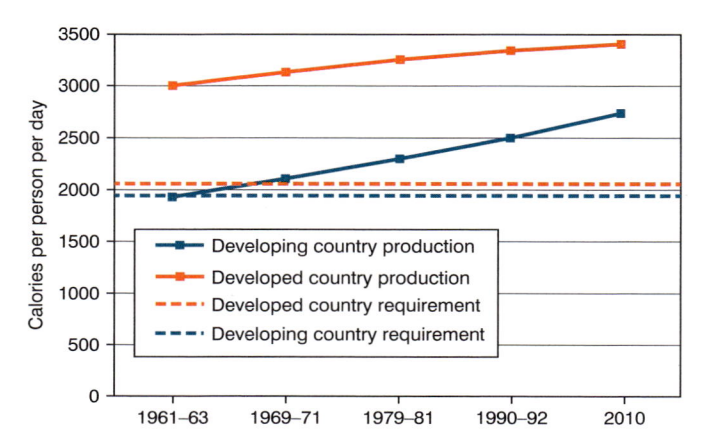

△ Fig. 2.71 World food production from 1961 to 2010.

EXTENDED

Problems with world food supplies

Although food production is increasing, especially as new technologies become more available to farmers in less industrialised countries, not everyone in the world has enough food to eat. According to the FAO (Food and Agriculture Organisation of the United Nations), about 14% of people in the world do not get enough food, and starvation kills around 6 million children every year. Most people who do not have enough food live in very poor countries where there is not enough money to buy the new technologies.

World food production does produce more than enough food for everyone. However, one problem is that the food is not distributed to everyone equally. People in industrialised countries have access to far more food, and eat more every day, than people in poor areas.

Famine (widespread hunger) occurs in areas where crops fail year after year due to drought, as in parts of Eastern Africa. It may also occur after major floods, such as happen in low-lying areas of Pakistan and Bangladesh. One major cause of famine is war, which not only destroys crops and animals, it also displaces large numbers of people to refugee camps when they try to escape from the fighting.

As the human population increases, particularly in areas where there is not enough food, there is an increasing risk of famine.

△ Fig. 2.72 Food distribution during a famine in Congo.

Using microorganisms to produce food

Microorganisms have been used for many centuries to produce food such as bread and cheese.

Yoghurt is made by adding a culture of a bacterium called *Lactobacillus* to warm milk and keeping the culture warm for several hours. The bacteria in the culture break down a protein in the milk, which makes the yoghurt easier to digest than milk. The bacteria also produce lactic acid, which gives yoghurt its characteristic sour taste.

A more recent use of microorganisms in food production has been to make **single-cell protein** (SCP), also called mycoprotein. This is made using a fungus that is mixed with carbohydrate and kept in warm conditions so that it grows rapidly. The fungus is separated and dried. SCP has a very high level of protein and its production is very efficient. Compared with common sources of protein such as meat from cattle, SCP uses far less space and is much quicker to produce. The process can also be used to produce more protein from a food that is usually thrown away: for example, SCP can be added to the whey in cheese production to produce a meat-free protein.

QUESTIONS

1. Describe four ways in which modern technology has increased food supply.
2. **EXTENDED** Give reasons why people in some areas of the world suffer from famine.
3. **EXTENDED** Explain how microorganisms are used to produce a) yoghurt, b) single-cell protein.

FOOD ADDITIVES

Additives are substances added to foods in small quantities for a variety of reasons. They may be used to enhance the flavour, improve the texture and appearance, or extend the shelf life of food. Often they do not have any nutritional value, and many people think that additives are sometimes used when there is no real need for them: for example, food colouring. However, many additives have a useful role, preventing food from spoiling so that it can be stored safely for longer. Not all additives are 'modern'. Sugar, salt and vinegar have been used as preservatives for centuries.

Function	Examples	Advantages	Disadvantages
Flavourings	Monosodium glutamate (MSG) (an amino acid)	Improves taste	High doses can cause sweating and hyperactivity
	Salt	Improves taste	Contributes to high blood pressure
	Chocolate	Improves taste	Contains caffeine, which can be addictive
	Sugar	Improves taste	Expensive and fattening
Additives to improve nutritional quality	Lysine is added to flours in places where diet is poor. Vitamin D is added to white flours.	Enriches poor diets	
Colourings	Annatto	Improves appearance by replacing colour lost in processing	Some people are allergic to annatto
Sweeteners	Nutrasweet	Food can taste sweet without being so fattening	Unsuitable for some people. Gives a bitter aftertaste
Preservatives	Vinegar, salt, sugar	Reduces bacterial spoilage, in ready-made meals, pickles and jams for example	Vitamin content of food declines with storage
Anti-oxidants	Sodium ascorbate	Stops fat spoiling and becoming rancid in pre-cooked meats	
Stabilisers	Carragheen	Stops food texture separating	
Thickeners	Guar gum	Gives a thicker consistency and better sensation in the mouth	

△ Table 2.6 Some common food additives.

Some modern additives are produced chemically, such as the yellow colouring tartrazine (E102) and orange colouring sunset yellow (E110). During the late 20th century, scientific evidence was collected which showed that these and other artificial colourings contributed to hyperactivity in young children. As the bright colourings were used frequently in sweets, this increased the chance that children would be affected. Since 2009, these colourings have been banned from foods in the UK.

All additives must now be tested for safety and licensed before they can be used in foods. An 'E number' (such as E102) indicates that a substance is an additive and has been licensed.

QUESTIONS

1. **EXTENDED** Give one advantage and one disadvantage of using colourings as food additives.

END OF EXTENDED

End of topic checklist

Key terms

absorption, alimentary canal, amino acid, amylase, artificial selection, assimilation, balanced diet, Benedict's reagent, bile, biuret test, canine tooth, carbohydrate, catalyst, cement, chemical digestion, chlorophyll, chloroplast, colon, constipation, cuticle, deamination, dentine, digestion, digestive enzyme, egestion, emulsify, enamel, enzyme, epidermis, eutrophication, faeces, famine, fatty acid, fibre, fluoride, food additive, glycerol, glycogen, herbicide, incisor, ingestion, insoluble, limiting factor, lipase, lipid, malnutrition, mechanical digestion, microvilli, mineral (mineral ion), molar, nutrition, osmosis, palisade cell, peristalsis, pesticide, phloem , photosynthesis, physical digestion, premolar, protease, protein, pulp cavity, secretion, simple sugar, single cell protein, soluble, spongy mesophyll cell, starch, starvation, stomata, sucrose, urea, vascular bundle, villus, vitamin, waste product, xylem, yield, yoghurt

During your study of this topic you should have learned:

○ How to define *nutrition* as the taking in of nutrients for growth and tissue repair.

○ About the elements in carbohydrates, fats and proteins.

○ How to describe the synthesis of large molecules from smaller basic units.

○ How to describe tests for starch, reducing sugars, protein, fats.

○ How to list the main sources of, and describe the importance of, carbohydrates, fats, proteins, vitamins C and D, calcium and iron, fibre, water.

○ How to describe the deficiency symptoms for vitamins C and D and the mineral salts calcium and iron.

○ **EXTENDED** About the use of microorganisms in the food industry.

○ **EXTENDED** About the uses, benefits and health hazards associated with food additives.

○ The definition of *photosynthesis* as the fundamental process by which plants manufacture carbohydrates from raw materials using energy from light.

○ The word equation for the production of simple sugars and oxygen.

○ How to investigate the necessity for chlorophyll, light and carbon dioxide for photosynthesis, using appropriate controls.

○ How to describe the intake of carbon dioxide and water by plants.

End of topic checklist continued

- How to explain that chlorophyll traps light energy and converts it into chemical energy for the formation of carbohydrates and their subsequent storage.
- **EXTENDED** The balanced equation for photosynthesis in symbols.
- **EXTENDED** How to investigate and state the effect of varying light intensity, carbon dioxide concentration and temperature on the rate of photosynthesis.
- **EXTENDED** The definition of the term *limiting factor* as something present in the environment in such short supply that it restricts life processes.
- **EXTENDED** How to explain the concept of limiting factors in photosynthesis.
- **EXTENDED** How to explain the use of carbon dioxide enrichment, optimum light and optimum temperatures in glasshouse systems.
- How to identify the structures in a leaf and describe their roles in photosynthesis, gas exchange, transport and support.
- How to describe the importance of nitrate and magnesium ions in plants.
- How to describe the uses, and the dangers of overuse, of nitrogen fertilisers.
- **EXTENDED** How to explain the effects of nitrate ion and magnesium ion deficiency on plant growth.
- What is meant by the term *balanced diet* and describe a balanced diet related to age, sex and activity of an individual.
- How to describe the effects of malnutrition in relation to starvation, coronary heart disease, constipation and obesity.
- About ways in which food production has increased due to modern technology.
- **EXTENDED** About the problems of world food supplies.
- **EXTENDED** About the problems that contribute to famine.
- How to define the terms *ingestion* and *egestion*.
- How to identify the main regions of human alimentary canal and associated organs.
- How to describe the functions of the regions of the alimentary canal, in relation to *ingestion*, *digestion*, *absorption*, *assimilation* and *egestion*.
- How to define *digestion* as the breakdown of large insoluble food molecules into small, water-soluble molecules using mechanical and chemical processes.

- How to identify the types of human teeth and describe their structure and functions.
- About the causes of dental decay and describe the proper care of teeth.
- **EXTENDED** How to describe how fluoride reduces tooth decay and explain arguments for and against adding fluoride to public water supplies.
- How to describe the process of chewing.
- How to describe the role of longitudinal and circular muscles in peristalsis.
- An outline of the role of enzymes and bile in digestion.
- About the significance of chemical digestion in the alimentary canal in producing small, soluble molecules that can be absorbed.
- Where, in the alimentary canal, amylase, protease and lipase enzymes are secreted.
- About the functions of a typical amylase, a protease and lipase, listing the substrate and end-products.
- How to define *absorption* as movement of digested food molecules through the wall of the intestine into the blood or lymph.
- How to identify the small intestine as the region for the absorption of digested food.
- How to describe the significance of villi in increasing the internal surface area of the small intestine.
- **EXTENDED** How to describe the structure of a villus, including the role of capillaries and lacteals.
- **EXTENDED** About the role of the hepatic portal vein in the transport of absorbed food to the liver.
- **EXTENDED** How to identify the role of the small intestine and colon in absorption of water.
- How to define *assimilation* as movement of digested food molecules into the cells of the body where they are used, becoming part of the cells.
- How to describe the role of the liver in the metabolism of glucose.

End of topic checklist continued

○ How to describe the role of fat as an energy storage substance.

○ **EXTENDED** How to define *deamination* as removal of the nitrogen-containing part of amino acids to form urea, followed by release of energy from the remainder of the amino acid.

○ **EXTENDED** That the liver is the site of breakdown of alcohol and other toxins.

End of topic questions

Note: The marks awarded for these questions indicate the level of detail required in the answers. In the examination, the number of marks awarded to questions like these may be different.

1. **a)** Explain why carbon, hydrogen and oxygen are the most common elements found in a human body. **(1 mark)**

    b) Why does the body need other elements, in addition to those in a)? **(1 mark)**

2. A sample of bread was ground up. Some of the breadcrumbs were tested with Benedict's reagent and some with iodine solution. The rest of the crumbs were mixed with Substance A. After 20 minutes, some of the mixture was tested with Benedict's reagent and some with iodine solution. The results of the tests are shown in the table.

	Test with Benedict's solution	Test with iodine solution
Before adding Substance A	No precipitate	Change to blue-black colour
After 20 mins with Substance A	Orange-red precipitate	No colour change

    a) Describe what the results show. **(2 marks)**

    b) What was Substance A? Explain your answer. **(4 marks)**

3. Using what have you learned about the effect of concentration gradient and surface area to volume ratio, explain the adaptations of a leaf for photosynthesis. **(4 marks)**

4. **EXTENDED** Sketch the axes of a graph with time of day along the *x*-axis and rate of photosynthesis on the *y*-axis. The units on the *x*-axis should start at midnight on one day and end at midnight on the following day. Add an arrowhead at the top of the *y*-axis to show that the units are arbitrary (they have no values) but increase as you go up the axis.

    a) Draw a line on your axes to show how the rate of photosynthesis might change during the day for a large tree. **(1 mark)**

    b) Annotate your graph to explain which factor (or factors) may be limiting the rate of photosynthesis at different times of day. **(3 marks)**

5. Describe the importance of the following in a healthy diet:

    a) vitamins C and D **(2 marks)**

    b) the minerals calcium and iron **(2 marks)**

    c) water **(1 mark)**

    d) dietary fibre **(2 marks)**

End of topic questions continued

6. There is an old saying that you should chew your food 100 times before swallowing to help look after your stomach. Explain why chewing food well helps digestion. **(3 marks)**

7. Identify the organs of the digestive system involved, and their role, in each of the following processes:

   a) ingestion (2 marks)

   b) digestion (5 marks)

   c) absorption (2 marks)

   d) assimilation (2 marks)

   e) egestion (2 marks)

8. In a greenhouse a grower is growing tomato plants. Explain as fully as you can why she might do the following:

   a) leave lights on in the greenhouse all night (2 marks)

   b) close the greenhouse windows at night but open them during the day (4 marks)

   c) add a liquid feed containing nitrogen and magnesium ions to the water for the plants. (2 marks)

9. This is the diet schedule for a male Olympic athlete training for a competition, not including drinks during training.

Breakfast	large bowl of cereal, such as porridge or muesli
	half pint semi-skimmed milk plus chopped banana
	1–2 thick slices wholegrain bread with olive oil or sunflower spread and honey or jam
	glass of fruit juice + 1 litre fruit squash
Post-training 2nd breakfast	portion of scrambled eggs
	portion of baked beans
	1–2 rashers grilled lean bacon
	portion of grilled mushrooms or tomatoes
	2 thick slices wholegrain bread with olive oil spread
	1 litre fruit squash
Lunch	pasta with bolognese or chicken and mushroom sauce
	mixed side salad
	fruit
	1 litre fruit squash

Post-training snack	4 slices toast with olive oil or sunflower spread and jam
	large glass of semi-skimmed milk
	fruit
	500ml water
Dinner	grilled lean meat or fish
	6-7 boiled new potatoes, large sweet potato or boiled rice
	large portion of vegetables, such as broccoli, carrots, corn or peas
	1 bagel
	1 low-fat yoghurt and 1 banana or other fruit
	750ml water and squash
Bedtime snack	low-fat hot chocolate with 1 cereal bar

a) Identify the foods that contribute to each of these food types:

   i) carbohydrates (5 marks)

   ii) proteins (3 marks)

   iii) lipids (2 marks)

   iv) vitamins and minerals (2 marks)

   v) dietary fibre (1 mark)

b) Which food type is most represented in this diet? (1 mark)

c) Explain why this food type is so important in this diet. (1 mark)

d) Which food group would you expect to be more represented in an athlete's diet in the early stages of training? Explain your answer. (2 marks)

e) Explain why this diet is not suitable for everyone. (2 marks)

10. Using the graph of world food production (Fig. 2.71), describe and suggest explanations for the changes in food production in developing countries since 1961. (2 marks)

11. **EXTENDED** Explain as fully as you can why, although there is enough food for every human on earth, some people are still starving. (4 marks)

12. **EXTENDED** Should additives be used in food? Give examples to support your argument. (3 marks)

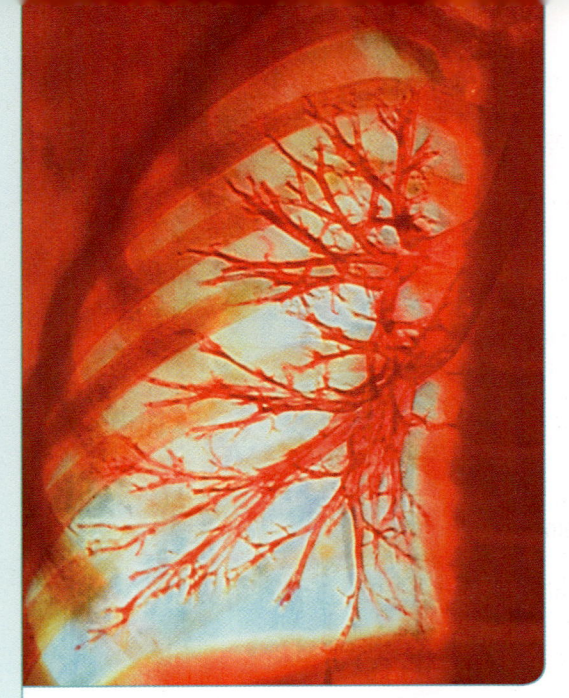

△ Fig. 2.73 Network of arteries in a lung. The arteries branch from the pulmonary artery, which supplies blood to the lungs.

Transportation

INTRODUCTION

Almost every cell in your body is less than 20 μm (0.02 mm) from a blood vessel. This is because every cell needs a constant supply of oxygen and glucose for respiration, which are supplied by the blood. So it's not surprising that, no matter where you cut yourself, you will bleed. Many of the blood vessels in the tissues are extremely narrow – about 5 to 10 μm wide, which is about the width of one red blood cell. It has been calculated that if you placed the blood vessels of one adult in a line, it would wrap four times around the Earth's equator.

KNOWLEDGE CHECK

- ✓ Know that cells in a plant leaf make glucose by photosynthesis, which is converted to sucrose and transported to other parts of the plant in phloem cells.
- ✓ Know that xylem vessels transport water and mineral ions from the roots of a plant through the stem to the leaves.
- ✓ Know that the heart and blood vessels form the human circulatory system.
- ✓ Know that for respiration, cells need a continuous supply of oxygen and glucose, supplied by the blood in a human body.

LEARNING OBJECTIVES

- ✓ Be able to state the functions of xylem and phloem and identify their positions as seen in transverse sections of roots, stems and leaves.
- ✓ Be able to identify root hair cells as seen under the light microscope and state their functions.
- ✓ Be able to state the pathway taken by water through root, stem and leaf.
- ✓ Be able to investigate, using a suitable stain, the pathway of water through the above-ground parts of a plant.
- ✓ Be able to define *transpiration* as evaporation of water at the surfaces of the mesophyll cells followed by loss of water vapour from plant leaves, through the stomata.
- ✓ Be able to describe how water vapour loss is related to cell surfaces, air spaces and stomata.
- ✓ Be able to describe the effects of variation of temperature, humidity and light intensity on transpiration rate.
- ✓ Be able to describe how wilting occurs.
- ✓ Be able to define translocation in terms of the movement of sucrose and amino acids in phloem.

- ✓ Be able to describe the circulatory system as a system of tubes with a pump and valves to ensure one-way flow of blood.
- ✓ Be able to describe the double circulation and relate the differences in the two circuits to their functions.
- ✓ Be able to describe the structure of the heart, including muscular wall and septum, chambers, valves and blood vessels.
- ✓ Be able to describe the function of the heart in terms of muscular contraction and the working of the valves.
- ✓ Be able to investigate, state and explain the effect of physical activity on pulse rate.
- ✓ Be able to describe coronary heart disease in terms of blockage of coronary arteries and state the possible causes and preventive measures.
- ✓ Be able to name the main blood vessels to and from the heart, lungs, liver and kidney.
- ✓ Be able to describe the structure and functions of arteries, veins and capillaries.
- ✓ Be able to identify red and white blood cells.
- ✓ Be able to list the components of blood as red blood cells, white blood cells, platelets and plasma, and state their functions.
- ✓ **EXTENDED** Be able to relate the structure and functions of root hairs to their surface area and to water and ion uptake.
- ✓ **EXTENDED** Be able to explain the mechanism of water uptake and movement in terms of transpiration.
- ✓ **EXTENDED** Be able to discuss the adaptations of the leaf, stem and root to three contrasting environments.
- ✓ **EXTENDED** Be able to describe translocation throughout the plant of applied chemicals, including systemic pesticides.
- ✓ **EXTENDED** Be able to compare the role of transpiration and translocation in the transport of materials from sources to sinks, within plants at different seasons.
- ✓ **EXTENDED** Be able to explain how structure and function are related in arteries, veins and capillaries.
- ✓ **EXTENDED** Be able to describe the transfer of materials between capillaries and tissue fluid.
- ✓ **EXTENDED** Be able to describe the immune system in terms of antibody production, tissue rejection and phagocytes.
- ✓ **EXTENDED** Be able to describe the function of the lymphatic system in circulation of body fluids, and the production of lymphocytes.
- ✓ **EXTENDED** Be able to describe the process of clotting.

TRANSPORT IN FLOWERING PLANTS

Transport tissues in plants

In plants, water and dissolved substances are transported throughout the plant in a series of tubes or vessels. In plants, there are two types of transport vessel, called **xylem** and **phloem**.

- Xylem tissue contains long, hollow xylem cells that form long tubes through the plant. The tubes are the hollow remains of dead cells. The thick strong cell walls help to support the plant. Xylem tubes are important for carrying water and dissolved mineral ions, which have entered the plant through the roots, to all the parts of the plant that need them. They are particularly important for supplying the water that the leaf cells need for photosynthesis.

- Phloem cells are living cells that are linked together to form continuous phloem tissue. Dissolved food materials, particularly sucrose and amino acids that have been formed in the leaf, are transported all over the plant from the leaves. For example, sucrose will be carried to any cell that needs glucose for respiration. Sucrose is less reactive than glucose and therefore is easier to transport without causing problems for other cells. Sucrose may also be carried to parts of the plants where it will be stored, often as another carbohydrate, such as starch, which is stored in seeds and root tubers. This transport of sucrose and other materials is called **translocation**.

In roots the xylem and phloem vessels are usually grouped together separately, but in the stem and leaves they are found together as **vascular bundles** or **veins**.

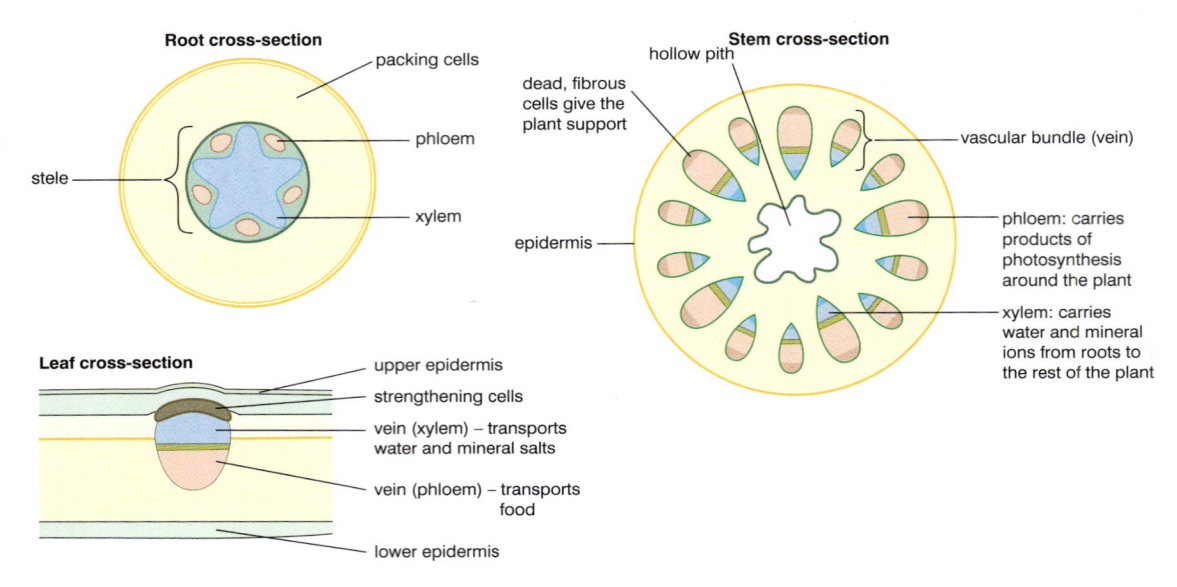

△ Fig. 2.74 The positions of xylem and phloem tissue in a root, stem and leaf.

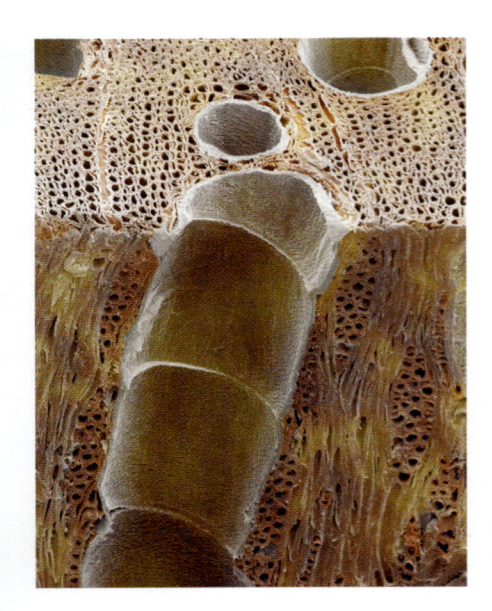

◁ Fig. 2.75 Xylem vessels are long, thick-walled tubes that run through the veins of a plant.

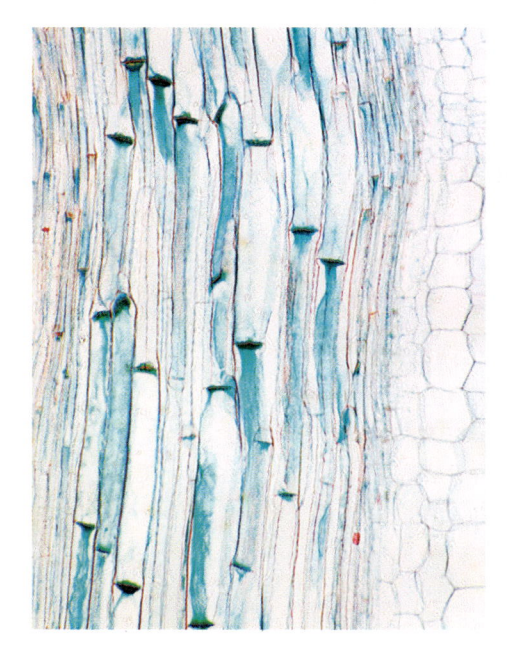

◁ Fig. 2.76 Phloem cells in a longitudinal section of a stem. Some cells still show the green-stained cytoplasm of the phloem cells.

REMEMBER

Remember that substances enter and leave xylem and phloem tissue by diffusion. For the highest marks, be prepared to describe the concentration gradient for each substance and try to explain how the gradient is set up, both into and out of the transport tissues, and is maintained as a result of photosynthesis and translocation.

SCIENCE IN CONTEXT

GROWTH RINGS IN TREES

The wood of a tree is mostly xylem tissue. Every year, new xylem cells are produced from a ring of cells just inside the bark of the tree. When the tree is growing rapidly, the new xylem cells are large. In temperate regions, such as the UK, the rate of growth and the size of new cells decrease as autumn approaches, and stop during winter. The difference in size of cells produced over one year gives the tree its 'rings' and makes it possible to estimate the age of the tree.

△ Fig. 2.77 Growth rings occur in temperate climates when new xylem cells alternately grow (in spring and summer) and stop growing (in winter).

QUESTIONS

1. Where would you find xylem and phloem tissue in a plant?
2. Describe the structure and function of xylem tissue.
3. Describe the structure and function of phloem tissue.

Water uptake in a plant

Plants absorb water and dissolved mineral ions from the soil through **root hair cells**. Root hair cells are found in a short region just behind the growing tip of every root. They are very delicate, and easily damaged. As the root grows, the hairs of the cells are lost, and new root hair cells are produced near the tip of the root.

◁ Fig. 2.78 The root of this germinating seed has many fine root hair cells that greatly increase its surface area.

Water enters the root hair cells, then passes across the root from cortical cell to cortical cell by osmosis. It then enters the xylem tissue in the root and can move from there to all other parts of the plant, including the leaves.

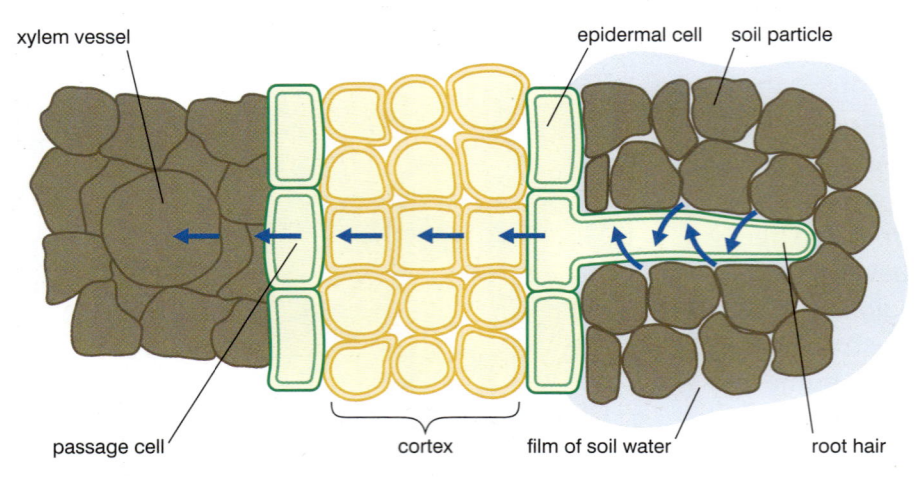

△ Fig. 2.79 The passage of water across a root.

In the leaves, water moves out of the xylem cells in the vascular bundle, into the cells of the spongy mesophyll by osmosis.

EXTENDED

Root hair cells are specially adapted for absorption of substances. They have a fine extension that sticks out into the soil and greatly increases their surface area for absorption. Water enters the root hair cell by **osmosis**, because the concentration of water molecules is higher outside the root, in the soil water, than inside the cytoplasm of the root hair cells. This means that the **water potential** outside the root is higher than inside, so water molecules move down their **water potential gradient**.

Soil water contains minute amounts of dissolved mineral ions. So dissolved mineral ions are usually in higher concentration inside root cells than in the soil water. This means that essential mineral ions cannot usually enter the root by **diffusion**, because that would be against their concentration gradient. Instead, the cell membranes of root hair cells are adapted to take in mineral ions such as nitrates and magnesium ions by **active transport**.

END OF EXTENDED

Investigating water movement through a plant

The movement of water through the above-ground parts of a plant can be investigated by adding food colouring to the water added to the plant. Food colouring is soluble and is carried through the plant with the water in the xylem. After a day or two in coloured water, the veins of the leaves and flowers of a plant will show the colour.

△ Fig. 2.80 A section across a celery stalk that has been standing in coloured water for a day will show colour mainly within the veins (vascular bundles) of the stalk.

△ Fig. 2.81 A carnation that has been standing in coloured water shows the colour vividly in the petals.

QUESTIONS

1. Describe the route that water takes as it moves through a plant.
2. How could you investigate the above-ground route that water takes as it moves through a plant? Explain your answer.
3. Which process is used in a root to absorb the following from soil water?
   a) water
   b) mineral ions
4. **EXTENDED** Copy the diagram of water movement across a root and annotate it to explain how water potential controls the movement of water in a root.

Losing water: transpiration

Water is a small molecule that easily crosses cell membranes. Inside the leaf, water molecules cross the cell membranes and walls of the spongy mesophyll cells into the air spaces. This process is called **evaporation** because the liquid water in the cells becomes water vapour in the air spaces. Whenever the **stomata** in a leaf are open, water molecules diffuse from the air spaces out into the air (where there are usually fewer water molecules). So, in addition to using water in the process of photosynthesis, plants lose water by evaporation from the leaf. This loss of water from the leaves is called **transpiration**.

If the rate of transpiration of water from the leaves of a plant is greater than the rate of water absorption by root hair cells, the plant will start to **wilt** (see Fig. 2.23). This is when all the cells of the plant are not full of water, so the strength of the cell walls cannot support the plant and it starts to collapse.

Factors that affect the rate of transpiration

The rate of transpiration from a leaf will be affected by anything that changes the concentration gradient of water molecules between the leaf and the air. The steeper the concentration gradient, the faster will be the rate of transpiration. Several factors can affect the rate of transpiration.

- **Temperature** – Increased temperature gives particles more heat energy, which results in faster movement of the particles. The faster particles move, the easier it is for them to evaporate from cell surfaces into the air spaces, diffuse out of the leaf and move away. So increased temperature increases the rate of transpiration.

- **Humidity** – This is a measure of the concentration of water vapour in the air. When the air is very humid, it feels damp because there is a high concentration of water vapour in the air. When the air feels dry, the humidity is low. The concentration of water molecules inside the air spaces in the leaf is high. The higher the humidity of the air, the lower the concentration gradient between the air outside and inside the leaf.
- **Light intensity** – The higher the light intensity, the more photosynthesis is taking place in the palisade cells. So the stomata are usually opened at their widest in order to exchange carbon dioxide and oxygen as quickly as possible with the cells inside the leaf. Open stomata also make it possible for water molecules to diffuse out of the air spaces into the air more quickly. So a higher light intensity increases the rate of transpiration, up to a maximum when the stomata are fully open.

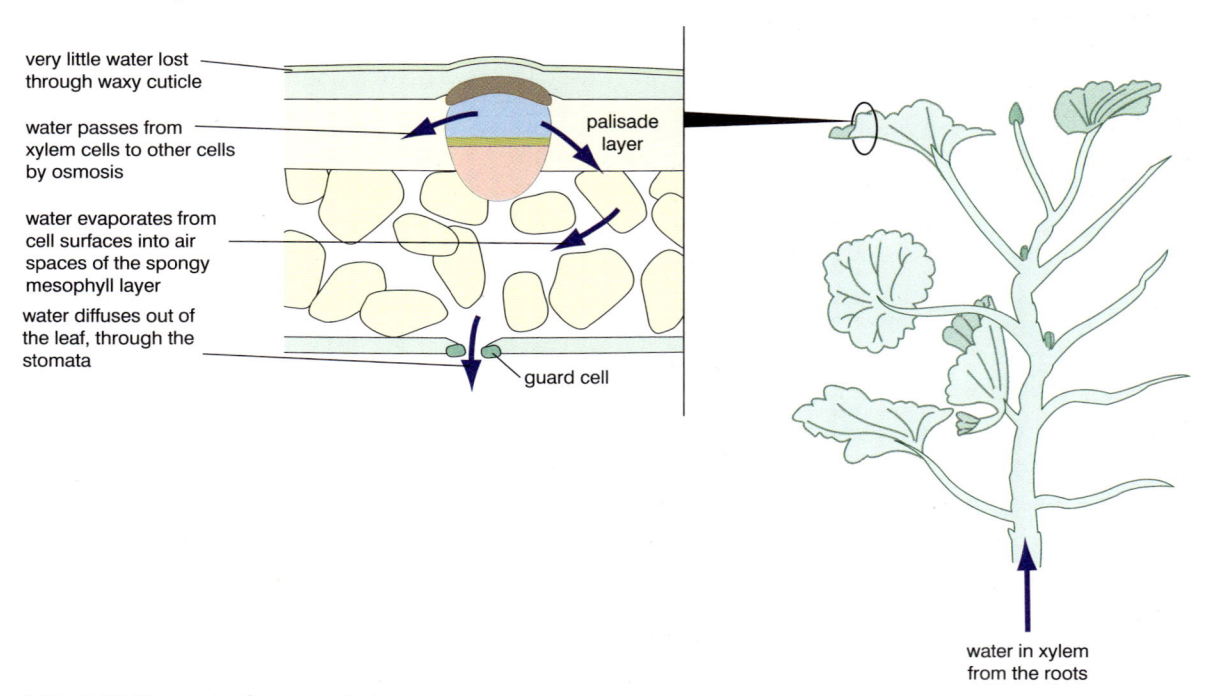

△ Fig. 2.82 How water leaves a plant.

Developing investigative skills

The diagram shows apparatus called a potometer that can be used to investigate the effect of a range of factors on the rate of transpiration. As water evaporates from the leaf surface, the bubble of air in the potometer moves nearer to the leafy twig.

Using and organising techniques, apparatus and materials

❶ Suggest how you could use a potometer to measure the effect of the following factors on transpiration: a) temperature, b) light intensity, c) wind speed, d) humidity.

Observing, measuring and recording

The table below shows the results of an investigation, using a potometer in five different sets of conditions.

Conditions	Time for water bubble to move 5cm/seconds
Still air, sunlight	135
Moving air, sunlight	75
Still air, dark cupboard	257
Moving air, dark cupboard	122
Hot, moving air, sunlight	54

❷ For each of the following factors, identify which data in the table should be compared to show the effect of the factor, and explain why those are the right data to compare.

  a) temperature
  b) light intensity
  c) wind speed

❸ Using the data you have identified, draw a conclusion about the effect of the following on the rate of transpiration.

  a) temperature
  b) light intensity
  c) wind speed

Handling experimental observations and data

❹ Explain why the time taken for the bubble to move 5 cm is a measure of transpiration.

❺ What else could make the bubble move?

❻ Explain how could you improve the reliability of the conclusions you drew in Q3.

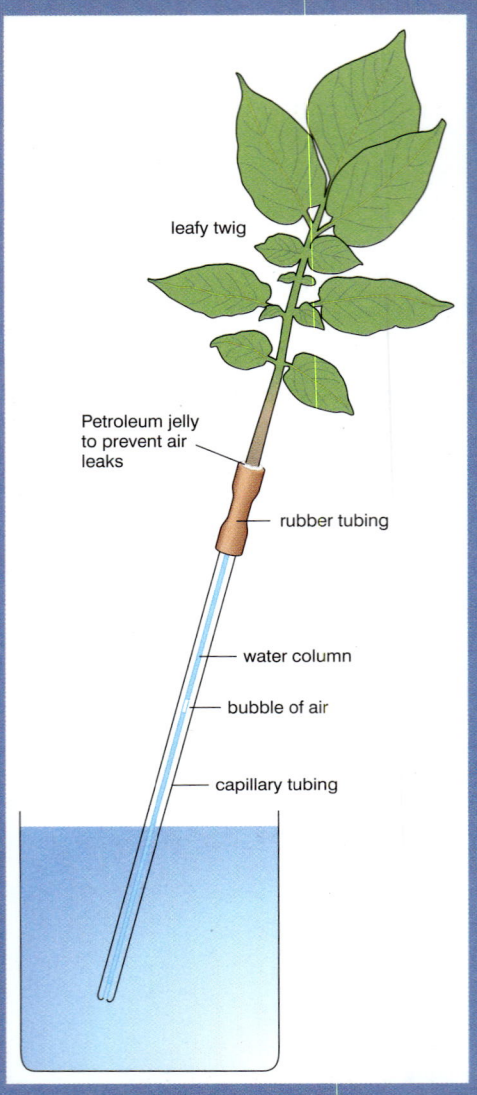

△ Fig. 2.83 Apparatus needed for the investigation into the rate of transpiration.

EXTENDED

Transpiration and water potential

The loss of water from spongy mesophyll cells to the air spaces in a leaf increases the water potential of the cell cytoplasm. So water molecules will move from surrounding cells into these mesophyll cells by osmosis. That creates a water potential gradient between those cells and the ones further into the leaf, so water molecules move into them by osmosis, and so on, all the way back to the xylem in the vascular bundles. Water molecules will move out of the xylem into surrounding cells by osmosis.

Water molecules are **cohesive**, meaning that they have a tendency to stick together. So, as water molecules move out of the xylem cells in a leaf vascular bundle, they create a **tension** (or 'pull') at the top of a xylem tube that connects all the way through the plant to a root (rather like when you suck on a straw). This tension pulls water molecules up the xylem into the leaf, which pulls water molecules out of the root cortical cells into xylem tubes in a root. The water potential gradient is continued through the root cortical cells to the root hair cells and into the soil water.

END OF EXTENDED

QUESTIONS

1. Define the word *transpiration*.
2. Copy the diagram of the plant and leaf section and add your own annotations to explain how water moves through the plant. Include the following words in your labels: evaporation, osmosis, diffusion, transpiration.
3. Explain the advantage to plants of closing their stomata at night in terms of water loss.
4. Explain, in terms of the movement of water molecules, why transpiration rate is faster when:

    a) the temperature is higher

    b) the humidity of the air is lower.
5. **EXTENDED** Explain the relationship between cohesion between water molecules and a continual water potential gradient between leaf and root cells.

EXTENDED

Plant adaptations to water loss

Plants live in many different conditions, where the supply of water may vary greatly. To survive well, plants need to be well adapted to the conditions in which they live.

Garden plants

Many of the plants found in our gardens are adapted to conditions in which there is rarely too much or too little water. The descriptions of plants in the text above apply to many garden plants, because they can usually get sufficient water and nutrients for healthy growth.

Pond plants

Plants that live in water have no problem getting sufficient water for photosynthesis and transpiration. So their stomata may remain open all the time. Floating plants usually have their stomata on the upper surface, where they can exchange gases more easily with the air.

Pond plants may have very small roots because they can extract nutrients from the surrounding water through other tissues also. The water will also support the plant, so roots may do little more than help keep a plant in place. Many pond plants have large air spaces in their leaves, to help keep their leaves near the surface of the water where there is more light for photosynthesis.

Desert plants

In deserts, water is usually very limited and light is very intense during the day. The air temperature may range from beyond 40 °C during the day to below zero at night. It is not surprising therefore, that many plants cannot survive in these conditions. Those that are well adapted may have some or all of these adaptations:

- large root systems that penetrate a large volume of soil and may be very deep to reach deep soil water, to capture as much water from the soil as possible
- reduced or no leaves, to reduce the rate of transpiration (leaves may be reduced to spines to deter herbivores)
- green stems for photosynthesis to replace leaves
- stomata sunk deep into pits in the leaf or stem surface, to reduce the rate of transpiration
- thickened leaves or stems (**succulent**) that contain cells that store water
- hairy surfaces to reduce air flow across stomata, reducing transpiration during the day, and acting as insulation at night.

△ Fig. 2.84 Desert plants such as the cactus have large root systems and succulent stems to take in and store as much water as possible.

QUESTIONS

1. **EXTENDED** Explain why the stomata of pond plants may remain open all the time, but those of a garden plant close at night.
2. **EXTENDED** Sketch a diagram of a desert plant such as a cactus, and annotate it to show how it is adapted to survive in the desert.

END OF EXTENDED

Translocation

Translocation is the transport or movement of materials, including sucrose and amino acids, in the phloem of a plant. Sucrose is made directly from glucose that is produced in photosynthesis. Sucrose is a less active molecule than glucose, and so safer to transport around the plant. Amino acids are produced by cells using glucose from photosynthesis and nitrogen ions taken from the soil.

Sucrose is transported in the phloem to other parts of the plant:

- for conversion back to glucose and use in respiration
- for conversion back to glucose and used to produce other molecules in the cell for cell growth
- for conversion to starch to be stored until needed.

Amino acids are used to produce proteins for the formation of new plant tissue during growth, and to produce enzymes needed to control cell reactions.

EXTENDED

Pests, such as insects, can damage crop plants and reduce the yield. Pesticides are often used to kill the pests and protect the plants. Some pesticides only kill the pests they touch, and so only have an effect on pests on the surfaces of the plant that the pesticide can reach. These pesticides may also be washed off easily by rain and so no longer protect the plant.

Systemic pesticides are absorbed through the leaves of the plant and spread through the phloem to all parts of the plant. This means that if a pest eats any part of the plant, not just the parts that were sprayed, they will be killed by the pesticide. Systemic pesticides also last longer in the plant because they cannot be washed off by rain.

Sources and sinks

Transpiration may be described in terms of sources and sinks.

- A **source** is where a substance is produced or is supplied to the plant.
- A **sink** is where the substance is used or converted to another substance.

In transpiration, the soil water is the *source* of water and dissolved mineral ions such as nitrogen ions that enter the plant through root hair cells. The air is the source of carbon dioxide. The photosynthesising cells are the *sink* for carbon dioxide and water, when they use them to produce glucose.

In translocation, the *source* of glucose is the photosynthesising cells of the leaf. The main *sink* for glucose may differ at different times of the year. Early in the year, when the plant is growing rapidly, the root and shoot tips will be the main sinks for glucose, for use in producing new plant tissue. Later in the year, a plant may direct more glucose to flowers for reproduction, and then to producing seed. Some plants may also direct glucose to storage organs, such as root and stem tubers (such as potato), to form starch that can be used at the start of the next growing season for rapid new growth.

END OF EXTENDED

QUESTIONS

1. In which tissue does translocation occur in a plant?
2. Give two examples of substances that are translocated.
3. **EXTENDED** Explain the advantages of using pesticides that are systemic.
4. **EXTENDED** Explain what is meant by a source and a sink in terms of a) transpiration and b) translocation.

TRANSPORT IN HUMANS

Human circulatory system

The human circulatory system is a system of continuous tubes that carry blood around the body. The tubes are connected to a pump, the heart, which forces the blood through the circulation. Valves in the heart and in the veins make sure that blood only circulates in one direction.

The human circulatory system is described as a **double circulation** because the blood passes through the heart twice for each time that it passes through the body tissues. This is because when blood leaves the right side of the heart, it passes through the tissues of the lungs before returning to the heart for pumping around the rest of the body.

Separating the circulation to the lungs from the circulation to the body means that the blood in the two circulations can be at different pressures.

- Blood leaving the right side of the heart is normally below 4 kPa (about 30 mmHg). Blood does not travel far to the lung tissue, so there is little loss of pressure before it reaches the capillaries

surrounding the alveoli. This lower pressure prevents damage to the delicate capillaries that pass through lung tissue.
- Blood leaving the left side of the heart has to travel all round the body and back to the heart. So it needs to start at a much higher pressure, at about 16 kPa (c. 120 mmHg) as it leaves the heart. By the time it reaches the capillaries within body tissues, the pressure has dropped to below 3 kPa (below 23 mmHg) and so will not damage them.

SCIENCE IN CONTEXT: MEASURING BLOOD PRESSURE

Blood pressure may be measured in units of kilopascals (kPa), which is the standard unit of pressure.

A nurse or doctor usually measures blood pressure in units of millimetres of mercury (mmHg). This unit relates to the mercury sphygmomanometers that were used before digital pressure monitors were introduced.

Medical practitioners continue to use this unit because they recognise values that are normal or abnormal.

1 mmHg = 0.13 kPa

△ Fig. 2.85 A patient having her blood pressure measured.

QUESTIONS

1. What is the role of the heart in the circulatory system?
2. What prevents blood flowing the wrong way through the circulatory system?
3. Explain what is meant by a *double circulatory system*.
4. Explain the advantage of a double circulatory system.

The heart

The heart is a muscular organ that pumps blood by expanding in size as it fills with blood, and then contracting, forcing the blood back out and around the body.

The heart is two pumps in one. The right side and left side are separated by a layer of tissue called the septum. The right side of the heart pumps blood to the lungs to collect oxygen. The left side pumps **oxygenated** blood around the rest of the body. The **deoxygenated** (without oxygen) blood returns to the right side to be sent to the lungs again.

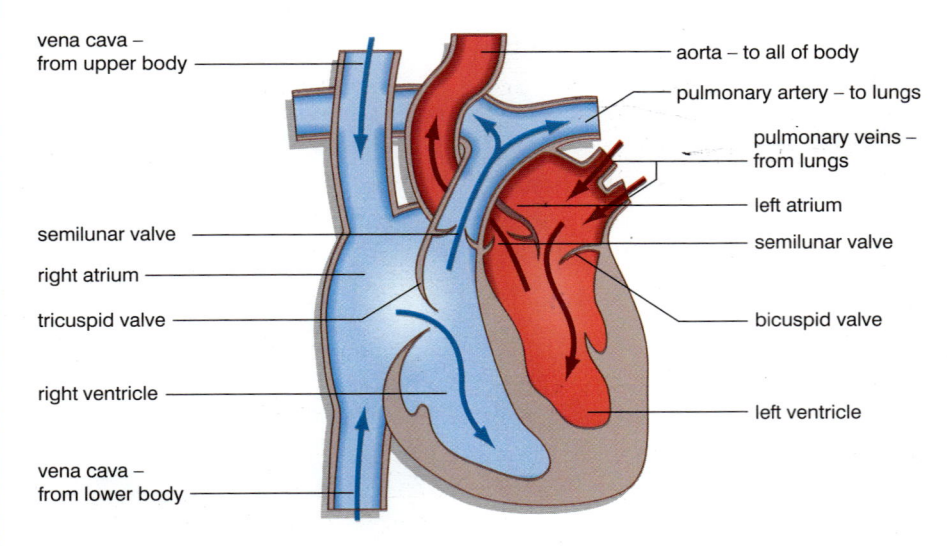

△ Fig. 2.86 Oxygenated blood (shown in red) and deoxygenated blood (blue).

> **REMEMBER**
>
> Diagrams of the circulatory system and heart are always drawn as if looking in a mirror, or at another person. So in the diagram the 'left' side of the heart/circulation in a body is drawn on the right side of the diagram.

The heart consists of four chambers: two **atria** (single: atrium) and two **ventricles**. The walls of the chambers are formed from thick muscle. Blood passes through the chambers of the heart in a particular sequence as the walls of the chambers contract. First the atria contract at the same time, then the ventricles both contract at the same time, to move the blood through the heart.

- Blood from the body arrives at the heart via the vena cava, and enters the right atrium.
- Contraction of the right atrium passes blood to the right ventricle.
- Contraction of the right ventricle forces blood out through the pulmonary artery to the lungs.
- Blood enters the left atrium from the lungs through the pulmonary vein.
- Contraction of the left atrium passes blood to the left ventricle.
- Contraction of the left ventricle forces blood out through the aorta towards the rest of the body. (Note that the muscular wall of the left ventricle is thicker than that of the right ventricle, because it produces a greater force.)

To make sure that blood only flows in one direction through the heart, there are **valves** at the points where blood vessels enter and leave the heart, and between the atria and ventricles. These close when the heart contracts, to prevent backflow of blood.

Changing heart rate

Heart rate is the measure of how frequently the heart beats, generally given as beats per minute. Heart rate is usually measured by feeling for a 'pulse point' where the blood flows through an artery near to the skin, such as in the wrist or at the temple.

Taking a pulse is actually measuring the expansion and relaxation of the artery wall as the blood passes through it. However, as each pulse of blood is created by one contraction of the ventricles, we say that we are measuring heart beats.

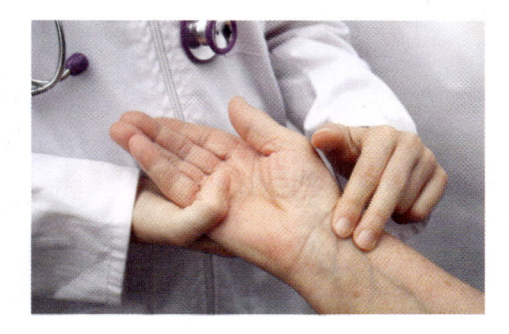

△ Fig. 2.87 Taking a wrist pulse.

Resting heart rate is the rate at which the heart beats when the person is at rest. On average it is between 60 and 80 beats per minute for an adult human, but this range is very variable. Resting heart rate may vary as a result of:

- age – children usually have a faster average than adults
- fitness – an athlete who is trained for endurance sports (such as marathon running) may have a resting heart rate as low as 40 beats per minute because their heart contains more muscle and can pump out more blood on each contraction
- illness – infection can raise resting heart rate, but some diseases of the circulatory system can slow resting heart rate.

Heart rate increases during activity in order to pump blood more rapidly around the body. This supplies oxygen and glucose more rapidly to respiring cells particularly in the muscles, and removes waste products more rapidly.

△ Fig. 2.88 How heart rate changes with exercise.

Developing investigative skills

The effect of exercise on heart rate can be measured by taking pulse measurements after different levels of exercise.

Using and organising techniques, apparatus and materials

In an investigation of the effect of exercise on the heart rate, a student was asked to exercise at different levels for 4 minutes, at which point the pulse rate was measured. The student was then allowed to rest for 5 minutes and then continue to exercise at the next level of activity.

1 a) Explain why the pulse rate was taken after 4 minutes of exercise and not sooner.

  b) Explain why they rested for 5 minutes before starting the next level of exercise.

Observing, measuring and recording

The table shows the results of an investigation into the effect of exercise on heart rate of one student.

	Resting	Walking	Jogging	Running
Heart rate/beats per minute	72	81	96	122

2 Describe the pattern shown in the data.
3 Use the data to draw a conclusion for the investigation.
4 Explain why heart rate responds like this to different levels of exercise.

Handling experimental observations and data

5 How reliable is this conclusion, and what could have been done during the investigation to improve the reliability?

QUESTIONS

1. Starting in the vena cava, list the chambers and blood vessels in the order that blood passes through them until it reaches the aorta.

2. Explain why resting heart rate in an adult is given as a range of values and not a single value.

3. Describe and explain the effect of activity on heart rate.

Coronary heart disease

The muscle of the heart needs its own blood supply to provide the oxygen and sugars it needs for respiration. It cannot get these materials from the blood that flows through it, so there are **coronary arteries** and **veins** that supply the heart muscle. If the blood flow through these coronary blood vessels is reduced, it can reduce the amount of oxygen and sugars getting to the muscle cells, and so reduce the amount of energy they can release through respiration.

Blockage of the coronary arteries can occur when layers of cholesterol and other fatty material form thick layers on the inner lining of the blood vessel. This makes the blood vessel narrower or may completely block it. Partial blockage can cause health problems, such as angina (heart pains) or high blood pressure. A total blockage will cause a heart attack, which may result in death.

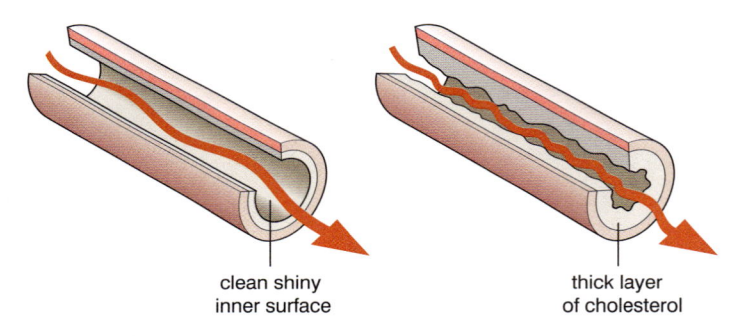

△ Fig. 2.89 Deposits of cholesterol inside arteries makes it more difficult for blood to flow through freely, increasing the risk of diseases of the circulatory system.

Some factors can increase the risk of a blockage of the coronary arteries:

- High levels of saturated fats (such as found in red meats) in the diet can cause increased deposits of cholesterol.
- Chemicals in tobacco smoke that pass into the blood can damage the delicate lining of arteries, which increases the chance that deposits of cholesterol are laid down at these points.

Some health care professionals consider stress a direct cause of coronary heart disease; others dispute this. However, some people respond to stress by smoking or eating too much. Over a long time, these habits can increase the risk of heart disease.

QUESTIONS

1. Explain why the heart needs its own blood supply.
2. Describe the effects of partial and complete blockage of a coronary artery.
3. Explain how a person can reduce their risk of coronary heart disease.

Blood vessels

Fig. 2.90 shows a simplified layout of the human circulatory system, including the major blood vessels. The name of a major blood vessel is often related to the organ it supplies: *hepatic* for liver (from the Greek *hepatos* meaning 'liver'), *renal* for kidneys (from the Latin *renes* meaning 'kidneys'), *pulmonary* for lungs (from the Latin *pulmonis* meaning 'lungs'). Learn the names of the blood vessels that are associated with the heart, the lungs, liver and kidneys.

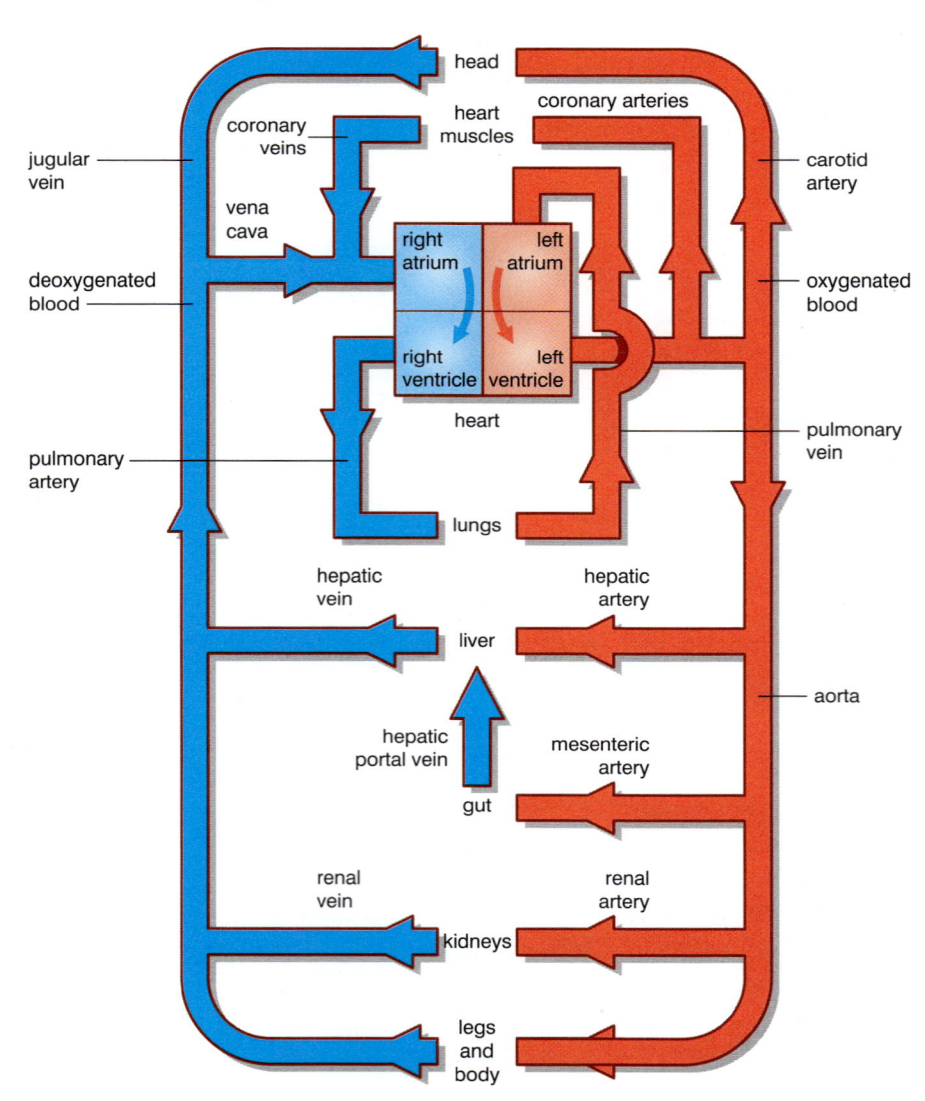

△ Fig. 2.90 Plan of the human circulatory system.

The blood vessels are grouped into three different types: arteries, capillaries and veins.

REMEMBER

A for **a**rteries that travel **a**way from the heart. **V**eins carry blood into the heart and contain **v**alves.

- **Arteries** are large blood vessels that carry blood that is flowing away from the heart. Arteries have thick muscular and elastic walls, with a narrow central space (lumen) through which the blood flows.
- **Capillaries** are the tiny blood vessels that flow through every tissue and connect arteries to veins. Capillaries have very thin walls that are only one cell thick and contain no muscle. All the exchange of substances between the blood and tissues happens in the capillaries.
- **Veins** are large blood vessels that carry blood that is flowing back towards the heart. Veins have a large lumen through which blood flows. **Valves** in the veins prevent backflow.

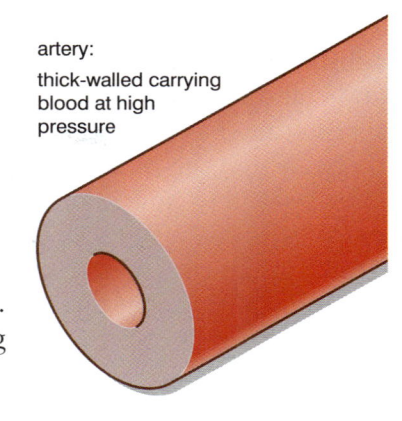

artery: thick-walled carrying blood at high pressure

△ Fig. 2.91 Arteries vary in diameter from about 10 to 25 mm.

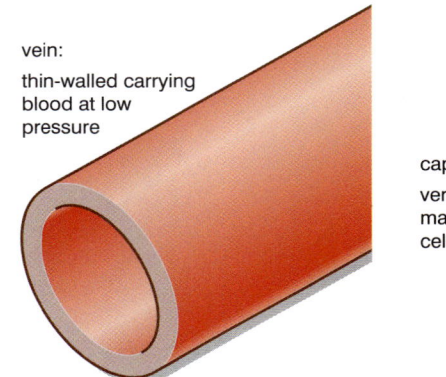

vein: thin-walled carrying blood at low pressure

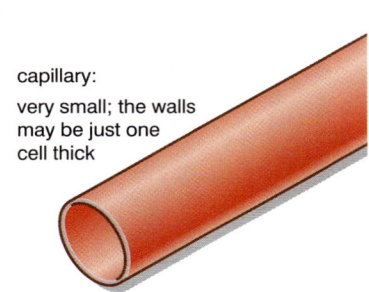

capillary: very small; the walls may be just one cell thick

△ Fig. 2.92 Veins vary in diameter from about 5 to 15 mm. Capillaries are very small, with a diameter of around 0.01 mm.

EXTENDED

Structure related to function

The structure of different blood vessels enables them to carry out their function most efficiently.

- Blood carried in the arteries is at higher pressure than in the other vessels. The highest pressure is in the aorta, the blood vessel that leaves the left ventricle. The thick walls of arteries help to protect them from bursting when the pressure increases as the pulse of blood enters them. The recoil of the elastic wall after the pulse of blood has passed through helps to maintain the blood pressure and even out the pulses. By the time the blood enters the fine capillaries, the change in pressure during and after a pulse has been greatly reduced.

Blood vessels	Blood pressure/kPa
Aorta	>13
Arteries	13–5.3
Capillaries	3.3–1.6
Veins	1.3–0.7
Vena cava	0.3

△ Table 2.8 Blood pressure in different vessels.

- The thin walls of capillaries help to increase the rate of diffusion of substances by keeping the distance for diffusion between the blood and cell cytoplasm to a minimum.
- By the time blood leaves the capillaries and enters the veins, there is no pulse and the blood pressure is very low. The large lumen (centre) allows blood to flow easily back to the heart. The contraction of body muscles, such as in the legs, helps to push the blood back toward the heart against the force of gravity. The valves make sure that blood can flow only in the right direction, back towards the heart.

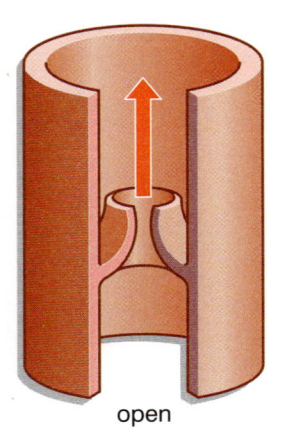

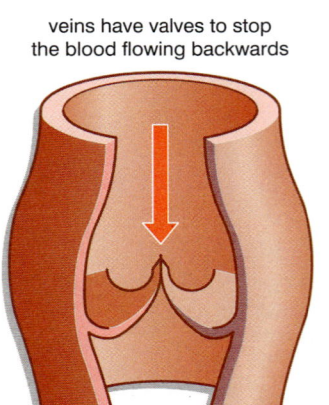

△ Fig. 2.93 Valves in the veins make sure that blood can only move in one direction, towards the heart.

END OF EXTENDED

QUESTIONS

1. Name the following blood vessels:
   a) the vessels that carry blood to the kidneys
   b) the vessel that carries blood from the heart towards the body
   c) the vessels that carry blood from the liver back towards the heart
2. Describe the differences in structure of arteries, capillaries and veins.
3. **EXTENDED** Explain how the structure of arteries helps to reduce and even out the blood pulses from the heart.

Blood

The human circulatory system carries substances around the body. Table 2.9 shows some of the important substances transported around the human body. These substances are carried within the blood, in different forms. Blood also helps to distribute heat around the body.

Substance	Carried from	Carried to
Food (glucose, amino acids, fat)	Small intestine	All parts of the body
Water	Intestines	All parts of the body
Oxygen	Lungs	All parts of the body
Carbon dioxide	All parts of the body	Lungs
Urea (waste)	Liver	Kidneys
Hormones	Glands	All parts of the body (different hormones affect different parts)

△ Table 2.9 Substances carried by the blood around the body.

Blood is made from plasma, red blood cells, white blood cells and platelets. Each of these has a particular function in the body.

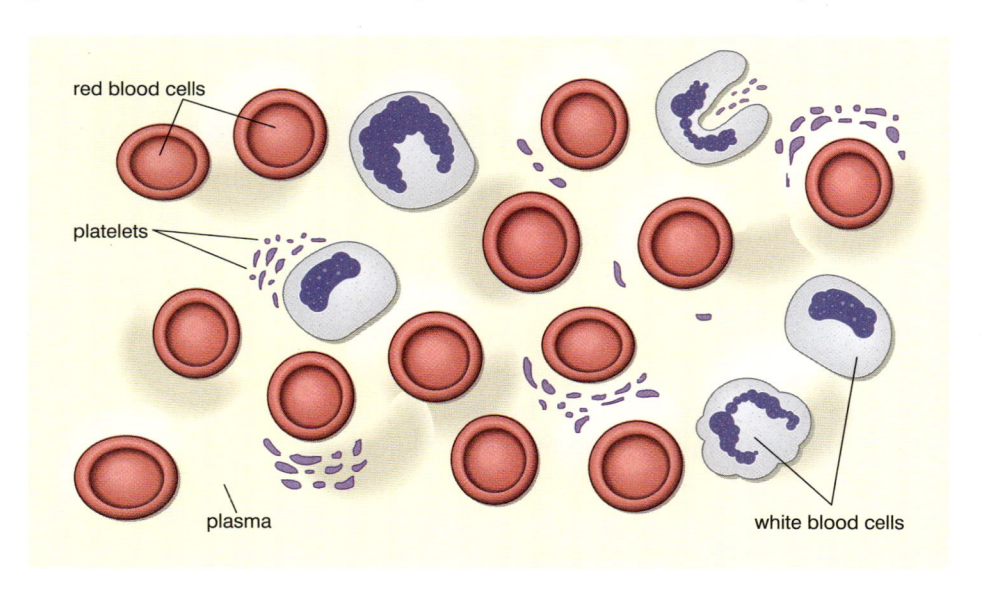

△ Fig. 2.94 Blood is mostly water, containing cells and many dissolved substances.

Plasma

Plasma is the straw-coloured liquid part of blood. It mainly consists of water, which makes it a good solvent for many substances. Digested food molecules, such as glucose and amino acids, easily dissolve in plasma. Urea, which is formed by the liver from excess amino acids, is also soluble in plasma. Many hormones (see Coordination and response) are also soluble and are carried round the body dissolved in plasma. Carbon dioxide dissolves in water to form carbonic acid (H_2CO_3), and most carbon dioxide is carried in the blood in this form.

Plasma proteins are large molecules that remain in the plasma and have a range of functions.

- Some carry lipids or hormones that are not soluble in plasma.
- Some are enzymes that control processes such as blood clotting.
- Some are involved with the immune system.

Red blood cells

Red blood cells are the most common cell type in blood. They have several adaptations that make them efficient at transporting oxygen.

- They have a **biconcave** disc shape (a disc that is thinner in the middle than at the edges) that increases their surface area to volume ratio compared with other cells. This increases the rate at which diffusion into and out of the cell can take place.
- They contain **haemoglobin**, which is the red chemical that carries oxygen in the cell. In areas of high oxygen concentration (such as in capillary next to alveoli), haemoglobin binds with oxygen. This reaction is reversible, because in areas of low oxygen concentration (such as in a capillary running through respiring tissue) the oxygen is released from the haemoglobin.
- They have no nucleus. This maximises the volume that can be filled with haemoglobin, but means that the cells cannot divide to make new cells as they get older. They are cleaned out of the blood by the liver, and new ones are released into the blood from the bone marrow in long bones such as the femur.
- They are small and flexible, which makes it possible for them to get through the smallest blood vessels (capillaries) that are sometimes no wider than a single red blood cell.

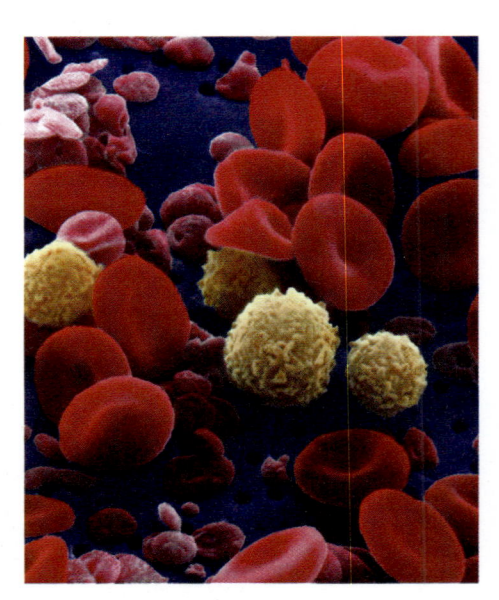

△ Fig. 2.95 Red blood cells (shown in red), white blood cells (yellow) and platelets (pink).

In the lungs, haemoglobin combines with oxygen to form oxyhaemoglobin. In other organs and tissues, oxyhaemoglobin releases the oxygen to form haemoglobin again.

EXTENDED

One of the factors measured in a blood test is red blood cell count. This is the number of red blood cells in a given volume of blood. In a normal adult male at around sea level it is about 4.7–6.1 million cells per mm^3.

If someone who lives at around sea level travels to a place at high altitude, their red blood cell count gradually increases by up to 50% over about two weeks. At high altitude the air is often described as being 'thinner', meaning that the oxygen concentration is lower than nearer sea level.

1. Describe how haemoglobin is adapted for exchanging oxygen with the air and with respiring cells.

2. Suggest how a person's breathing might change if they travelled from low to high altitude quickly. Explain your answer.

3. What advantage is there to the body of producing more red blood cells at high altitude? Explain your answer as fully as you can.

4. Some elite athletes train at high altitude for a few weeks, just before a competition even if the race is held at low altitude. Explain as fully as you can what advantage high-altitude training would give over low-altitude training.

END OF EXTENDED

White blood cells

There are several different types of white blood cell, but they all play an important role in defending the body against disease. They are part of the **immune system** that responds to infection by trying to kill the **pathogen** (the disease-causing organism).

- **Phagocytes:** Several types of white blood cell belong to this group, but they all kill pathogens by engulfing (flowing around the pathogen until it is completely enclosed) and then digesting them. Different types of phagocytes target different pathogens, such as bacteria, fungi and protoctist parasites.

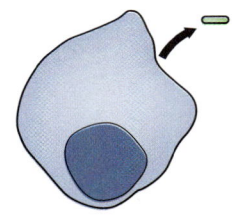

1 A phagocyte moves towards a bacterium

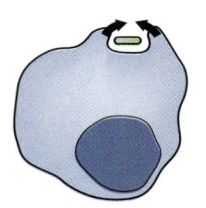

2 The phagocyte pushes a sleeve of cytoplasm outwards to surround the bacterium

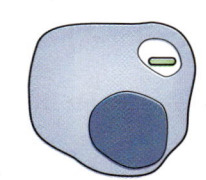

3 The bacterium is now enclosed in a vacuole inside the cell. It is then killed and digested by enzymes

∆ Fig. 2.96 Phagocytosis of a bacterium by a phagocyte, a type of white blood cell.

- **Lymphocytes:** This type of white blood cell has a very large nucleus, and is responsible for producing chemicals called **antibodies**. Antibodies are proteins that attack pathogens (see The immune system).

Platelets

Platelets are small fragments of much larger cells that are also important in protecting us from infection by causing blood to clot where there is damage to a blood vessel. This helps to seal the cut and prevent blood from leaking out and pathogens from getting in.

EXTENDED

The liver makes a protein called **fibrinogen** which is released into the plasma. When there is damage to a blood vessel, the platelets respond by releasing an enzyme that causes the soluble fibrinogen to change and become a fibrous protein called **fibrin**. The fibres of this protein trap blood cells to produce the clot.

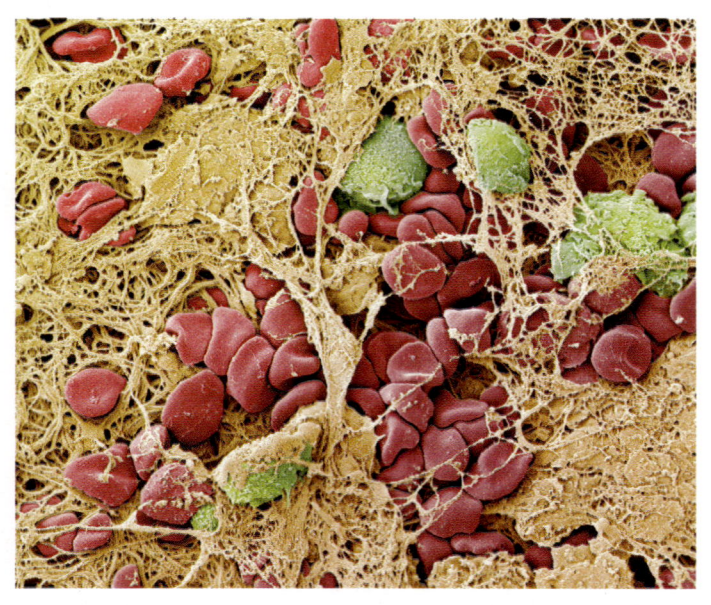

△ Fig. 2.97 A blood clot is made of protein fibres that trap blood cells to control damage to a blood vessel.

END OF EXTENDED

SCIENCE IN CONTEXT

BLOOD TESTS

Doctors often take blood samples for testing when a patient is unwell. These tests not only check the concentration of different substances in the blood, such as glucose; they also count the number of different kinds of cells. Taking a blood sample is quick and easy to do, and the tests help in diagnosing what is wrong with the patient.

- An abnormal concentration of glucose may indicate diabetes.
- Too few white blood cells may indicate liver or bone marrow disease.
- Too few red blood cells can make the blood look paler than normal, causing anaemia, and usually results in the patient feeling more easily exhausted than usual.

Any abnormal results need following up with other tests to confirm diagnosis.

QUESTIONS

1. Draw up a table to show the components of blood and the role that they play in the body.
2. Explain how the structure of a red blood cell is adapted to its function.
3. Describe the role of white blood cells in the immune system.
4. Explain how platelets can protect us from infection.

EXTENDED

THE IMMUNE SYSTEM

The **immune system** coordinates the body's response to infection. This includes the production of phagocyte cells that directly attack the invading pathogen, and **lymphocyte** cells that produce antibodies.

Lymphocytes produce antibodies that specifically match the invading pathogen. The antibodies attach to the pathogen and either attract phagocytes to engulf the pathogen or cause the pathogen to break open and die. Lymphocytes also produce **memory cells**, which remain in the blood after the pathogen has been destroyed. If the same pathogen attacks again, the memory cells rapidly respond causing a rapid increase in the production of antibodies. These antibodies attack and destroy the pathogen often before you are aware of another infection.

Antibodies are a group of proteins that work by shape. Molecules on the surface of a pathogen have a particular shape, which differs depending on the pathogen. An antibody for a particular pathogen has a matching shape, so that it can attach to the surface of the pathogen and either attract phagocytes or cause the pathogen to break up. This is why the vaccination for one infection, such as measles, cannot protect you from other infections, such as chicken pox.

This also explains why you can catch a common cold or influenza more than once. The pathogens that cause these infections mutate and change the shape of their surface molecules frequently, so memory cells do not recognise them when the pathogen infects another time.

> ### SCIENCE IN CONTEXT
>
> ### VACCINATION
>
> **Vaccination** is a way of making the body respond as if it has been infected and helping to protect it from illness in the future. The vaccine is prepared from small amounts of material from the pathogen, which is either dead or weakened so that it cannot cause disease.
>
> The vaccine is put into the body, either through the mouth (for polio vaccination) or by injection. Lymphocytes in the immune system respond by making antibodies to the pathogen, and also memory cells. These memory cells remain in the blood, sometimes for the rest of your life, after the pathogen has been destroyed.
>
> △ Fig. 2.98 An injection of vaccine can be a painful experience, but childhood vaccinations can give life-long protection from dangerous infections.
>
> If you are ever infected by the live pathogen, your memory cells recognise it very rapidly and stimulate lymphocytes to produce huge quantities of antibodies very quickly. This response kills off the pathogen rapidly, often before you develop symptoms and realise you have been infected. This rapid response makes you immune to that pathogen.

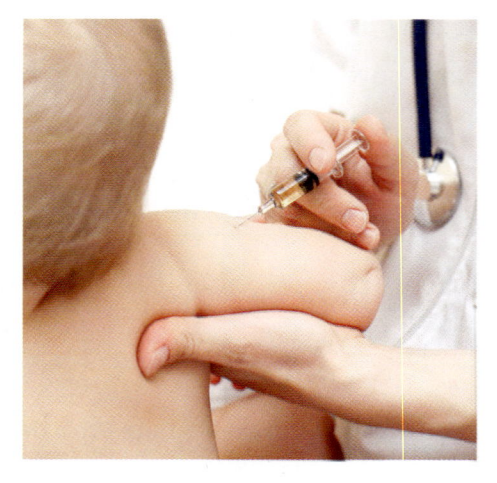

Tissue rejection

Antibodies are produced whenever cells that are 'foreign' (not from your body) are introduced. This happens after an **organ transplant**, such as when a person's defective heart or kidney is replaced with a heart or kidney from someone else. The immune system of the person receiving the transplant recognises the tissue as 'foreign' and reacts against it. This reaction is called **tissue rejection**. To prevent tissue rejection after a transplant, the patient needs to be treated with drugs that suppress the immune system.

The lymphatic system

The walls of capillaries are so thin that water, dissolved solutes (such as glucose and mineral ions) and dissolved gases (such as oxygen) can easily pass through the wall from the plasma into the **tissue fluid** surrounding the cells. Cells exchange materials across their cell membrane with the tissue fluid.

More fluid is pushed out of the capillaries by the pressure of the blood inside them, but only some of this returns to the capillaries. The rest passes into the **lymphatic system** and becomes a fluid called **lymph**. The lymphatic system is formed from a series of connected tubes that flow from tissue back towards the heart. It connects with the blood system near to the heart, where the lymph is returned to the blood plasma.

Tissues associated with the lymphatic system, such as the bone marrow in the centre of long bones of the legs and arms, produce the white blood cells called lymphocytes. These cells are carried through the lymphatic system and enter the blood system with the lymph near to the heart.

QUESTIONS

1. **EXTENDED** Explain how antibodies help to protect the body from infection.
2. **EXTENDED** Explain the problem of antibodies in transplants.
3. **EXTENDED** Describe the role of the lymphatic system.

END OF EXTENDED

End of topic checklist

Key terms

active transport, atrium, biconcave, cohesion, coronary arteries/veins, deoxygenated, diffusion double circulation, evaporation, fibrin, fibrinogen, haemoglobin, heart, heart rate, humidity, immune system, lymph, lymphatic system, lymphocyte, memory cell, organ transplant, osmosis oxygenated, pathogen, phagocyte, phloem, plasma, platelet, root hair cell, sink, source, stomata succulent, systemic pesticide, tension, tissue rejection, translocation, transpiration, vaccination valve, vascular bundle, vein (*animal*), vein (*plant*), ventricle, water potential, water potential gradient, wilting, xylem

During your study of this topic you should have learned:

○ About the functions of xylem and phloem.
○ How to identify the positions of xylem and phloem as seen in transverse sections of roots, stems and leaves.
○ How to identify root hair cells as seen under the light microscope and state their functions.
○ About the pathway taken by water through root, stem and leaf (root hair, root cortex cells, xylem, mesophyll cells).
○ How to investigate, using a suitable stain, the pathway of water through the above-ground parts of a plant.
○ **EXTENDED** How to relate the structure and functions of root hairs to their surface area and to water and ion uptake.
○ The definition of *transpiration* as evaporation of water at the surfaces of the mesophyll cells followed by loss of water vapour from plant leaves, through the stomata.
○ How water vapour loss is related to cell surfaces, air spaces and stomata.
○ How to describe the effects of variation of temperature, humidity and light intensity on transpiration rate.
○ How wilting occurs.
○ **EXTENDED** How to explain the mechanism of water uptake and movement in terms of transpiration producing a tension ('pull') from above, creating a water potential gradient in the xylem, drawing cohesive water molecules up the plant.
○ **EXTENDED** About the adaptations of the leaf, stem and root to three contrasting environments (pond, garden and desert).
○ The definition of *translocation* in terms of the movement of sucrose and amino acids in phloem:

- from regions of production
- to regions of storage or to regions of utilisation in respiration or growth.

○ **EXTENDED** How to describe translocation throughout the plant of applied chemicals, including systemic pesticides.

○ **EXTENDED** How to compare the role of transpiration and translocation in the transport of materials from sources to sinks, within plants at different seasons.

○ How to describe the circulatory system as a system of tubes with a pump and valves to ensure one-way flow of blood.

○ How to describe the double circulation in terms of a low-pressure circulation to the lungs and a high-pressure circulation to the body tissues and relate the differences in the two circuits to their functions.

○ How to describe the structure of the heart, including muscular wall and septum, chambers, valves and blood vessels.

○ How to describe the function of the heart in terms of muscular contraction and the working of the valves.

○ How to investigate, state and explain the effect of physical activity on pulse rate.

○ How to describe coronary heart disease in terms of blockage of coronary arteries and state the possible causes (diet, stress and smoking) and preventive measures.

○ The names of the main blood vessels to and from the heart, lungs, liver and kidney.

○ How to describe the structure and functions of arteries, veins and capillaries.

○ **EXTENDED** Be able to explain how structure and function are related in arteries, veins and capillaries.

○ **EXTENDED** Be able to describe the transfer of materials between capillaries and tissue fluid.

○ How to identify red and white blood cells as seen under the light microscope on prepared slides, and in diagrams and photomicrographs.

○ About the components of blood and their functions:
- red blood cells – haemoglobin and oxygen transport
- white blood cells – phagocytosis and antibody formation
- platelets – causing clotting
- plasma – transport of blood cells, ions, soluble nutrients, hormones, carbon dioxide, urea and plasma proteins.

○ **EXTENDED** Be able to describe the immune system in terms of antibody production, tissue rejection and phagocytes.

○ **EXTENDED** Be able to describe the function of the lymphatic system in circulation of body fluids, and the production of lymphocytes.

○ **EXTENDED** Be able to describe the process of clotting.

End of topic questions

Note: The marks awarded for these questions indicate the level of detail required in the answers. In the examination, the number of marks awarded to questions like these may be different.

1. Flower sellers sometimes produce flowers of unusual colours for sale, starting with white flowers. Explain how they may do this and why it is possible. **(4 marks)**

2. A potted plant sitting on a sunny windowsill wilts more quickly than an identical plant on a shaded shelf in the room. Explain why. **(4 marks)**

3. EXTENDED This is a section through a cactus stem. Sketch the main features of the cactus stem and label it to show how it is adapted to the conditions in which it lives. **(6 marks)**

4. EXTENDED Explain how and why different parts of a plant are sinks for the translocation of sucrose at different times of the year in a country where there are seasons, such as the UK. **(7 marks)**

5. The blood pressure of blood leaving the right ventricle of the human heart is 3 kPa, and blood from the left ventricle is around 16 kPa. Explain how the heart can produce these different pressures, and why this difference is important for the body. **(4 marks)**

6. People who suffer from anaemia often have a low red blood cell count (fewer blood cells per mm^3 blood) than usual. One of the symptoms of anaemia is becoming tired more easily than usual. Explain why these symptoms occur. **(4 marks)**

7. People who have suffered a thrombosis, where the blood unnecessarily clots and blocks a blood vessel, may be given aspirin every day to help protect them from it happening again. Aspirin interferes with the way platelets function.

    a) Describe the role of platelets in the blood. **(2 marks)**

    b) Explain why this normal function helps us to remain healthy. **(2 marks)**

    c) What damage might be caused if a blood vessel that supplies heart muscle is blocked by a blood clot? **(3 marks)**

    d) Explain why aspirin is effective in reducing the risk of another thrombosis. **(1 mark)**

8. EXTENDED Explain why the immune response to an infection of the pathogen that causes measles:

    a) will protect you from a future infection by the measles pathogen **(2 marks)**

    b) will not protect you against infection by the pathogen that causes chickenpox. **(2 marks)**

Respiration

INTRODUCTION

Respiration and photosynthesis need substances that are gases in air (or in water for aquatic organisms), and they produce gases that need to be returned to the environment. These gases must get into and out of the body fast enough to support the rate at which the processes work. For single-celled organisms, this isn't a problem. They have a large surface area-to-volume ratio, and diffusion across the cell membrane can supply the gases quickly. Larger organisms cannot do this. Not only do they have a much smaller surface area-to-volume ratio, which slows the rate of diffusion, but many of them live on land, where the delicate surface for gas exchange would dry out. Different groups of organisms have different solutions to these problems. Plants exchange gases inside the leaf; insects have internal tubes where they exchange gases; fish have gills; and many vertebrates have lungs.

△ Fig. 2.99 The axolotl is a relative of frogs and newts. It has external gills throughout its life, through which it exchanges gases with the water.

KNOWLEDGE CHECK

✓ Know that organisms need energy for all the life processes that keep them alive.
✓ Know that plants get this energy from the sugars they make in photosynthesis.
✓ Know that animals get this energy from their food.
✓ Know that animals breathe in oxygen from the air and breathe out carbon dioxide.
✓ Know that humans use lungs for breathing.
✓ Know that plants take in oxygen and give out carbon dioxide as a result of respiration.

LEARNING OBJECTIVES

✓ Be able to define *respiration* as the chemical reactions that break down nutrient molecules in living cells to release energy.
✓ Be able to state the uses of energy in the human body: muscle contraction, protein synthesis, cell division, active transport, growth, the passage of nerve impulses, the maintenance of a constant body temperature.
✓ Be able to define *aerobic respiration* as the release of a relatively large amount of energy in cells by the breakdown of food substances in the presence of oxygen.
✓ Be able to state the word equation for aerobic respiration.
✓ Be able to define *anaerobic respiration* as the release of a relatively small amount of energy by the breakdown of food substances in the absence of oxygen.

- ✓ Be able to state the word equation for anaerobic respiration in muscles during hard exercise.
- ✓ Be able to state the word equation for anaerobic respiration in the microorganism yeast.
- ✓ Be able to describe the role of anaerobic respiration in yeast during brewing and bread-making.
- ✓ Be able to compare aerobic respiration and anaerobic respiration in terms of relative amounts of energy released.
- ✓ Be able to list the features of gas exchange surfaces in animals.
- ✓ Be able to identify on diagrams and name the larynx, trachea, bronchi, bronchioles, alveoli and associated capillaries.
- ✓ Be able to state the differences in composition between inspired and expired air.
- ✓ Be able to use lime water as a test for carbon dioxide to investigate the differences in composition between inspired and expired air.
- ✓ Be able to investigate and describe the effects of physical activity on rate and depth of breathing.
- ✓ **EXTENDED** Be able to state the symbol equation for aerobic respiration.
- ✓ **EXTENDED** Be able to state the symbol equation for anaerobic respiration in muscles.
- ✓ **EXTENDED** Be able to state the symbol equation for anaerobic respiration in yeast.
- ✓ **EXTENDED** How to describe the effect of lactic acid in muscles during exercise.
- ✓ **EXTENDED** Be able to describe the role of the ribs, internal and external intercostal muscles and diaphragm in producing volume and pressure changes leading to the ventilation of the lungs.
- ✓ **EXTENDED** Be able to explain the role of mucus and cilia in protecting the gas exchange system from pathogens and particles.
- ✓ **EXTENDED** Be able to explain the link between physical activity and rate and depth of breathing in terms of changes in the rate at which tissues respire and therefore of carbon dioxide concentration and pH in tissues and in the blood.

RESPIRATION OUTLINE

When most people talk about respiration, they usually mean breathing (or ventilation), when gases are exchanged across a respiratory surface. However, this is not the scientific meaning of respiration. In the first part of this section we will look at **cellular respiration**, which is the release of energy from the chemical bonds in food molecules such as glucose. This only takes place inside cells, and every living cell carries out cellular respiration.

> **REMEMBER**
>
> Be clear in your answers that you are using the term *respiration* to mean *cellular respiration*, and to use *ventilation* not *respiration* when talking about breathing.

Every cell in a living organism requires energy, and this energy comes from respiration which is the breakdown of chemical bonds in food molecules such as glucose to release energy in a form that can be used in cells.

In human cells, this energy is used in many ways, for example:
- to produce the contraction of muscle cells
- to produce new chemical bonds during the synthesis (formation) of new protein molecules
- to produce new chemical bonds in other molecules needed for growth and reproduction of cells
- for the active transport of molecules across cell membranes
- to produce the movement of nerve impulses along nerve cells
- to maintain a constant core body temperature.

Glucose is usually the molecule that is meant when people talk about *nutrient molecules* or *food molecules* broken down in respiration. This is because glucose is commonly used in this reaction. If glucose is in short supply, then other molecules from the breakdown of fats or proteins may be used instead.

AEROBIC RESPIRATION

Most plant and animal cells use oxygen during cellular respiration. This kind of respiration is called **aerobic respiration**. Water and carbon dioxide are produced as waste products. This is similar in some ways to burning fuel except that in our bodies enzymes control the process.

Aerobic respiration can be summarised by a word equation:

glucose + oxygen → water + carbon dioxide (+ energy)

EXTENDED

It can also be written as a symbol equation:

$$C_6H_{12}O_6 + 6O_2 \rightarrow 6H_2O + 6CO_2 \text{ (+ energy)}$$

END OF EXTENDED

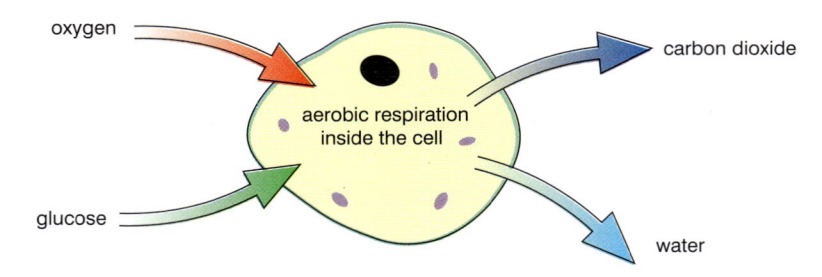

△ Fig. 2.100 Aerobic respiration in a cell.

The oxygen needed for respiration comes from the air (except for a small proportion in photosynthesising plants, which comes from photosynthesis). The carbon dioxide from cellular respiration is released to the air, and the water is either used in the body or excreted through the kidneys.

During aerobic respiration, many of the chemical bonds in the glucose molecule are broken down. This releases a lot of energy: around 2900 kJ of energy are released for each mole of glucose molecules used in aerobic respiration.

> ### SCIENCE IN CONTEXT
> ### WATER FROM RESPIRATION
>
> A camel can survive for many days without drinking liquid water, which means it survives well in desert conditions. The camel's hump is not a store of water, but a store of fat. Over a long period without food, the fat breaks down to release substances for aerobic respiration. Since water is one of the products of aerobic respiration, this also helps the camel to survive longer without drinking water.
>
> Going entirely without food and water would kill a human in a few days, because we don't metabolise fat as well as the camel does, or retain water as well. So, before setting out into the desert for a long trip, make sure your camel has a large hump (and you have plenty of food and water).
>
> △ Fig. 2.101 An Arabian camel (dromedary).

Developing investigative skills

You can use this apparatus to investigate what happens to seeds as they germinate.

Using and organising techniques, apparatus and materials

❶ What changes will this apparatus measure? Explain your answer.

Observing, measuring and recording

❷ The results from this investigation are shown in the table. (Lime water was used in the tube.)

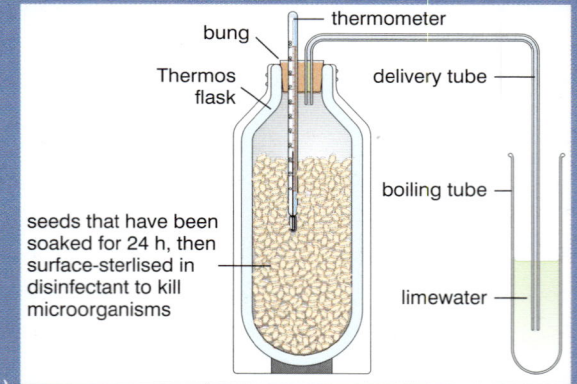

△ Fig. 2.102 Apparatus needed for the investigation.

Start	Limewater	Clear
	Temperature/°C	15
After 2 days	Limewater	Milky
	Temperature/°C	19

a) Describe the changes over 2 days.

b) Explain, as fully as you can, the changes over 2 days.

Handling experimental observations and data

❸ What control would you need to carry out to show that it was the seeds that caused these changes? Explain your answer.

QUESTIONS

1. **a)** Write out the word equation for aerobic respiration.

   b) Annotate your equation to show where the reactants come from.

   c) Annotate your equation to show what happens to the products of the reaction in a human.

   d) Describe how your answer to part c) might differ for a camel on a long journey without water, and explain your answer.

2. Where does respiration take place in the body?

3. Give three examples of the use of energy from respiration in the human body.

4. **EXTENDED** Write the balanced symbol equation for aerobic respiration.

ANAEROBIC RESPIRATION

Aerobic respiration provides most of the energy that plant cells and animal cells need most of the time. However, there are times when not enough oxygen is available for aerobic respiration to occur fast enough to deliver all the energy that is needed, for example:

- in diving animals, such as whales and seals
- in muscle cells, when vigorous exercise requires more energy than can be provided by an increased supply of oxygen from deeper faster breathing and a faster heart rate.

In these cases, the additional energy needed is supplied by **anaerobic respiration**. This kind of respiration also releases energy from glucose molecules, but without the need for oxygen.

Some organisms, such as yeast (a unicellular fungus) and some bacteria, respire anaerobically in preference to aerobically even when oxygen is present.

In anaerobic respiration, the glucose molecule is only partly broken down, so far less energy is released from each glucose in anaerobic respiration than in aerobic respiration. Only about 150 kJ is produced from every mole of glucose molecules respired anaerobically in a muscle cell.

△ Fig. 2.103 Diving animals use anaerobic respiration when they have been under water for some time.

Anaerobic respiration in muscle cells

When animal cells respire anaerobically, such as muscle cells during vigorous exercise, the glucose is broken down to lactic acid:

glucose → lactic acid (+ energy)

EXTENDED

The balanced symbol equation for this reaction is:

$C_6H_{12}O_6 \rightarrow 2C_3H_6O_3$

END OF EXTENDED

Note that even when a muscle cell is respiring anaerobically, aerobic respiration is also taking place and using all the oxygen that is available. Where aerobic respiration cannot supply all the energy needed, only the additional energy needed comes from anaerobic respiration.

During anaerobic respiration, the concentration of lactic acid builds up in the cells until a time when oxygen is available again (such as at the end of exercise). It is then converted back to glucose and then glycogen in muscle and liver cells and stored for use later, or broken down fully using oxygen to carbon dioxide and water.

The additional oxygen needed after exercise used to be called the *oxygen debt*, but it is now understood that additional oxygen is needed after any prolonged exercise, even if it is completely aerobic, to return many processes in the body back to their resting state. This need, which you can feel when you breathe more deeply after exercise, is now called *EPOC* (excess post-exercise oxygen consumption).

SCIENCE IN CONTEXT — RESPIRATION IN ATHLETICS

If you watch carefully, you will not see a sprint athlete breathe during a race. At the start of the race there will be some oxygen in their muscle cells, but this is rapidly used up as they start running. Anaerobic respiration provides virtually all of the energy used in a 100 m sprint by a well-trained athlete.

Sprinting cannot be maintained for long, because the muscle cells also need a rapid supply of glucose for respiration. So longer distance races use a combination of aerobic and anaerobic respiration. In marathons, most of the race is run aerobically, with only the last stretch being managed as a sprint using anaerobic respiration.

△ Fig. 2.104 A fit athlete in the middle of a sprint is using almost entirely anaerobic respiration

Anaerobic respiration in yeast cells

Yeast cells break down glucose without using oxygen, even if oxygen is present. The word equation for this reaction is:

glucose → ethanol + carbon dioxide (+ energy)

EXTENDED

The balanced symbol equation for anaerobic respiration in yeast cells is:

$C_6H_{12}O_6 \rightarrow 2C_2H_5OH + 2CO_2$ (+ energy)

END OF EXTENDED

This reaction occurs when yeast is used in brewing to make beer and wine. The ethanol is formed when sugars in the barley or grapes is broken down. Ethanol causes the alcoholic effects that occur when people drink these drinks. Yeast is also used in making bread because the carbon dioxide released forms bubbles in the dough, making the bread light and spongy.

▷ Fig. 2.105 The froth on top of this yeast culture is carbon dioxide from anaerobic respiration escaping from the mixture.

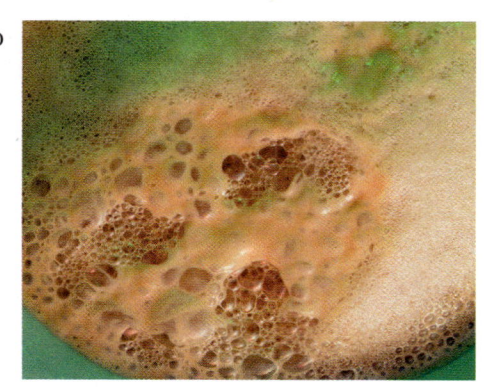

QUESTIONS

1. Explain why muscle cells sometimes need to respire anaerobically.
2. Describe the similarities and differences between anaerobic respiration in yeast cells and in animal cells.
3. Compare the amount of energy released during aerobic respiration and anaerobic respiration of glucose in a muscle cell.
4. Give two examples of products we make that use the anaerobic respiration of yeast cells.

GAS EXCHANGE IN ANIMALS

Animals need to exchange gases with the environment, to supply oxygen for respiration in cells and to remove the waste product of respiration, carbon dioxide. These gases are exchanged at surfaces by diffusion, so gas exchange surfaces need adaptations to maximise the rate at which diffusion occurs.

The rate of diffusion increases with:

- surface area – most gas exchange surfaces have features that increase the surface area as much as possible

- decreasing diffusion distance – gas exhange surfaces are usually closely associated with a large network of capillaries, separated by as short a distance as possible
- concentration gradient – there is constant movement of air (or water in the case of aquatic animals) across the gas exchange surface, and constant movement of blood through the associated capillaries, which makes sure that the concentration gradient across the surface is maintained.

Different animals have different gas exchange surfaces:

- fish and immature amphibians have gills that exchange gases with water
- adult amphibians may use their skin surface for gas exchange, or a lung
- reptiles, birds and mammals use lungs to exchange gases with air
- insects and other invertebrates have breathing tubes to exchange gases with air.

REMEMBER

You may be asked to apply your knowledge of the features of gas exchange surfaces to an example that you do not already know. Make sure you identify an example of each of the features mentioned in this list.

GAS EXCHANGE IN HUMANS

Breathing is the way that oxygen is taken into our bodies and carbon dioxide removed. Sometimes it is called **ventilation**.

When we breathe, air is moved into and out of our lungs. This involves different parts of the respiratory system within the thorax (the chest cavity).

When we breathe in, air enters though the nose and mouth. In the nose the air is moistened and warmed. The air passes over the **larynx** where it may be used to make sounds, such as when we talk. The air travels down the **trachea** (windpipe) to the **lungs**.

The air enters the lungs through the **bronchi** (singular: bronchus), which branch and divide to form a network of **bronchioles**.

EXTENDED

Cells in the lining of the trachea, bronchi and bronchioles secrete **mucus,** which is a sticky liquid. This traps microorganisms and dust particles that are breathed in. The lining of the trachea and bronchi are covered in tiny hairs called **cilia.** These sweep in a combined motion to move the mucus up from the lungs, up the trachea to the back of the mouth where it can be swallowed. The combined action of mucus and cilia helps to prevent dirt and microorganisms entering the lungs and causing damage and infection.

END OF EXTENDED

At the end of the bronchioles are air sacs. The bulges on an air sac are called **alveoli** (singular: alveolus). The alveoli are covered in tiny blood capillaries. This is where oxygen and carbon dioxide are exchanged between the blood and the air in the lungs. This is called **gas exchange**.

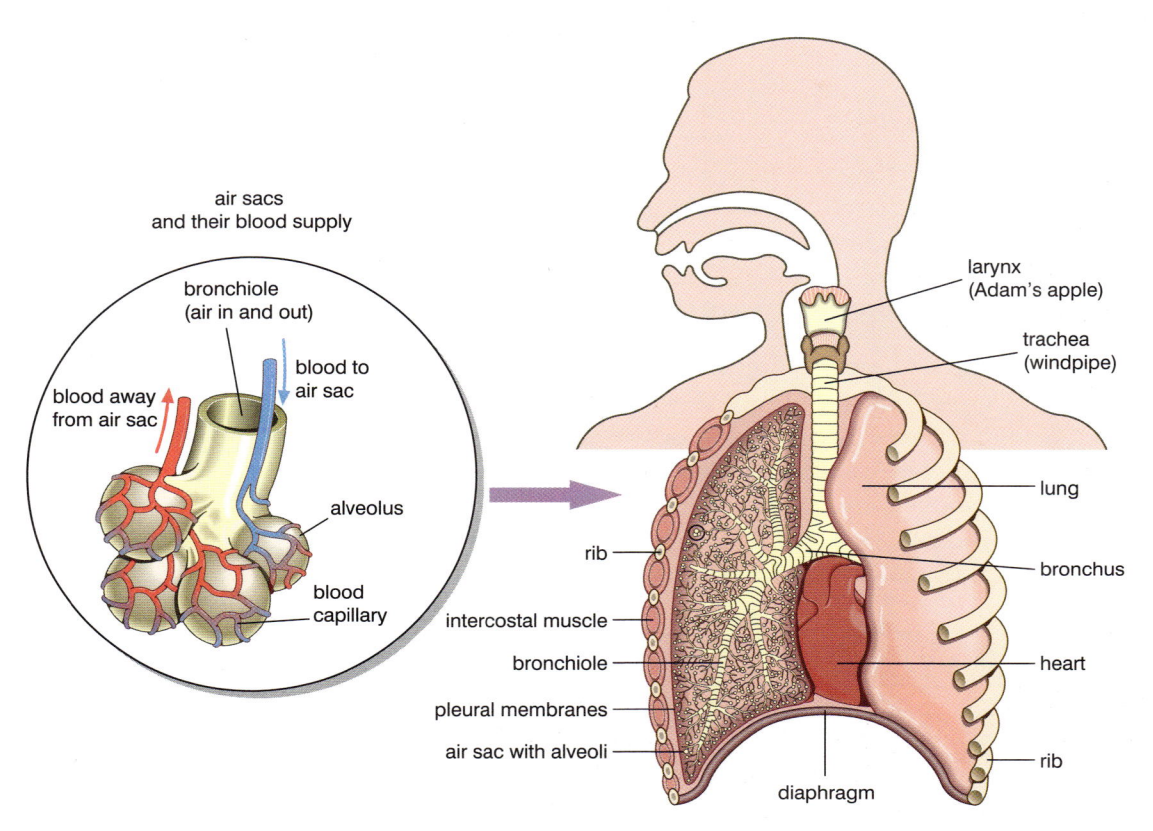

△ Fig. 2.106 The respiratory system.

The **alveoli** are where oxygen and carbon dioxide diffuse into and out of the blood. For this reason the alveoli are described as the site of gas exchange or the respiratory surface.

The alveoli are adapted for efficiency in exchanging gases by diffusion. They have:

- thin permeable walls which keeps the distance over which diffusion of gases takes place, between the air and blood, to a minimum
- a moist lining, in which the gases dissolve before they diffuse across the cell membranes
- a large surface area because there are hundreds of millions of alveoli in a human lung, giving a surface area of around 70 m^2 for diffusion
- high concentration gradients for the gases, because the blood is continually flowing past the air sacs, delivering excess carbon dioxide and taking on additional oxygen, and because of ventilation of the lungs which refreshes the air in the air sacs.

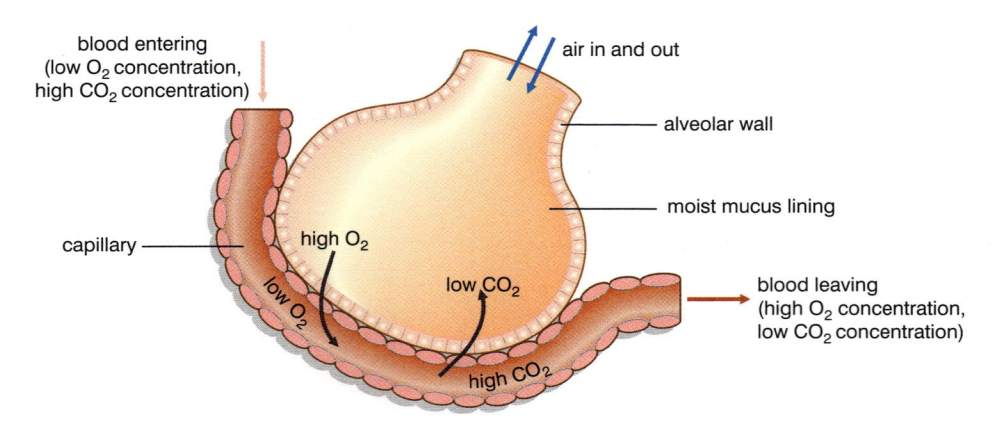

△ Fig. 2.107 Gas exchange in an air-filled alveolus.

REMEMBER

For the highest marks, be careful how you describe the process of gas exchange between the air in the lungs and the blood. Remember that diffusion is a passive process, so that it only occurs while there is a concentration gradient. Avoid answering in simple terms that imply the movement of oxygen is only from the air to the blood, and that the movement of carbon dioxide is only from the blood to the air.

QUESTIONS

1. Explain as fully as you can why gas exchange systems show adaptations for a rapid rate of diffusion.
2. List the structures of the human respiratory system. For each structure, explain its role in breathing.
3. Sketch a diagram of an alveolus and annotate it to show how it is adapted for efficient gas exchange. (Hint: remember to refer to diffusion.)
4. **EXTENDED** What is the role of the cilia and mucus in the human respiratory system?

EXTENDED

BREATHING IN AND OUT

Breathing in is known as inhalation or **inspiration**, and breathing out as exhalation or **expiration**.

Both happen because of changes in the volume of the thorax. The change in volume causes pressure changes, which in turn cause air to enter or leave the lungs.

The ribs surrounding the thorax are joined together by **intercostal muscles**, and below the lungs is the **diaphragm**. The diaphragm is a domed sheet of tough tissue surrounded by muscle that attaches to the thorax. There are two sets of intercostals: the internal intercostal muscles and the external intercostal muscles. The ribs, intercostal muscles and diaphragm work together to bring about breathing or ventilation of the lungs.

In gentle breathing, only the diaphragm may be involved. In deeper breathing, such as during exercise, or if we think about it, the ribs and intercostal muscles become involved.

Breathing in

Air is breathed into the lungs in large breaths as follows.

1. The diaphragm contracts and flattens in shape.

2. The external intercostal muscles contract, making the ribs move upwards and outwards.

3. These changes cause the volume of the thorax to increase.

4. This causes the air pressure in the thorax to decrease.

5. This decrease in pressure draws air into the lungs.

Rings of cartilage in the trachea and bronchi keep the air passages open and prevent them from collapsing when the air pressure decreases and bursting when air pressure increases.

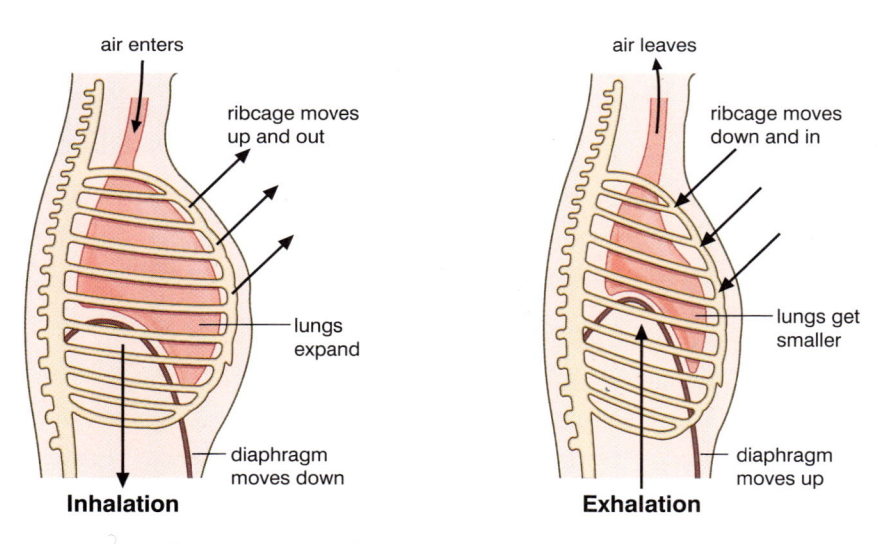

△ Fig. 2.108 Left: breathing in. Right: breathing out.

Breathing out

Air is breathed out from the lungs as follows.

1. The diaphragm relaxes and returns to its domed shape, pushed up by the liver and stomach. This means it pushes up on the lungs.

2. The external intercostal muscles relax, allowing the ribs to drop back down. This also presses on the lungs. If you are breathing hard, the internal intercostal muscles also contract, helping the ribs to move down.

3. These changes cause the volume of the thorax to decrease.

4. This causes the air pressure in the thorax to increase.

5. This causes air to leave the lungs.

> **REMEMBER**
>
> Be prepared to answer questions on breathing by comparing the pressure inside the lungs and external air pressure. Air moves from an area of higher pressure to an area of lower pressure.
> - During inhalation, air enters because the air pressure inside the lungs is lower than the air pressure outside the body.
> - During exhalation, air leaves the lungs because the air pressure inside is higher than the air pressure outside the body.

END OF EXTENDED

> **SCIENCE IN CONTEXT**
>
> **ARTIFICIAL RESPIRATION**
>
> Sometimes accident or illness can damage a person's ability to breathe. Since exchange of gases is essential for respiration, and so for life, it is critical that this process is continued artificially until the patient is sufficiently recovered to be able to do it for themselves again.
>
> △ Fig. 2.109 A ventilator mask forces air into the patient's lungs by increasing air pressure, and allows air out of the patient's lungs by decreasing the air pressure in the mask.
>
> In the past, the patient was placed inside a large machine, sometimes called an iron lung. The machine was sealed from the air, and changes in pressure inside the machine caused changes in thoracic volume, which resulted in air being drawn in or forced out of the patient's lungs. Today, a sealed mask is placed over the patient's mouth and nose, and air is forced into the lungs by increasing the air pressure. The air is naturally breathed out again as the stretched muscles relax.

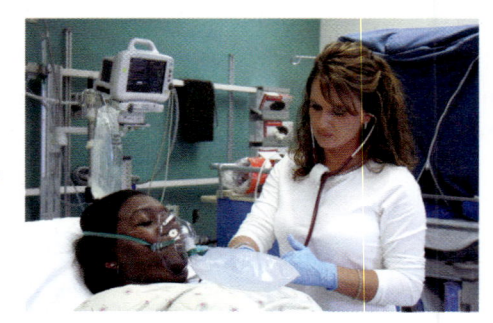

Inspired air and expired air

The air we breathe in and out contains many gases. Oxygen is taken into the blood from the air we breathe in. Carbon dioxide and water vapour are added to the air we breathe out. The other gases in the air we breathe in are breathed out almost unchanged, except for being warmer.

	In inhaled air	In exhaled air
Oxygen	21%	16%
Carbon dioxide	0.04%	4.5%
Nitrogen and other gases	79%	79%
Water	variable	high
Temperature	variable	high

△ Table 2.10 Characteristics of inspired and expired air.

You can compare the carbon dioxide in inspired air and expired air using the apparatus shown in the diagram. Lime water reacts with carbon dioxide and turns cloudy, so this is a test for carbon dioxide.

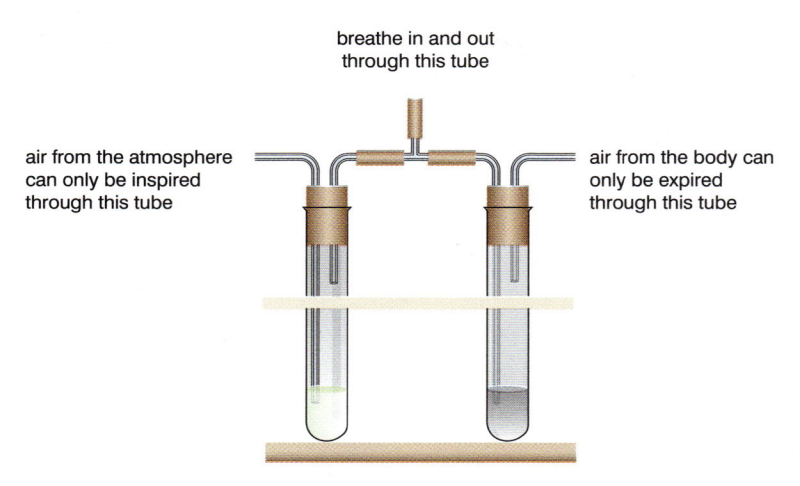

◁ Fig 2.110 This apparatus can be used to investigate the production of carbon dioxide when breathing. Note: one-way valves should be used when setting up this apparatus.

Investigating the effect of exercise on breathing

There are two aspects of breathing that can change during exercise: the rate of breathing and the volume of breath.

- Rate of breathing is usually counted as number per minute.
- The volume of a breath can be measured in dm^3 using a spirometer. A simple spirometer can be made using a 2-litre plastic bottle that has been marked down the side with volumes of water. (This can be done by adding 500 cm^3 of water at a time, and marking the volume on the side of the bottle with a waterproof marker.) When the bottle is full of water, turn it upside-down into a water trough without allowing any air into the bottle. Insert a flexible plastic tube into the neck of the bottle and secure the bottle and tube in position. Clean the other end of the tubing with antiseptic solution. (Alternatively, add a mouthpiece to the end of the tubing that can easily be removed and sterilised after each test.) To measure the volume of a breath, ask another person to wear a noseclip and then breathe out a normal breath into the tube. The scale on the bottle can be used to measure the volume of air breathed out.

[Note: this apparatus must only be used for measuring one breath. The bottle must be set up again before measuring another breath to ensure a fair test.]

Developing investigative skills

Using and organising techniques, apparatus and materials

❶ Design an investigation into the effect of exercise on breathing rate or volume. (Hint: think carefully about how many people to test, and how to test them, in order to get reliable results.)

Observing, measuring and recording

❷ This investigation could involve vigorous exercise. What risks will you need to prepare for, and how should they be minimised?

Handling experimental observations and data

The data in the table are the results from an investigation into the effect of exercise on breathing in 4 people. They were first tested at rest and then after 2 minutes of running on a treadmill set at the same speed.

Person	A		B		C		D	
	Rate/ breaths/ minute	Breath volume/ dm^3	Rate/ breaths/ minute	Breath volume/ dm^3	Rate/ breaths/ minute	Breath volume/ dm^3	Rate/ breaths/ minute	Breath volume/ dm^3
At rest	13	0.5	15	0.4	12	1.2	18	0.6
After exercise	19	1.3	23	0.9	18	1.3	26	1.5

❸ Explain how these data should be adjusted before they can give a reliable answer to the question 'How does exercise affect breathing?'

Note: The results of an investigation like this one should show that both the rate of breathing and depth of breathing increase with the level of activity. However, the rate and depth of breathing of a trained athlete will change less than those of an untrained person.

EXTENDED

The rate and depth of breathing increase with level of activity because, as the muscles contract faster, they respire faster and so make carbon dioxide more quickly. Carbon dioxide is an acidic gas that dissolves easily in water-based solutions, such as the cytoplasm of a cell and blood plasma. The more carbon dioxide there is in solution the more acidic the solution. A change in pH can affect the activity of many cell enzymes (see Enzymes), so it is important that carbon dioxide is removed from the cells and the body as quickly as possible.

END OF EXTENDED

> **SCIENCE IN CONTEXT**
>
> **MONITORING CARBON DIOXIDE IN THE BLOOD**
>
> Rate of breathing is controlled by a part of the brain that measures the carbon dioxide concentration of the blood, not the oxygen concentration. This is because a small increase in carbon dioxide concentration in body fluids could have a much more damaging effect on the body than a small decrease in oxygen concentration.

QUESTIONS

1. **EXTENDED** Describe what happens during a) inhalation, b) exhalation, in terms of movement of structures in the respiratory system and in terms of pressure and volume changes in the thoracic cavity.

2. Look at Table 2.10 of the gases found in inhaled and exhaled air. Explain the changes between inhaled and exhaled air shown.

3. a) Describe the effects of exercise on the rate and depth of breathing.

   b) **EXTENDED** Explain what would happen to cells if rate and depth of breathing did not change during exercise.

End of topic checklist

Key terms

aerobic respiration, alveoli, anaerobic respiration, bronchus, bronchioles, cellular respiration, cilia, diaphragm, expiration, inspiration, intercostal muscles, larynx, lungs, lymphocyte, mucus, phagocyte, synthesis, trachea, ventilation

During your study of this topic you should have learned:

- ○ The definition of *respiration* as the chemical reactions that break down nutrient molecules in living cells to release energy.
- ○ About the uses of energy in the human body:
  - muscle contraction
  - protein synthesis
  - cell division
  - active transport
  - growth
  - the passage of nerve impulses
  - the maintenance of a constant body temperature.
- ○ The definition of aerobic respiration as the release of a relatively large amount of energy in cells by the breakdown of food substances in the presence of oxygen.
- ○ The word equation for aerobic respiration:

  glucose + oxygen → carbon dioxide + water (+ energy)

- ○ **EXTENDED** The symbol equation for aerobic respiration:

  $C_6H_{12}O_6 + 6O_2 \rightarrow 6CO_2 + 6H_2O$

- ○ The definition of *anaerobic respiration* as the release of a relatively small amount of energy by the breakdown of food substances in the absence of oxygen.
- ○ The word equation for anaerobic respiration in muscles during hard exercise:

  glucose → lactic acid

- ○ The word equation for anaerobic respiration in the microorganism yeast:

  glucose → ethanol + carbon dioxide

- **EXTENDED** The symbol equation for anaerobic respiration in muscles:

  $C_6H_{12}O_6 \rightarrow 2C_3H_6O_3$

- **EXTENDED** The symbol equation for anaerobic respiration in yeast:

  $C_6H_{12}O_6 \rightarrow 2C_2H_5OH + 2CO_2$

- About the role of anaerobic respiration in yeast during brewing and bread-making.
- How to compare aerobic respiration and anaerobic respiration in terms of relative amounts of energy released.
- **EXTENDED** How to describe the effect of lactic acid in muscles during exercise.
- About the features of gas exchange surfaces in animals.
- How to identify on diagrams and name the larynx, trachea, bronchi, bronchioles, alveoli and associated capillaries.
- About the differences in composition between inspired and expired air.
- How to use lime water as a test for carbon dioxide to investigate the differences in composition between inspired and expired air.
- How to investigate and describe the effects of physical activity on rate and depth of breathing.
- **EXTENDED** About the role of the ribs, internal and external intercostal muscles and diaphragm in producing volume and pressure changes leading to the ventilation of the lungs.
- **EXTENDED** How to explain the role of mucus and cilia in protecting the gas exchange system from pathogens and particles.
- **EXTENDED** How to explain the link between physical activity and rate and depth of breathing in terms of changes in the rate at which tissues respire and therefore of carbon dioxide concentration and pH in tissues and in the blood.

End of topic questions

Note: The marks awarded for these questions indicate the level of detail required in the answers. In the examination, the number of marks awarded to questions like these may be different.

1. **a)** List the body systems in a human that are involved in supplying the reactants of cellular respiration. **(3 marks)**

   b) List the body systems in a human that are involved in removing the products of cellular respiration. **(3 marks)**

2. Draw up a table to summarise the similarities and differences between aerobic and anaerobic respiration. **(8 marks)**

3. Students were studying the results of respiration in some woodlice. They set up two identical sets of apparatus, of a boiling tube fitted with a bung and linked to a tube of lime water through a delivery tube. They placed some woodlice in one boiling tube, and no woodlice in the other tube. The boiling tubes were fitted with their bungs so that no additional air could enter the apparatus and then left overnight.

   a) What was the role of the second set of apparatus? Explain your answer. **(2 marks)**

   b) Suggest what happened to the lime water in the two sets of apparatus. **(2 marks)**

   c) Explain as fully as you can your answer to part **b)**. **(3 marks)**

4. A whale takes a deep breath of air and then dives for half an hour. Suggest how energy is generated in the whale's muscles over the period of the dive. **(4 marks)**

5. **a)** Define the terms *diffusion* and *gas exchange*. **(2 marks)**

   b) Using a suitable example, describe the role of diffusion in gas exchange in organisms, and explain how the tissues and organs of the gas exchange surfaces are adapted to maximise the rate of gas exchange. **(5 marks)**

6. **EXTENDED** The diagram shows a model that can be used to demonstrate the role of the diaphragm in breathing.

   a) Describe and explain what will happen to the balloon 'lungs' when the rubber diaphragm is

   i) pulled down

   ii) pushed up. **(8 marks)**

   b) Which parts of the body that can be involved in breathing are not included in this model? Explain their role in breathing. **(7 marks)**

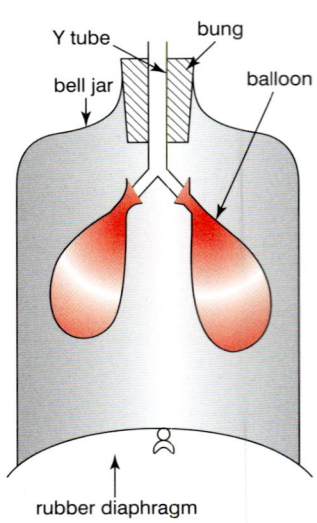

Excretion in humans

INTRODUCTION

More blood flows through the kidneys every hour than through the blood vessels of any other organ in the body. The kidneys filter out waste and toxic substances from the blood as quickly as possible and excrete them in urine. In the past doctors used to check the colour of urine, smell it and even taste it, to help diagnose what was wrong with a patient. Today, doctors and nurses use different diagnostic urine tests as a quick way to identify if a patient is healthy or not.

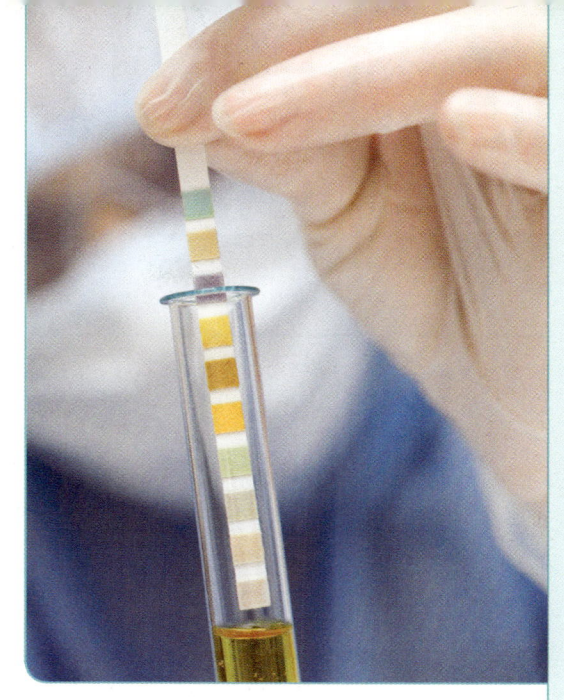

△ Fig. 2.111 A dipstick test like this can quickly test urine for a number of excreted substances.

KNOWLEDGE CHECK

✓ Know that humans breathe out carbon dioxide produced from respiration.
✓ Know that excess amino acids from digestion are broken down in the liver.

LEARNING OBJECTIVES

✓ Be able to define *excretion* as the removal from organisms of toxic materials, the waste products of metabolism and substances in excess of requirements.
✓ Be able to describe function of the kidney in terms of the removal of urea and excess water and the reabsorption of glucose and some salts.
✓ Be able to state the relative positions of ureters, bladder and urethra in the body.
✓ Be able to state that urea is formed in the liver from excess amino acids.
✓ Be able to state that alcohol, drugs and hormones are broken down in the liver.
✓ **EXTENDED** Be able to outline the structure of a kidney and the structure and functioning of a kidney tubule.
✓ **EXTENDED** Be able to explain that dialysis can be used to treat patients with kidney failure, by cleaning the blood on a regular basis using a dialysis machine.
✓ **EXTENDED** Be able to discuss the application of dialysis in kidney machines.
✓ **EXTENDED** Be able to discuss the advantages and disadvantages of kidney transplants, compared with dialysis.

EXCRETION

Excretion is the process or processes by which an organism gets rid of materials that it doesn't need. These materials include:

- toxic (poisonous) materials that could harm the body
- waste products of metabolism – from the chemical reactions inside cells, including respiration

- substances that are in excess – there is more of them in the body than is needed.

> **REMEMBER**
> Excretion is different from egestion.

The activities in human cells produce many substances that need to be excreted.

- Carbon dioxide is the waste product from respiration, which if it remained in cells would change their pH and affect the activity of enzymes. It diffuses from respiring cells into the plasma of the blood, and is carried around the body until it reaches the lungs. Here it diffuses through the capillary and alveoli walls and is breathed out.
- Waste products of many cell processes, including urea, dissolve in the blood and are carried to the kidneys, where they are excreted.
- Food and drink can add more water and salts to the body than the cell needs. So the excess is excreted by the kidneys.

The liver plays an important role in preparing some substances for excretion:

- **urea** is produced in liver cells from the breakdown of excess amino acids
- some **hormones** such as adrenaline are broken down by liver cells when they are no longer needed by the body
- toxic substances, such as alcohol and drugs, absorbed into the body through the alimentary canal wall are carried by the hepatic portal vein directly to the liver (so they have little effect on the rest of the body) where they are broken down to less toxic substances.

The products of these breakdown processes, such as urea, are carried in the blood to the kidneys where they are excreted in urine.

QUESTIONS

1. Define the term *excretion* in your own words.
2. Describe the role of the lungs in excretion in humans.
3. Name three groups of substances that the liver prepares for excretion.

The human urinary system

Humans have two kidneys situated near the spine at the back of the body, about halfway down the the abdomen. The kidneys are well supplied with blood by the renal arteries and veins. Inside the kidneys:

- the blood is filtered to remove waste substances, including urea, and excess substances, such as water, that are not needed by the body

- substances that the body does need, such as glucose and some mineral salts, are reabsorbed from the filtered solution
- this leaves excess water, urea and mineral ions, which together form **urine**.

Urine flows out of the kidneys down the **ureters**, and into the **bladder**. The urine is stored in the bladder until a ring of muscle at the base is released (usually when you go to the toilet). The urine then flows out of the bladder, through the **urethra** to the environment.

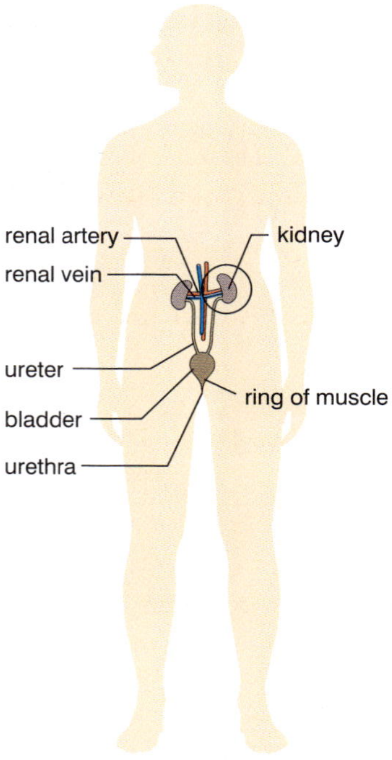

△ Fig. 2.112 The human urinary system.

QUESTIONS

1. Draw up a table to list the main structures of the urinary system and their functions.
2. Name two processes that occur in the kidney to produce urine.
3. State three substances found in urine.

EXTENDED

Structure of the kidney

Each kidney consists of around one million tiny tubules called **nephrons**. Each nephron is associated with a blood capillary. The exchange of substances between the nephron and capillary forms urine, which passes into the ureter.

All the nephrons pass between the outer area of the kidney, called the **cortex**, and the inner area, called the **medulla**. The cortex is brighter red than the medulla because of the large number of capillaries flowing through this region. The innermost part of the kidney is the pelvis, into which the nephrons drain and where the urine collects before it flows down the ureter.

Different parts of the nephron are involved in urine formation and excretion, and in osmoregulation (water control).

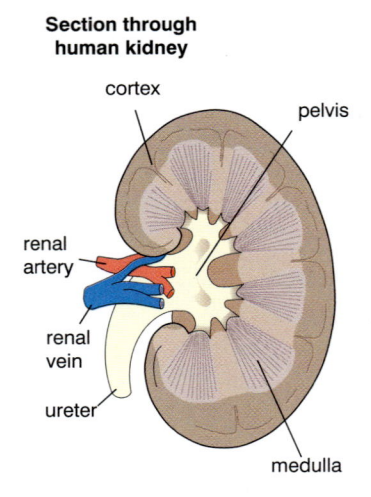

Section through human kidney
- cortex
- pelvis
- renal artery
- renal vein
- ureter
- medulla

△ Fig. 2.113 The structure of the kidney.

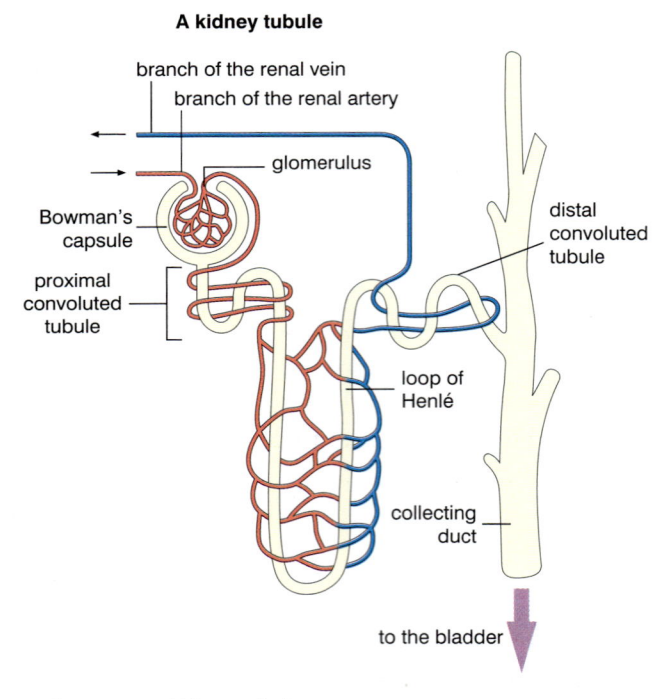

A kidney tubule
- branch of the renal vein
- branch of the renal artery
- glomerulus
- Bowman's capsule
- proximal convoluted tubule
- distal convoluted tubule
- loop of Henlé
- collecting duct
- to the bladder

△ Fig. 2.114 A kidney tubule.

EXCRETION BY THE KIDNEYS

The formation and excretion by the kidneys involves two separate processes: filtration and reabsorption.

Filtration

The walls of the renal capsule and glomerulus are only one cell thick, and there are tiny gaps between the cells. Blood enters the glomerulus from the renal artery. The blood vessel that enters the glomerulus is wider than the blood vessel that leaves it. This produces a high blood pressure in the glomerulus. This pressure squeezes many small molecules out from the blood and into the capsule. These molecules include water, glucose, mineral ions (such as sodium and chloride), hormones, vitamins and urea, which together form the filtrate. Large molecules such as blood proteins and blood cells are too large to be filtered through into the capsule, so they remain in the blood.

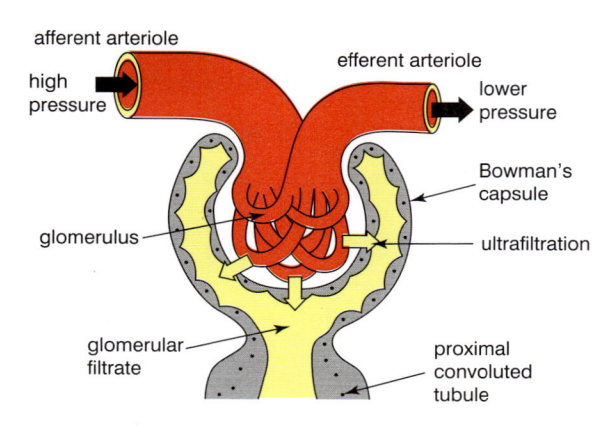

△ Fig. 2.115 The glomerulus and renal capsule.

> **REMEMBER**
>
> In order to get the best marks, remember to describe the movement of particles in terms of their concentration gradients.

Reabsorption

As the filtrate passes through the proximal convoluted tubule, some substances are reabsorbed from the filtrate back into the blood in the capillary that runs close by. Some substances are reabsorbed more than others, so this is called **selective reabsorption**. For example, most of the sodium ions and all of the glucose which were filtered into the tubule are reabsorbed by active transport. These substances are very important in the body, so it is essential that they are not lost in urine.

A lot of water is also reabsorbed from the convoluted tubule by osmosis, as the concentration of water molecules is higher in the filtrate than in the blood. Many other substances that the body needs, such as vitamins, other mineral ions and amino acids, are also reabsorbed here.

Additional water is reabsorbed into the blood from the collecting duct.

> **SCIENCE IN CONTEXT**
>
> **SURVIVING WITHOUT WATER**
>
> Australian hopping mice live in the deserts of Australia.
>
> These mice can survive without ever drinking liquid water. This is because most of the water that is released from food molecules during respiration is reabsorbed in their kidneys and kept within the body.
>
> The mice produce very small amounts of thick, syrupy urine.
>
> △ Fig. 2.116 An Australian hopping mouse.

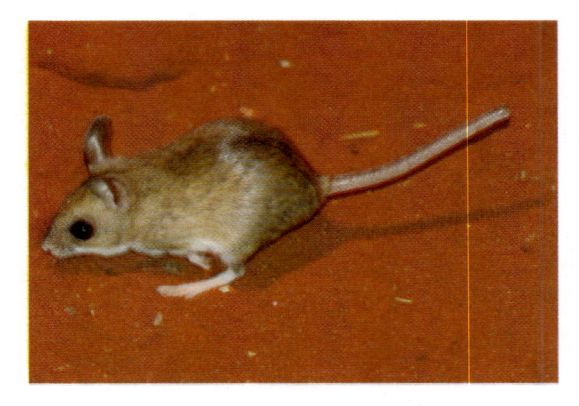

QUESTIONS

1. **EXTENDED** List the structures in the order that a urea molecule would pass through from the blood to the urethra.

2. **EXTENDED** In which parts of the nephron do the following take place?

    a) filtration

    b) reabsorption.

3. a) **EXTENDED** Name one substance that is selectively reabsorbed by the nephron.

    b) Which process is used to reabsorb this substance?

KIDNEY FAILURE

You have two kidneys but you can survive with only one. However, if both kidneys become damaged through accident or disease, then some means of replacing them must be found. There are two possibilities: kidney transplant and kidney dialysis.

Kidney dialysis

If the kidneys stop functioning, it takes only a few days for the concentration of urea and excess substances in the body to increase to a level that damages the functioning of cells. It is essential that these substances are removed from the blood to keep them at a safe level. This is done using an artificial method of cleaning the blood, called **dialysis**.

Blood from an artery in the patient's arm is pumped through the dialysis machine, checked for bubbles and then returned to the body through a vein.

The dialysis machine imitates the effect of the kidney. It passes the blood through a tube that is separated from the dialysis fluid by a partially permeable membrane. The dialysis fluid contains a balance of solutes that cause waste products such as urea and excess ions, protein and water to diffuse out of the blood into the fluid. The fluid in the machine is continually refreshed so that the concentration gradient is maintained between the dialysis fluid and the blood, and to make sure the blood is restored to the correct balance of substances.

Dialysis may take three or four hours to complete. It needs to be done several times a week to prevent the build-up of substances in the blood.

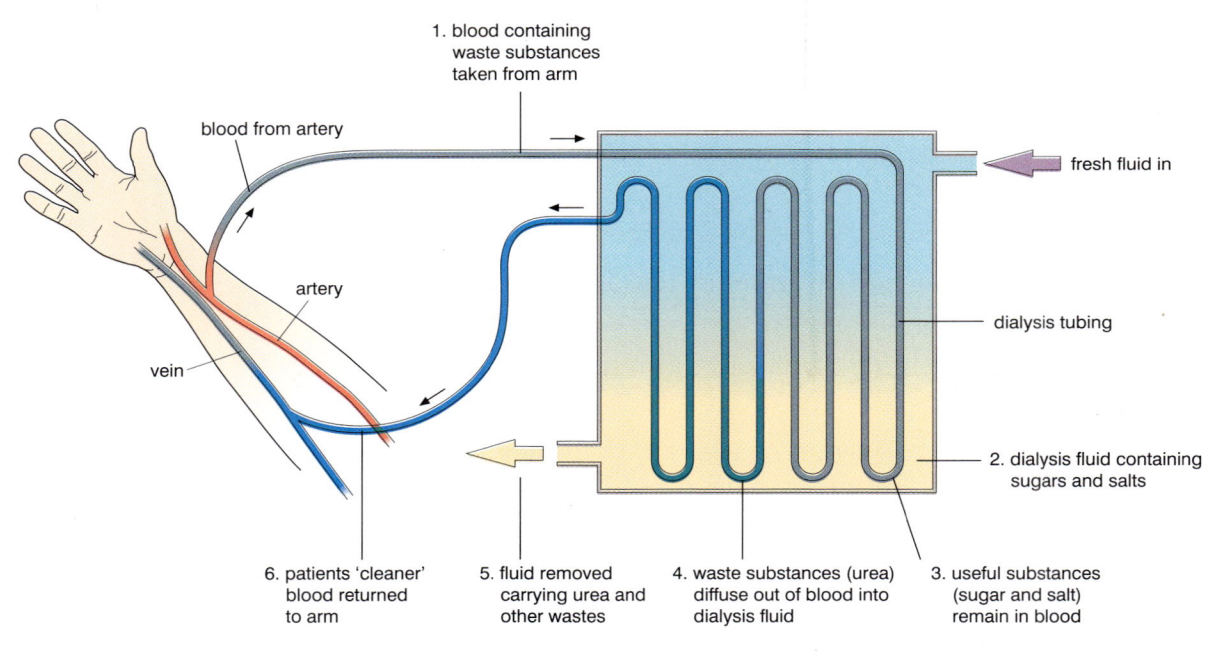

△ Fig. 2.117 How dialysis works.

Kidney transplant

Another solution to kidney failure is to **transplant** a healthy kidney into the patient. This is a better long-term solution because a healthy kidney works continually to maintain the right concentration of substances in the blood. The failed kidneys are only removed if they are diseased, and the new kidney is attached to the blood system lower in the body. The transplant kidney, which must be healthy, may come from someone who has just died, such as in a car accident, or from a living person who has two healthy kidneys and has donated one.

Successful organ transplant is difficult because of the immune system, which attacks any cells in the body that it does not recognise. This is useful for fighting infection, but a serious problem for transplants. If the immune system attacks the transplanted organ, the tissue will be destroyed. This is called **tissue rejection**.

The immune system identifies cells by some of the molecules in their cell membranes that act as 'markers'. Different people have different markers of their cell surfaces, and people who are closely related usually have markers that are more alike than those of people who are not related. The risk of rejection can be minimised by matching the markers on the tissue of the patient and the new kidney, so that they are as similar as possible. This is why a kidney from a close relative is more likely to be successfully transplanted than one from a stranger.

Even with a close tissue match, in order to prevent rejection, the patient will have to continually take drugs that suppress the immune system. This increases the risk of other infections, because the immune system does not fight them as well as usual. A transplanted kidney may last many years, or just a short time, depending on how well the body accepts the new tissue.

Until a kidney that is a close tissue match becomes available, the patient will have to continue with dialysis every few days.

QUESTIONS

1. **EXTENDED** Describe two treatments for kidney failure.
2. **EXTENDED** Explain why treatment for kidney failure is only needed if both kidneys fail.
3. **EXTENDED** Compare dialysis with normal kidney function to show its similarities and its differences.
4. **EXTENDED** Explain why one patient with kidney failure may need to wait years before they can have a kidney transplant, but a second patient may get a transplant much sooner.
5. **EXTENDED** Write down two disadvantages of kidney transplants.

END OF EXTENDED

End of topic checklist

Key terms

bladder, collecting duct, convoluted tubule, dialysis, excretion, glomerulus, hormone, nephron, renal capsule, selective reabsorption, tissue rejection, transplant, urea, ureter, urethra urine

During your study of this topic you should have learned:

○ The definition of *excretion* as the removal from organisms of toxic materials, the waste products of metabolism (chemical reactions in cells including respiration) and substances in excess of requirements:
  - carbon dioxide
  - urea
  - salts.

○ How to describe the function of the kidney in terms of the removal of urea and excess water and the reabsorption of glucose and some salts.

○ The relative positions of ureters, bladder and urethra in the body.

○ That urea is formed in the liver from excess amino acids.

○ That alcohol, drugs and hormones are broken down in the liver.

○ **EXTENDED** How to outline the structure of a kidney (cortex, medulla, and the start of the ureter) and the structure and functioning of a kidney tubule, including:
  - the role of the renal capsule in filtration from the blood of water, glucose, urea and salts
  - the role of the tubule in the reabsorption of glucose, most of the water and some salts back into the blood, leading to concentration of urea in the urine as well as loss of excess water and salts.

○ **EXTENDED** That dialysis can be used to treat patients with kidney failure, by cleaning the blood on a regular basis using a dialysis machine.

○ **EXTENDED** How to discuss the application of dialysis in kidney machines.

○ **EXTENDED** How to discuss the advantages and disadvantages of kidney transplants, compared with dialysis.

End of topic questions

Note: The marks awarded for these questions indicate the level of detail required in the answers. In the examination, the number of marks awarded to questions like these may be different.

1. **a)** Give two examples of organs of excretion in humans. **(2 marks)**

   b) For each organ name the substances they excrete. **(2 marks)**

2. Describe the role of the liver in controlling the concentration of amino acids in the body. **(4 marks)**

3. Name four organs of the human excretory system, and explain the role of each organ in the formation and excretion of urine. **(4 marks)**

4. **EXTENDED** Explain how the structure of the renal capsule is adapted to their function in filtration. **(2 marks)**

5. **a)** **EXTENDED** Explain what is meant by *selective reabsorption*. **(1 mark)**

   b) By which process is **i)** glucose and **ii)** water reabsorbed in the nephron? **(2 marks)**

   c) If some nephrons were treated with a poison that kills cells, would either or both of these processes from b) still be possible? Explain your answer. **(3 marks)**

   d) Explain as fully as you can why glucose is selectively reabsorbed in the kidneys. **(3 marks)**

6. **a)** **EXTENDED** Suggest which parts of a nephron in an Australian hopping mouse are longer than a human nephron. **(2 marks)**

   b) Explain your answer. **(2 marks)**

7. **EXTENDED** Explain why kidney failure must be treated rapidly and repeatedly. **(2 marks)**

8. **EXTENDED** Draw up a table comparing the processes of kidney dialysis and kidney transplant and identifying their advantages and disadvantages. **(9 marks)**

Coordination and response

INTRODUCTION

US tennis star Andy Roddick can hit a tennis ball so hard that it travels at 250 km per hour. This is only a little slower than the top speed of a Ferrari. In order to return the ball successfully, his opponents have only a fraction of a second to work out where to stand and how best to return the ball. Their response is built on years of training so that they can hit the ball back without thinking.

△ Fig. 2.118 Professional tennis players serve so fast that a radar gun is used to measure the speed of the ball.

KNOWLEDGE CHECK

- ✓ Know that plants and animals detect the environment with specialised sense organs.
- ✓ Know that animals respond to changes in the environment using nervous and hormonal systems.
- ✓ Know that plants respond to changes in the environment, such as by growth.
- ✓ Know that nerve cells are specialised cells adapted to the function of carrying electrical impulses.

LEARNING OBJECTIVES

- ✓ Be able to describe the human nervous system in terms of the central nervous system and the peripheral nervous system, which together serve to coordinate and regulate body functions.
- ✓ Be able to identify motor, relay and sensory neurones from diagrams.
- ✓ Be able to describe a simple reflex arc in terms of sensory, relay and motor neurones, and a reflex action.
- ✓ Be able to state that muscles and glands can act as effectors.
- ✓ Be able to describe the action of antagonistic muscles, including the biceps and triceps at the elbow joint.
- ✓ Be able to define *sense organs* as groups of receptor cells responding to specific stimuli: light, sound, touch, temperature and chemicals.
- ✓ Be able to describe the structure and function of the eye, including accommodation and pupil reflex.
- ✓ Be able to define a *hormone* as a chemical substance, produced by a gland, carried by the blood, which alters the activity of one or more specific target organs and is then destroyed by the liver.
- ✓ Be able to state the role of the hormone adrenaline in chemical control of metabolic activity, including increasing the blood glucose concentration and pulse rate.
- ✓ Be able to give examples of situations in which adrenaline secretion increases.
- ✓ Be able to compare nervous and hormonal control systems.
- ✓ Be able to define and investigate *geotropism* and *phototropism*.

- ✓ Be able to define *homeostasis* as the maintenance of a constant internal environment.
- ✓ Be able to identify, on a diagram of the skin, hairs, sweat glands, temperature receptors, blood vessels and fatty tissue.
- ✓ Be able to describe the maintenance of a constant body temperature in humans in terms of insulation and the role of temperature receptors in the skin, sweating, shivering, vasodilation and vasoconstriction of arterioles supplying skin-surface capillaries and the coordinating role of the brain.
- ✓ Be able to define a *drug* as any substance taken into the body that modifies or affects chemical reactions in the body.
- ✓ Be able to describe the use of antibiotics in medicine to treat bacterial infections.
- ✓ Be able to describe the effects of the abuse of heroin: a powerful depressant, problems of addiction, severe withdrawal symptoms and associated problems such as crime and infection with diseases such as HIV/AIDS through the use of shared needles.
- ✓ Be able to describe the effects of excessive alcohol consumption: reduced self-control, depressant, effect on reaction times, damage to liver and social implications.
- ✓ Be able to describe the effects of tobacco smoke and its major toxic components (tar, nicotine, carbon monoxide and smoke particles) on the gas exchange system.
- ✓ **EXTENDED** Be able to distinguish between voluntary and involuntary actions.
- ✓ **EXTENDED** Be able to distinguish between rods and cones, in terms of function and distribution.
- ✓ **EXTENDED** Be able to discuss the use of hormones in food production.
- ✓ **EXTENDED** Be able to explain the chemical control of plant growth by auxins including regulating differential growth and the effects of synthetic plant hormones used as weedkillers.
- ✓ **EXTENDED** Be able to explain the concept of control by negative feedback.
- ✓ **EXTENDED** Be able to describe the control of the glucose content of the blood by the liver, and by insulin and glucagons from the pancreas.
- ✓ **EXTENDED** Be able to explain why antibiotics kill bacteria but not viruses.

SENSITIVITY

Sensitivity is the ability to recognise and respond to changes in external and internal conditions, and is one of the key characteristics of living organisms.

A change in conditions is called a **stimulus**. To produce coordinated response to that stimulus there must be a **receptor** that can recognise the stimulus and an **effector**, a mechanism to carry out the response.

There are two systems involved in coordination and response in humans.

- One is the **nervous system**, which includes the brain, the spinal cord, the peripheral nerves and specialist sense organs such as the eye and the ear. Communication in the nervous system is in the form of electrical impulses and responses may be very rapid.
- The other is the **hormonal** (or endocrine) **system**, which uses chemical communication by means of hormones. Hormones are secreted by **endocrine glands** and act upon target cells in other

tissues and organs. The hormonal system helps to maintain basic body functions including metabolism and growth. Its response is usually slower than the nervous system, and it often has more long-term effects.

> **REMEMBER**
>
> For the top grades you must be able to compare and contrast the two systems of coordination in humans.
>
> Try also to identify the stimulus, receptor, effector and response in any example of the nervous or hormonal system.

Sensitivity in humans

In the human nervous system:

- specialised **sense organs** that contain receptor cells sense stimuli (changes in conditions)
- information about these stimuli is sent as electrical impulses from the receptor cells through nerves to the central nervous system
- the **central nervous system** (brain and spinal cord) processes the electrical impulses and coordinates the response
- the response passes as electrical impulses along **nerves** to effectors, which are often muscles but may be glands that produce hormones
- the effectors produce the response to the stimulus.

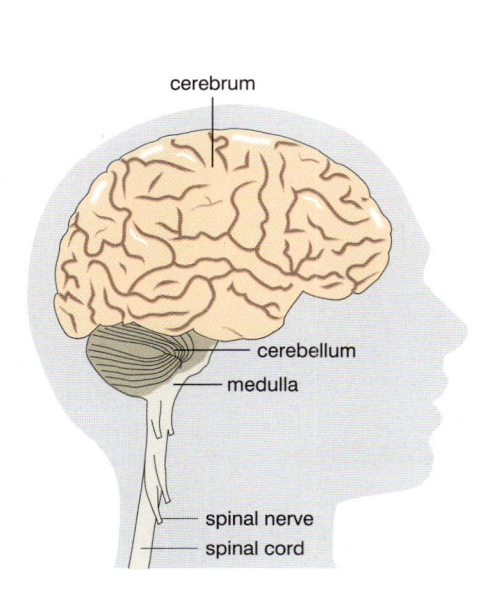

△ Fig. 2.119 The human central nervous system.

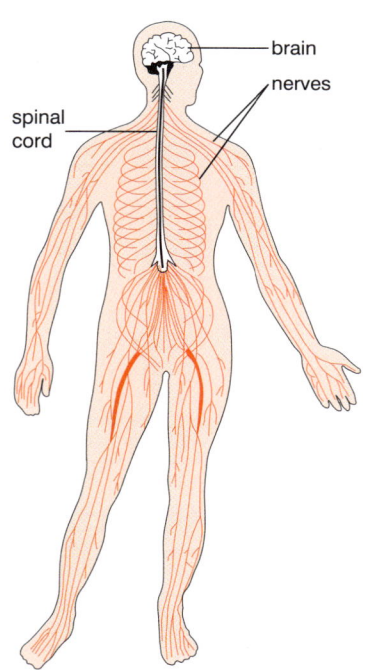

△ Fig. 2.120 The human nervous system.

QUESTIONS

1. What is sensitivity?
2. Explain the following terms:
   a) stimulus
   b) effector
   c) neurone
   d) central nervous system.
3. Which system in the human body is responsible for sensitivity?

Nerve cells

Nerves connect the sense organs to the central nervous system. Nerves, and the brain and spinal cord are made of specialised cells called **neurones**. These cells are specially adapted for their function. They have long cell extensions, called axons and dendrons, that carry the electrical impulses. They also have many endings that connect with other neurones, to pass on the electrical impulses.

- Neurones that link sense organs to the central nervous system are called **sensory neurones**.
- Neurones within the central nervous system may be very short, and are called **relay neurones** (sometimes intermediate or connecting neurones).
- Neurones that connect the central nervous system to an effector, such as a muscle, are called **motor neurones**.

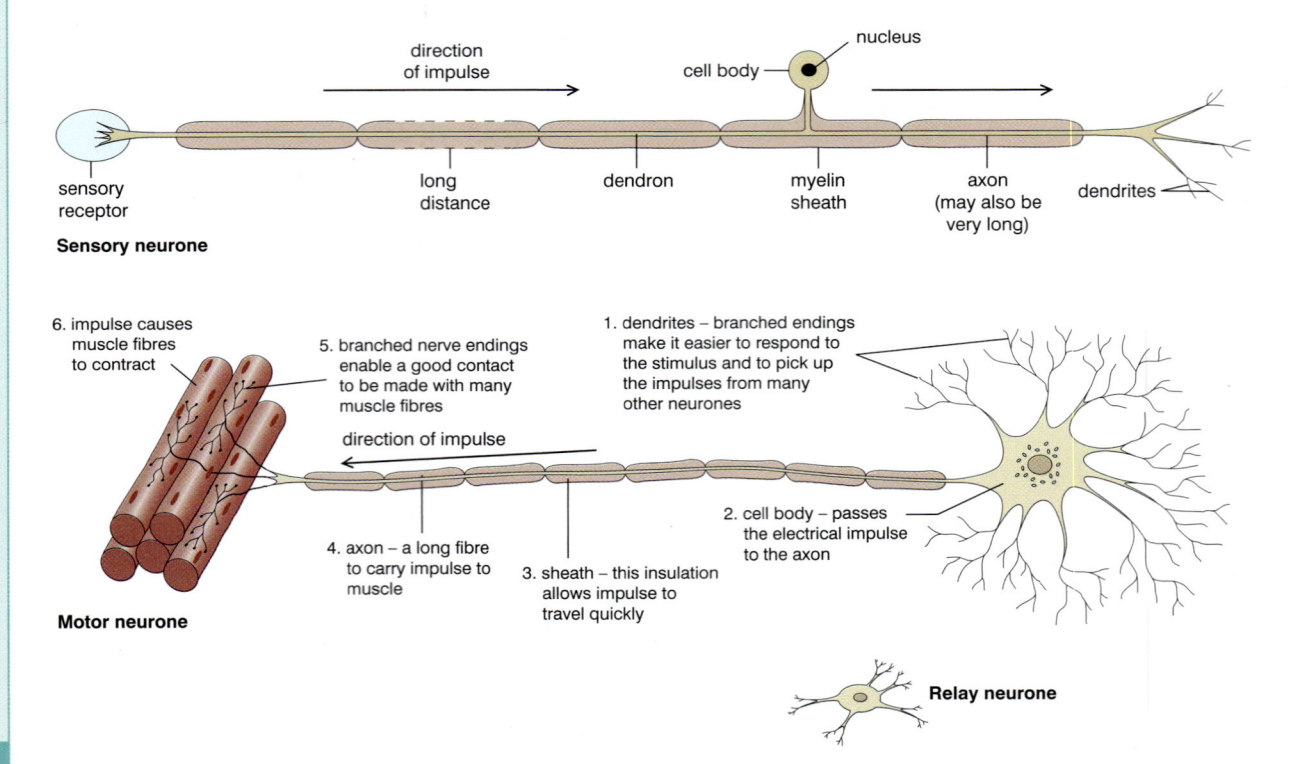

△ Fig. 2.121 Neurones.

Reflex responses

The simplest type of response to a stimulus is a **reflex**. Reflexes are rapid, automatic responses to a specific stimulus that usually protect you in some way, for example blinking if something gets in your eye or sneezing if you breathe in dust.

The pathway that nerve impulses travel along during a reflex is called a **reflex arc**:

stimulus → receptor → sensory neurone → relay neurone in CNS → motor neurone → effector → response

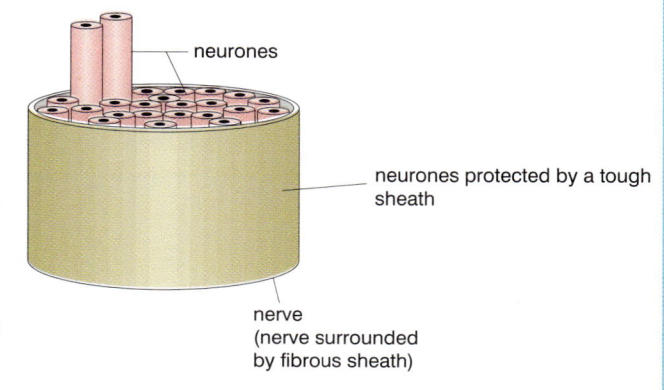

△ Fig. 2.122 Nerves are large bundles of many neurones.

(A few reflex arcs contain only a sensory neurone and a motor neurone, but most also contain a relay neurone.)

Simple reflexes are usually spinal reflexes, which means that the impulses are processed by the spinal cord, not the brain. The spinal cord sends an impulse back to the **effector**. Effectors are the parts of the body that respond, either muscles or glands. Examples of spinal reflexes include responses to stepping on a pin or touching a hot object.

stand on pin → nerve endings → sensory neurone → spinal cord → motor neurone → leg muscles → leg moves

When the spinal cord sends an impulse to an effector, other impulses are sent on to the brain so that it is aware of what is happening. This also allows the brain to over-ride the reflex response. For example, if you were holding a large bowl of hot food that you were looking forward to eating, you might look around quickly for somewhere to put it down rather than drop it immediately and risk breaking the bowl.

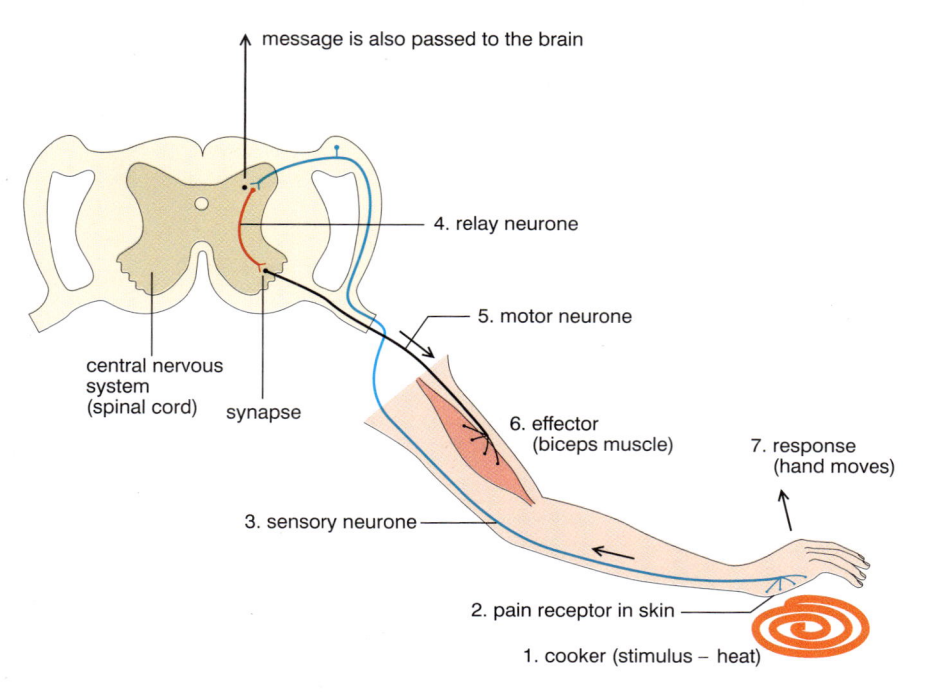

△ Fig. 2.123 A spinal reflex to touching a hot object.

REMEMBER

For higher marks you will need to understand the structure and functioning of the reflex arc, and be able to interpret diagrams and describe what happens at each step.

EXTENDED

Reactions that happen automatically, such as those caused by a reflex response, are usually **involuntary**, meaning that the body responds without the conscious control of the central nervous system. Many more of our responses are **voluntary**, meaning that we make a conscious decision, using the brain, to make a particular response. Involuntary actions are usually essential for survival and are immediate, while voluntary responses take longer.

END OF EXTENDED

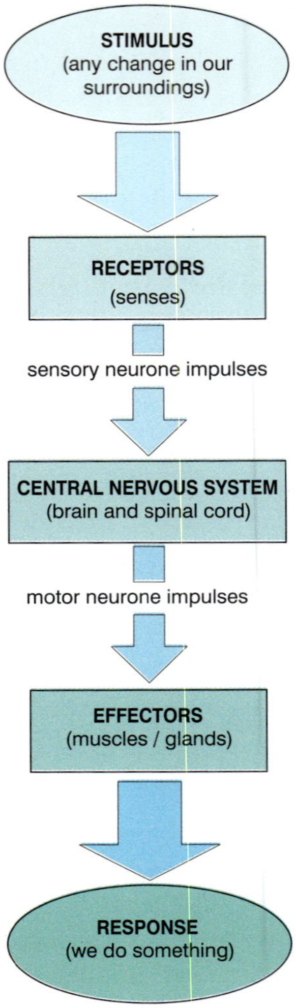

QUESTIONS

1. Describe three types of neurone, and the differences in structure and function in each type.
2. What is meant by the term *reflex response* and why are these responses important for survival?
3. **EXTENDED** With examples, distinguish between a *voluntary* and an *involuntary* reaction.

△ Fig. 2.124 A diagrammatic spinal reflex.

Sense organs

Different sense organs contain different specialised receptor cells that respond to different stimuli.

Sense organ	Sense	Stimulus
Skin	Touch	Pressure, pain, hot/cold temperatures
Tongue	Taste (chemicals)	Chemicals in food and drink
Nose	Smell (chemicals)	Chemicals in the air
Eyes	Sight	Light
Ears	Hearing	Sound
	Balance	Movement/position of head

Table 2.11. The different sense organs in humans.

The eye

In humans, the eye is the sense organ that responds to changes in light. The specialised light-sensitive receptor cells are found in the **retina** at the back of the eye. Light passing through the eye and reaching the retina causes changes in these cells, sending electrical impulses to the brain along the optic nerve. Other structures of the eye support this process.

- The cornea, lens, aqueous and vitreous humour are all transparent, to let light pass through without interference.
- The pupil is a hole in the iris that lets light through into the eye.
- The vitreous humour maintains the shape of the eyeball, so that the distance from front to back of the eye (the focusing distance) does not change.
- The retina is very dark, to absorb as much of the light that enters the eyes as possible. It also contains the cells that respond to light.

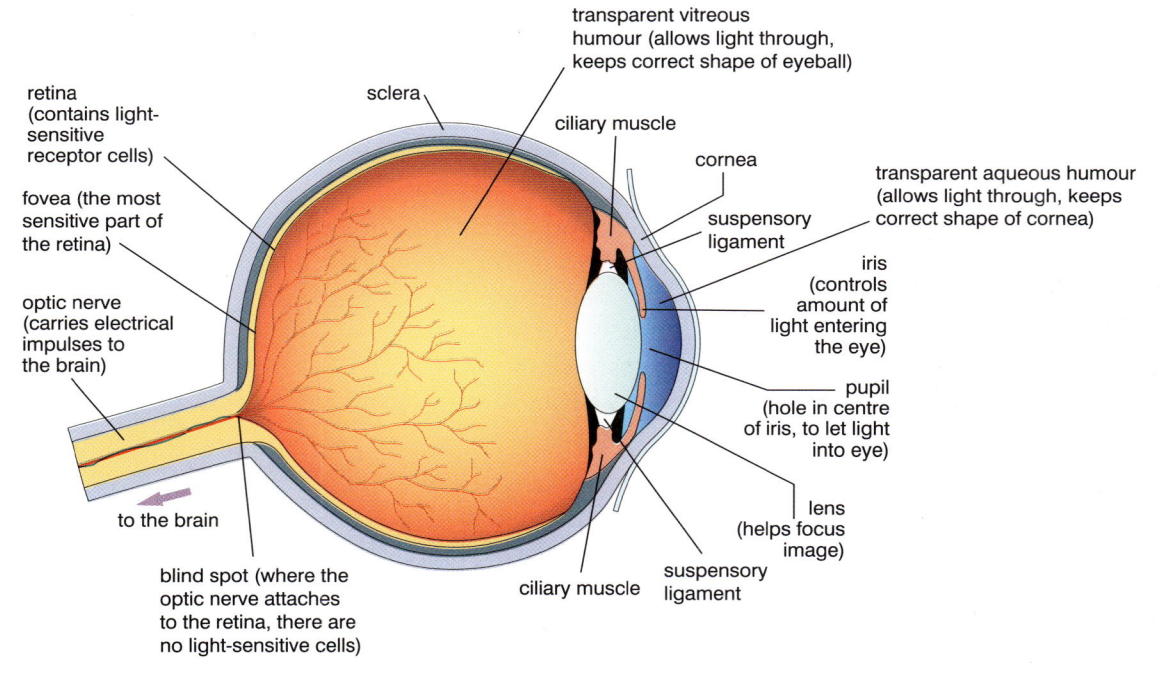

△ Fig. 2.125 Structure of the human eye.

EXTENDED

There are two different kinds of light-sensitive cells in the retina.

- **Cone** cells respond to light of a certain *wavelength*, and therefore colour. There are three types of cone cell, one type responding to red light, one to blue and one to green. The brain combines information from each type to interpret a much greater range of colours in what we see. Cone cells only work well in bright light, so we only see colour images when the light is bright enough. The cone cells are mostly clustered around the fovea on the retina, where most light falls in bright conditions.

- **Rod** cells respond to differences in *light intensity*, not wavelength. They are more sensitive than cone cells at low light intensities, so we use these mostly in low light conditions. The responses of rod cells give us images that are mainly black and white. Rod cells are found all over the retina.

> **SCIENCE IN CONTEXT**
>
> ### DARK ADAPTATION
>
> In bright light the rod cells lose their ability to respond. This is because the coloured pigment in the cells that respond to light is changed into another form. As the light becomes dimmer, the pigment slowly changes back into the form that detects light. However, it can take up to half an hour for the rods to fully recover. This is known as dark adaptation. It explains why, if you move from a bright place to a darker place, it takes a while for your eyes to adjust and see clearly again.

END OF EXTENDED

Changing light conditions

The light-sensitive cells in the retina only respond to the stimulus of light above a certain light intensity. When it is so dark that the cells are not stimulated, we cannot see. Since vision is an important sense to us, at low light intensities our eyes need to gather as much light as possible. However, rod and cone cells are easily damaged by high-intensity light (which is why you should NEVER look directly at the Sun or any other bright light source).

The iris (ring-shaped, coloured part of the eye) controls the amount of light entering the eye by controlling the size of the pupil, which is the hole in the centre of the iris. The iris contains circular and radial muscles. In bright light the circular muscles contract and the radial muscles relax, making the pupil smaller. This prevents damage by reducing the amount of light that enters the eye. The reverse happens in dim light, when the eye has to collect as much light as possible to see clearly.

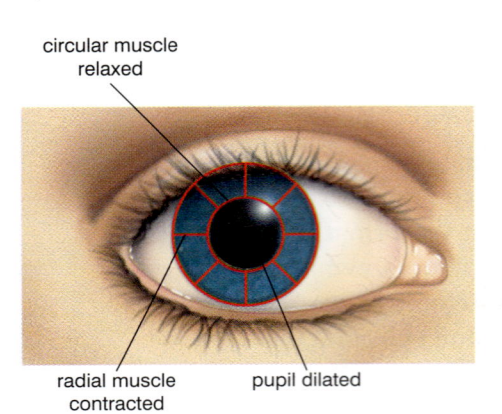

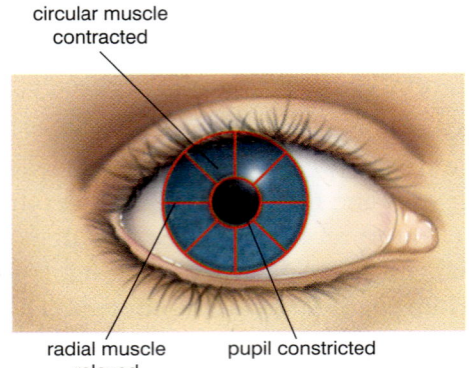

△ Fig. 2.126 The pupil response to light. Left: in dim light. Right: in bright light.

Focusing light

In order to see clear images of our surroundings, the light that enters our eyes needs to be focused properly on the retina. This is known as **accommodation**.

The thick clear cornea bends light rays as they enter the eye in order to focus them on the retina. The lens provides fine focus to sharpen the image. The shape of the cornea cannot change, so that focus is fixed. However, the lens shape can be changed. This makes it possible to focus clearly on objects at a distance and those that are close to us. The change of lens shape is controlled by the **ciliary muscle**, which forms a ring around the lens and is attached to the lens by the suspensory ligaments.

Rays of light from distant objects are almost parallel when they enter the eye. When the object is at a distance:

- the ciliary muscle relaxes and it becomes further from the lens
- this pulls more on the suspensory ligaments
- the suspensory ligaments pull more on the lens, and the lens is pulled into a thinner shape (Fig. 2.127)
- this reduces the focusing power of the lens, so that the light is focused properly on the retina.

Rays of light from an object nearby are more diverging than from objects at a distance. So they need a lens of greater focusing power to focus clearly on the retina. When the object is near:

- the ciliary muscle contracts, so it gets closer to the lens
- this reduces the pull of the suspensory ligaments on the lens (Fig. 2.127)
- the lens relaxes to form a more rounded shape
- this refracts (bends) the light more so that it focuses the image on the retina.

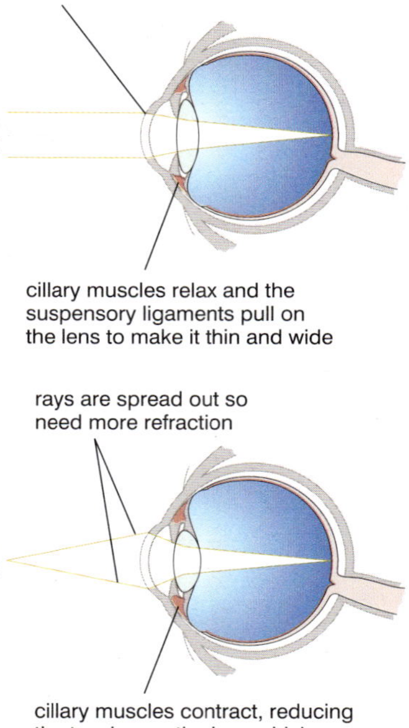

△ Fig. 2.127 How the eye focuses light from distant and near objects.

QUESTIONS

1. Explain how the following structures are adapted to support the role of the eye in sensing light:

    a) cornea

    b) pupil

    c) retina.

2. Explain how the eye responds to changing light intensity.

3. Explain how the eye produces a focused image of an object that is near to the person.

4. **EXTENDED** Explain the difference in distribution and role of rod and cone cells in the eye.

Antagonistic muscles

When you move a part of your body in response to a stimulus, you use your muscles to produce that movement. In this case the muscles are the effectors.

A muscle contracts in response to a stimulus. When muscles that are attached to the skeleton contract, they pull, causing movement at joints. Muscles can contract to pull, but they do not push when they relax. This means that you need a different muscle to move a joint the other way. A pair of muscles that work in opposition at a joint are called **antagonistic**.

The biceps and triceps are an example of a pair of antagonistic muscles. When the biceps contracts, it pulls on the forearm and flexes the elbow. To extend the elbow again, the biceps relaxes and the triceps contracts. It is the pull of the triceps that returns the forearm to its original position.

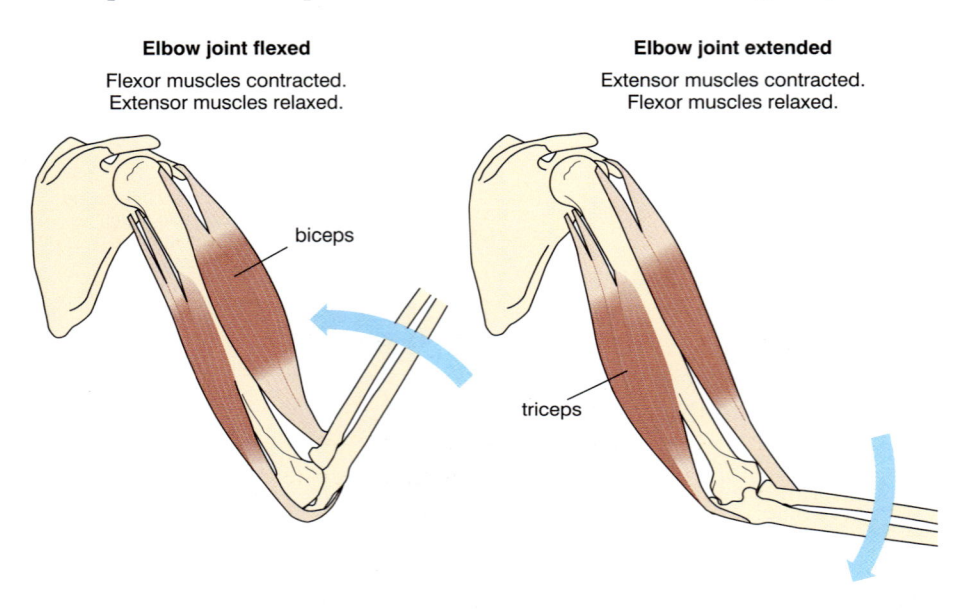

Elbow joint flexed
Flexor muscles contracted.
Extensor muscles relaxed.

Elbow joint extended
Extensor muscles contracted.
Flexor muscles relaxed.

△ Fig. 2.128 Extending and flexing the elbow.

QUESTIONS

1. Explain what is meant by *antagonistic muscles*.
2. Explain why we need antagonistic muscles.

Hormones

Hormones are chemical messengers. They are made in **endocrine glands**.

Endocrine glands do not have ducts (tubes) to carry away the hormones they make: the hormones are secreted directly into the blood to be carried around the body in the plasma.

Most hormones affect several specific parts of the body called the **target organs**; others may only affect one target organ. The changes caused by hormones are usually slower and longer-lasting than the changes brought about by the nervous system, and may continue until the hormone is broken down by liver cells.

Adrenaline

Adrenaline is a hormone produced in the adrenal glands just above the kidneys. This hormone is released in times of excitement, anger, fright or stress, and prepares the body for 'flight or fight': the crucial moments when an animal must instantly decide whether to attack or run for its life.

The effects of adrenaline are:

- increased heart (pulse) rate to circulate blood more rapidly around the body
- increased blood glucose concentration as glucose is released from liver and muscles, to provide more glucose for muscle cell respiration
- increased depth of breathing and breathing rate to deliver oxygen more rapidly to muscle cells and remove carbon dioxide more rapidly from the body
- increased sweating
- hair standing on end – this makes a furry animal look larger but only gives humans goose bumps
- dilated pupils for better vision
- paling of the skin as blood is redirected to muscles.

EXTENDED

Hormones in food production

Hormones have been used to increase the growth rate of livestock. For example, some hormones cause the body to develop more muscle tissue in cattle that are grown for meat. There are concerns that the hormones remain in the meat, milk or cheese from these animals, and that they may therefore affect people who eat the food. For that reason, many countries now ban the use of hormones to raise livestock.

Some countries, such as the US, continue to give hormones to animals used to produce our food, because they say that no health problem in humans has been proved. However, this makes it possible for hormones to pass into the environment, such as waterways, where they can affect and harm other organisms.

END OF EXTENDED

QUESTIONS

1. Explain the meaning of the following terms:
    a) hormone
    b) endocrine gland
    c) target organ.
2. In what conditions might adrenaline be released in the body?
3. Explain the advantages of adrenaline in preparing the body for action.
4. **EXTENDED** Describe one advantage and one disadvantage of using hormones in food production.

TROPIC RESPONSES IN PLANTS

Plants generally respond to changes in the environment by a change in the way they grow. For example, a shoot grows towards light, and in the opposite direction to the force of gravity; a root grows away from light, but towards moisture and in the direction of the force of gravity. These growth responses to a stimulus in plants are called **tropisms**. These responses help the plant produce leaves where there is the most light, and roots that can supply the water that the plant needs.

- Growth in response to the direction of light is called **phototropism**. If the growth is towards light, it is called *positive* phototropism, as in shoots. If the growth is away from light, it is called *negative* phototropism, as shown by roots.
- Growth in response to gravity is called **geotropism**. Plant shoots show negative geotropism, and plant roots show positive geotropism.

△ Fig. 2.129 Growing towards light helps a plant get more light for photosynthesis.

Developing investigative skills

The diagram shows apparatus that can be used to investigate the effect of light on the growth of seedlings.

Using and organising techniques, apparatus and materials

❶ Describe how the apparatus could be used for this investigation.

❷ Describe how you would set up a control for this investigation.

Observing, measuring and recording

❸ a) If this investigation were set up correctly, what result would you expect to see in the seedlings from the windowed box, compared with your control?

b) Explain your answer.

Handling experimental observations and data

❹ Suggest how this investigation could be extended to investigate whether roots also show a phototropic response.

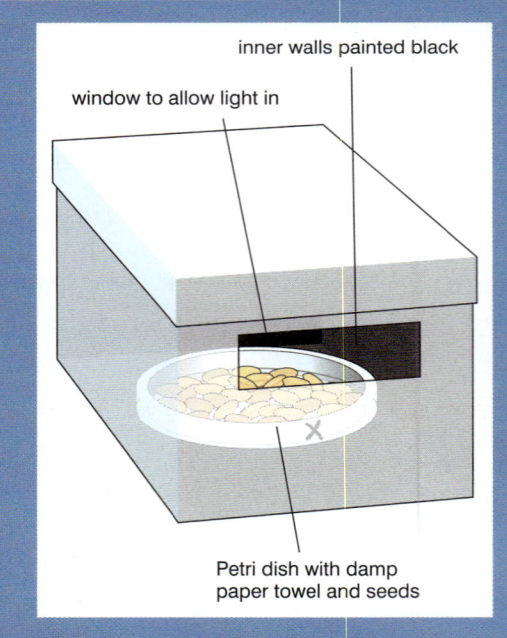

△ Fig. 2.130 Apparatus for investigating the effect of light on growing seeds.

EXTENDED

Control of tropic responses

Tropisms are controlled by plant hormones called **auxins**. Auxins are made in the tips of shoots and roots and diffuse away from the tip. Further back along a shoot, auxin *stimulates* cells to elongate so that the shoot longer. In a root, auxins *inhibit* elongation of cells.

The growth of shoots towards light can be explained by the response of auxin to light.

- When all sides of a shoot receive the same amount of light, equal amounts of auxin diffuse down all sides of the shoot. So cells all around the shoot are stimulated equally to grow longer. This means the shoot will grow straight up.
- When the light on the shoot comes mainly from the side, auxin on that side of the shoot moves across the shoot to the shaded side. The cells on the shaded side of the shoot will receive more auxin, and so grow longer, than those on the bright side. This causes the shoot to curve as it grows, so that it grows towards the light.

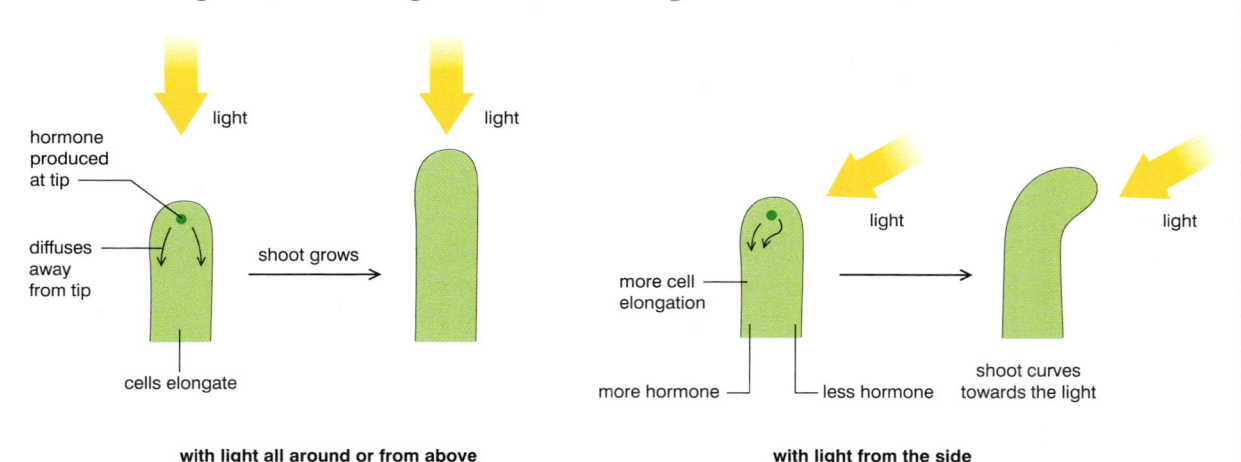

△ Fig. 2.131 The effect of light on the growth of shoots.

In roots, auxin has the opposite effect on cells, so that it reduces how much the cells elongate.

- When roots are pointing straight downwards, all sides of the root receive the same amount of auxin, so all cells elongate by the same amount.
- When the root is growing at an angle to the force of gravity, gravity causes the auxin to collect on the lower side. This reduces the amount of elongation of cells on the lower side of the root, so that the root starts to curve as it grows until it is in line with the force of gravity.

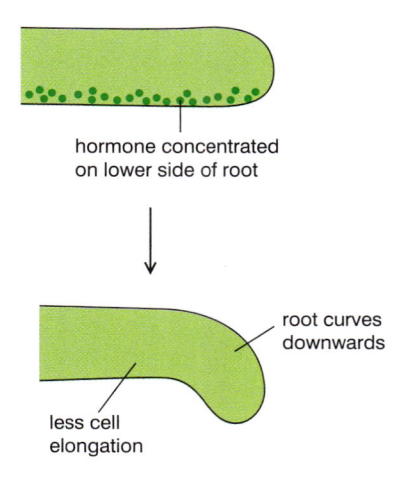

△ Fig. 2.132 The effect of gravity on the growth of roots.

REMEMBER

To gain higher marks, you need a full understanding of the phototropic responses in stems.

Remember that auxin causes shoots to curve by the elongation of existing cells, not by the production of more cells.

SCIENCE IN CONTEXT

GARDENER'S TIP

One effect of auxins is to inhibit the growth of side shoots. This is why a gardener who wants a plant to become bushy rather than tall will take off the shoot tip, so removing a source of auxin.

Using auxins

Auxins are now used to make synthetic weedkillers, or **herbicides**. One that is commonly used is 2,4-D. When it is sprayed on plants, the chemical is absorbed into the plant through its leaves. Inside the plant it encourages the stems and leaves to grow very rapidly. Such large concentrations of auxin are used that growth is too fast, and the plant is unable to supply all the nutrients that the new cells need for healthy growth. Eventually the plant dies.

The advantage of using synthetic weedkillers is that grasses are not affected by the herbicide; only broad-leaved plants are killed. This is useful because many of our crops, such as wheat and maize, are grasses, and many of the weed plants that we want to kill so they don't compete with crops are broad-leaved plants.

END OF EXTENDED

QUESTIONS

1. Define the term *tropism* in your own words.
2. Give one example of: a) positive phototropism, b) positive geotropism.
3. **EXTENDED** Describe the action of auxin in a shoot growing in one-sided light.
4. **EXTENDED** Explain why auxins may be used as weedkillers.

Homeostasis

For our cells to carry out life processes properly, they need the conditions in and around them, such as the temperature and amount of water and other substances, to stay within safe limits. Keeping conditions within these limits – that is, keeping the internal environment constant – is called **homeostasis**.

Temperature control

The temperature in the core of your body is about 37 °C, regardless of how hot or cold you may feel on the outside. This **core temperature** varies very little. A change of only 4 °C can cause death.

Heat energy is constantly released by cells as a result of respiration and other chemical reactions in the body, and is transferred to the surroundings outside the body. To maintain a constant body temperature, these must balance. The temperature of the blood from the core of the body is monitored by the hypothalamus in the brain. If the temperature varies too much from 37 °C, the hypothalamus causes the temperature to come back to about 37 °C. The hypothalamus also receives electrical impulses from heat sensor cells in the skin surface.

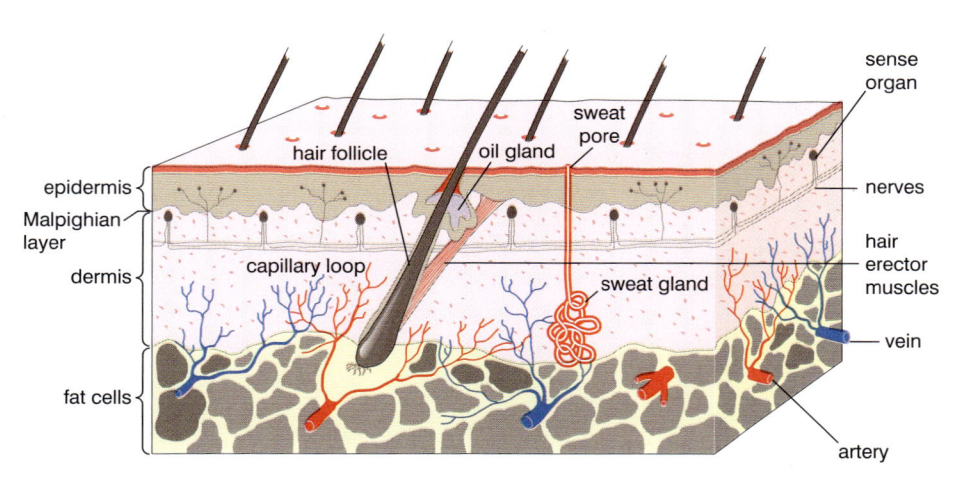

△ Fig. 2.133 Structure of the skin.

If core temperature rises too far:

- Sweat is released onto the surface of the skin from glands. Sweat is mostly water, and this water evaporates. Evaporation needs heat energy, so heat energy is removed from the skin surface as the sweat evaporates, cooling the skin.
- Blood vessels carrying blood near the surface of the skin dilate (get wider), which is known as **vasodilation**. This increases blood flow through the capillaries in the skin surface. Vasodilation makes it easier for heat energy to be transferred to the skin surface and from there to the environment by radiation and conduction. It is also what makes light-skinned people look pink when they are hot.

If core temperature falls too far:

- Blood vessels carrying blood near the surface of the skin constrict (get narrower), which is known as **vasoconstriction.** This reduces the amount of blood flowing through the skin capillaries. As the warm blood is kept deeper in the skin, this reduces the rate of heat transfer by conduction to the skin surface and from there to the environment.
- Body hair may be raised by muscles in the skin. This has little effect in humans (often called goose bumps) but is more effective in mammals with fur and in birds, because the fur or feathers trap air next to the skin. Air is not a good conductor of heat energy, so this still layer of air acts as insulation.
- Muscles may start to 'shiver'. This means they produce rapid, small contractions. Cellular respiration is used to produce these contractions, releasing heat energy at the same time which heats the blood flowing through the muscles.

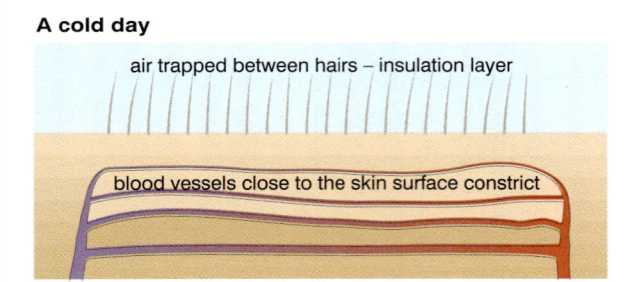

A cold day

air trapped between hairs – insulation layer

blood vessels close to the skin surface constrict

A hot day

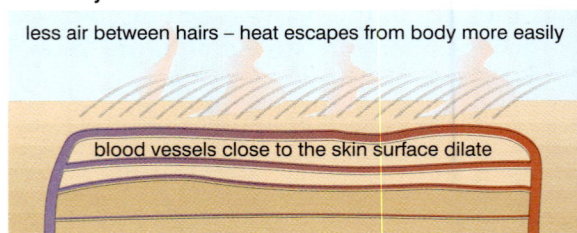

less air between hairs – heat escapes from body more easily

blood vessels close to the skin surface dilate

△ Fig. 2.134 Blood vessels respond to outside temperature.

Developing investigative skills

You can use a test tube of warm water wrapped in wet paper towel to investigate whether sweating really does cool the body, measuring how the temperature of the water changes over time.

Using and organising techniques, apparatus and materials

❶ Explain how the tube models sweating in a human.

❷ How would you set up the control for this investigation? Explain your answer.

Observing, measuring and recording

The table shows the results of an investigation like the one described above.

	Time/min	0	2	4	6	8	10	12	14	16
Temperature of water in tube/°C	Wet towel	56	50	46	42	39	36	34	32	31
	Dry towel	56	52	49	46	44	44	41	40	39

❸ Use the results to draw a suitable graph.

Handling experimental observations and data

❹ Describe any pattern shown in your graph.

❺ Draw a conclusion from the graph.

❻ Explain your conclusion using your scientific knowledge.

EXTENDED

NEGATIVE FEEDBACK

The control of core body temperature is an example of **negative feedback**. This is where a change in a stimulus causes a response that produces the opposite change. It depends on a monitoring control centre that detects changes and initiates responses, such as the hypothalamus in the brain. We can summarise a negative feedback response in a diagram like this.

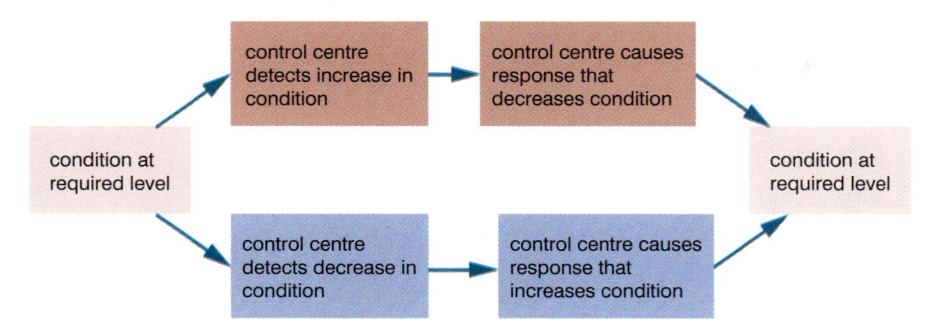

△ Fig. 2.135 A negative feedback response prevents large increases or decreases from the best conditions for the body. This is how the temperature and pH of the tissues is controlled so that enzymes work most effectively.

Control of blood glucose concentration

It is important that the blood glucose concentration remains within a small range. If it rises or falls too much, you can become very ill very quickly.

After a meal, the blood glucose concentration rises rapidly as glucose is absorbed by digested food in the small intestine. Cells in the pancreas detect this increase and respond by releasing the hormone **insulin**, which travels in the blood to the liver. Here it causes any excess glucose to be taken up by liver cells. Inside the liver cells, the glucose is converted to another carbohydrate, glycogen. This is insoluble and is stored in the liver.

Between meals, glucose in the blood is constantly diffusing into cells for use in cellular respiration. So the blood glucose concentration falls. When the pancreas detects a low level of glucose, the insulin-secreting cells stop secreting insulin and other cells start to secrete the hormone **glucagon** instead. Glucagon converts some of the stored glycogen back into glucose, which is released into the blood to raise the blood glucose concentration again.

END OF EXTENDED

QUESTIONS

1. Define the term *homeostasis* in your own words.
2. Give one example of homeostasis in the human body.
3. Explain the role of skin blood vessels in maintaining core body temperature.
4. **EXTENDED** Define the term *negative feedback*.
5. **EXTENDED** Use a negative feedback diagram like Fig. 2.135 to show how blood glucose concentration is kept at a safe level.

DRUGS

A **drug** is an substance that, when taken into the body, affects the way the body works by modifying or affecting chemical reactions.

Some drugs are **medicinal drugs** that are used to treat the symptoms or causes of disease. Many medicinal drugs may only be prescribed by a doctor because they are potentially harmful, or because their use should be restricted. For example, **antibiotics** are drugs that kill bacteria in the body or stop them from reproducing. This protects us from harmful bacterial diseases such as tetanus, pneumonia and some kinds of meningitis. However, over-use of antibiotics is resulting in the evolution of antibiotic-resistant bacteria (see Evolution of resistance to antibiotics).

EXTENDED

Antibiotics are specific to the type of bacteria they attack because of the way they work. Some cause the bacterial cell membrane to break open, killing the cell. Other antibiotics prevent some bacteria from making new cell walls and so prevent them from dividing and producing more bacteria.

The advantage is that bacterial cells and human cells are different in structure, so antibiotics don't harm human cells. The disadvantage is that viruses don't have cell membranes or cell walls, so antibiotics don't affect viruses. This means they cannot be used to treat viral diseases such as flu, colds and **HIV** (human immunodeficiency virus).

END OF EXTENDED

Heroin

Heroin is a powerful drug used sometimes by doctors to control high levels of pain in patients. However, it is illegally used by some people for its powerful effects on the brain, particularly as a **depressant**. Depressants can reduce feelings of fear or panic and so make the user feel better. The problem with heroin is that the body rapidly gets used to it, so the user needs increasingly large doses to have the same effect.

This is called **addiction**. It also means that if the user stops taking heroin, their body will suffer severe **withdrawal symptoms**, such as nausea, muscle cramps, sweating, anxiety and insomnia.

People who become addicted to heroin need more and more money to pay for their ever increasing dose. As the drug makes them less able to cope with everyday life, they may lose their job and turn to crime, such as burglary or theft, to get money.

Heroin can be taken into the body in different ways, one of which is injection with a syringe. To save money, heroin users may share needles for injection. This increases the risk of transmission of blood-borne diseases such as HIV, which causes the disease **AIDS** (acquired immune deficiency syndrome).

Alcohol

Alcohol also has a depressant effect on the body, which is why it is commonly used as a **recreational drug**. In limited amounts, it can produce a warm relaxing effect, although this can be dangerous as it slows reaction time, and so it takes longer to react to a situation. This is why most countries have laws to limit the amount of alcohol that can be drunk before driving a vehicle or operating machinery.

△ Fig. 2.136 Thousands of serious accidents – some of them fatal – are caused every year by people who have drunk alcohol before driving.

In larger quantities, alcohol can have many bad effects. In the short term, large quantities of alcohol may cause:

- vomiting, because alcohol is toxic in large amounts and this is the quickest way for the body to get rid of it
- violent behaviour because the user loses self-control – alcohol is a common factor in fights and domestic abuse
- unconsciousness – which increases the risk of death through choking on vomit.

Regularly drinking large amounts of alcohol can cause liver damage (**cirrhosis**), which can eventually lead to liver failure. Many liver transplants are carried out on patients who have been regular heavy drinkers of alcohol.

Smoking

When a person smokes tobacco, the chemicals in the smoke are taken into the lungs. Those chemicals that have small enough molecules can then diffuse into the blood and be carried around the body. Many of the chemicals in tobacco smoke have damaging effects, not only on the respiratory system, but also on other systems in the body.

- Nicotine in tobacco smoke is highly addictive, which makes it difficult for smokers to give up smoking. It also alters people's moods because it is a stimulant and a relaxant – smokers often say they feel more relaxed but alert after smoking. These feelings can also become addictive.
- The tar in tobacco smoke is a mixture of chemicals that form a black sticky substance in the lungs. This sticky layer can coat the tiny hair-like cilia lining the tubes of the lungs, making it more difficult for them to clear out dust and microorganisms. This can result in many lung infections and a thick cough as the smoker tries to clear sticky mucus from the lungs. The irritation and infection can cause a disease called bronchitis.
- Continued coughing, in order to clear tar and smoke particles from lungs, over a long time damages the alveoli, breaking down the divisions between them and so reducing their surface area. This causes a disease called emphysema where the patient has difficulty getting enough oxygen into their blood for any kind of activity. They may have to breathe pure oxygen in order to be able to do anything.

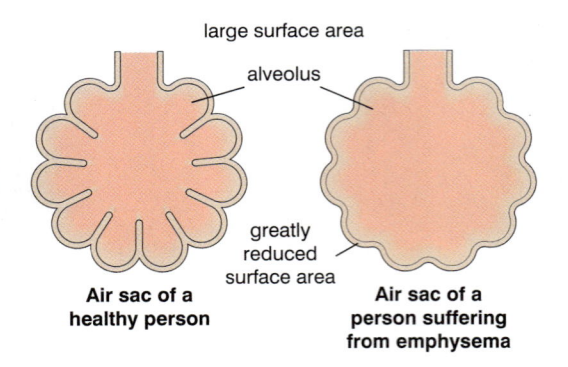

◁ Fig. 2.137 Repeated coughing over a long period breaks down the surface of each alveolus, reducing the surface area for exchange of gases. This condition is called emphysema.

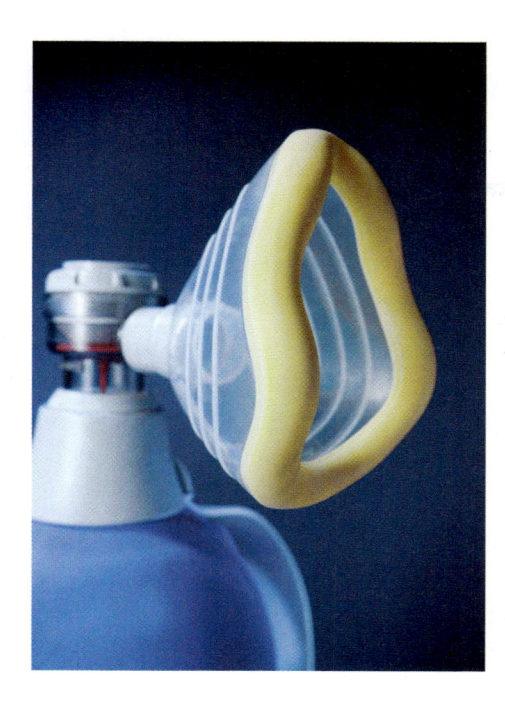

◁ Fig. 2.138 People with emphysema may have to breathe air containing a high concentration of oxygen, to make sure their damaged lungs can absorb enough oxygen into their bodies. Breathing masks attached to oxygen tanks such as this one shown can provide this for patients.

- Tobacco smoke also contains carbon monoxide, which is a toxic gas. It combines with haemoglobin in red blood cells and so prevents the cells from carrying oxygen. This reduces the amount of oxygen that gets to tissues, which in extreme cases can lead to cell death. In lower amounts it can result in breathlessness, when the body cannot get sufficient oxygen to cells for activity. During pregnancy, smoking passes through the placenta to the developing fetus. This can reduce the rate of growth of the fetus, resulting in a low birth weight which can cause complications during the birth and health problems through life.
- Many other gases in tobacco smoke are **carcinogenic**, meaning they cause cells to become cancerous and take over tissue. Smoking is the greatest cause of lung cancer.

QUESTIONS

1. Define the terms *drug* and *antibiotic*.
2. Describe the dangers of heroin use.
3. Describe the dangers of alcohol use.
4. Describe four constituents of tobacco smoke and explain how they damage health.
5. **EXTENDED** Explain why antibiotics can be used to treat pneumonia but not 'flu.

End of topic checklist

Key terms

addiction, AIDS, antibiotic, auxin, central nervous system, cirrhosis, cone, core temperature, depressant, drug, effector, endocrine gland, geotropism, glucagon, herbicide, heroin, homeostasis, hormonal system, hormone, insulin, involuntary medicinal drug, motor neurone, negative feedback, nerve, nervous system, neurone, recreational drug, reflex, reflex arc, relay neurone, retina, rod, sense organ, sensory neurone, stimulus, target organ, tropism, vasoconstriction, vasodilation, voluntary, withdrawal symptoms

During your study of this topic you should have learned:

○ How to describe the human nervous system in terms of the central nervous system (brain and spinal cord as areas of coordination) and the peripheral nervous system, which together serve to coordinate and regulate body functions.

○ How to identify motor (effector), relay (connector) and sensory neurones from diagrams.

○ How to describe a simple reflex arc in terms of sensory, relay and motor neurones, and a reflex action.

○ **EXTENDED** How to distinguish between voluntary and involuntary actions.

○ That muscles and glands can act as effectors.

○ How to describe the action of antagonistic muscles, including the biceps and triceps at the elbow joint.

○ The definition of *sense organs* as groups of receptor cells responding to specific stimuli: light, sound, touch, temperature and chemicals.

○ How to describe the structure and function of the eye, including accommodation and pupil reflex.

○ **EXTENDED** How to distinguish between rods and cones, in terms of function and distribution.

○ The definition of a *hormone* as a chemical substance, produced by a gland, carried by the blood, which alters the activity of one or more specific target organs and is then destroyed by the liver.

○ About the role of the hormone adrenaline in chemical control of metabolic activity, including increasing the blood glucose concentration and pulse rate.

○ About examples of situations in which adrenaline secretion increases.

- ◯ How to compare nervous and hormonal control systems.
- ◯ **EXTENDED** How to discuss the use of hormones in food production.
- ◯ How to define and investigate *geotropism* (as a response in which a plant grows towards or away from gravity) and *phototropism* (as a response in which a plant grows towards or away from the direction from which light is coming).
- ◯ **EXTENDED** How to explain the chemical control of plant growth by auxins including regulating differential growth and the effects of synthetic plant hormones used as weedkillers.
- ◯ The definition of *homeostasis* as the maintenance of a constant internal environment.
- ◯ How to identify, on a diagram of the skin, hairs, sweat glands, temperature receptors, blood vessels and fatty tissue.
- ◯ How to describe the maintenance of a constant body temperature in humans in terms of insulation and the role of temperature receptors in the skin, sweating, shivering, vasodilation and vasoconstriction of arterioles supplying skin-surface capillaries and the coordinating role of the brain.
- ◯ **EXTENDED** How to explain the concept of control by negative feedback.
- ◯ **EXTENDED** How to describe the control of the glucose content of the blood by the liver, and by insulin and glucagons from the pancreas.
- ◯ The definition of a drug as any substance taken into the body that modifies or affects chemical reactions in the body.
- ◯ About the use of antibiotics in medicine to treat bacterial infections.
- ◯ **EXTENDED** How to explain why antibiotics kill bacteria but not viruses.
- ◯ About the effects of the abuse of heroin: a powerful depressant, problems of addiction, severe withdrawal symptoms and associated problems such as crime and infection with diseases such as HIV/AIDS through the use of shared needles.
- ◯ About the effects of excessive alcohol consumption: reduced self-control, depressant, effect on reaction times, damage to liver and social implications.
- ◯ About the effects of tobacco smoke and its major toxic components (tar, nicotine, carbon monoxide and smoke particles) on the gas exchange system.

End of topic questions

Note: The marks awarded for these questions indicate the level of detail required in the answers. In the examination, the number of marks awarded to questions like these may be different.

1. Using the terms *stimulus*, *neurone*, *reflex arc*, *nerve impulse*, *effector* and *response*, describe the response of someone who touches something very hot. **(6 marks)**

2. **a)** Identify the similarities and differences between the nervous and hormonal systems in humans. **(5 marks)**

   b) Explain why it is advantageous to have both systems. **(3 marks)**

3. **a)** Describe the sequence of sensing and response in the nervous system of one of Andy Roddick's opponents who returns a serve successfully. **(4 marks)**

   b) Is this a reflex action? Explain your answer. **(2 marks)**

4. Explain the following.

   a) A student visiting a mine couldn't see anything in the mine when the lights were turned off. **(2 marks)**

   b) A cataract is a clouded lens in the eye, caused by many conditions. A patient with cataracts cannot see clear images. **(2 marks)**

   c) A person who is long-sighted needs to wear spectacles with converging lenses in order to read something near to them. **(2 marks)**

5. Some tomato seedlings are placed in a dimly lit room near to a brightly lit window.

   a) Which of these statements best describes the seedlings over the next few weeks? Explain your choice. **(2 marks)**

   i) The seedlings wilt and bend towards the light.

   ii) The seedlings grow towards the brightest light.

   iii) The seedlings grow straight up.

   iv) The seedlings bend towards the light.

   b) What is meant by the term *positive phototropism*? **(1 mark)**

   c) Explain the survival advantage to plants of having shoots that are positively phototropic. **(2 marks)**

   d) What is meant by *positive geotropism*? **(1 mark)**

   e) Explain the survival advantage to plants of having roots that are positively geotropic. **(2 marks)**

6. One example of homeostasis in humans is the control of core body temperature.

   a) Identify the receptors, monitoring area, and effectors in the response to a change in external temperature. **(6 marks)**

   b) Explain why changes in skin blood flow affect the rate of heat loss from the body. **(2 marks)**

   c) Explain why homeostasis of core body temperature is important for survival. **(4 marks)**

7. **EXTENDED** Using the example of blood glucose concentration, explain why negative feedback control is an essential feature of homeostasis. **(3 marks)**

8. Some people think that all drugs are bad. Explain, with examples, why this is only partly true. **(3 marks)**

Exam-style questions

Note: The questions, sample answers and marks in this section have been written by the authors as a guide only. The marks awarded for these questions indicate the level of detail required in the answers. In the examination, the number of marks awarded to questions like these may be different.

Sample student answer

EXTENDED Question 1

A Biology student set up an investigation in which three cubes of different sizes were cut from an agar jelly block. The agar jelly contained a red indicator that turns blue in the presence of alkali.

The cubes were placed in an alkali. The student measured the time taken for the cubes to turn completely blue.

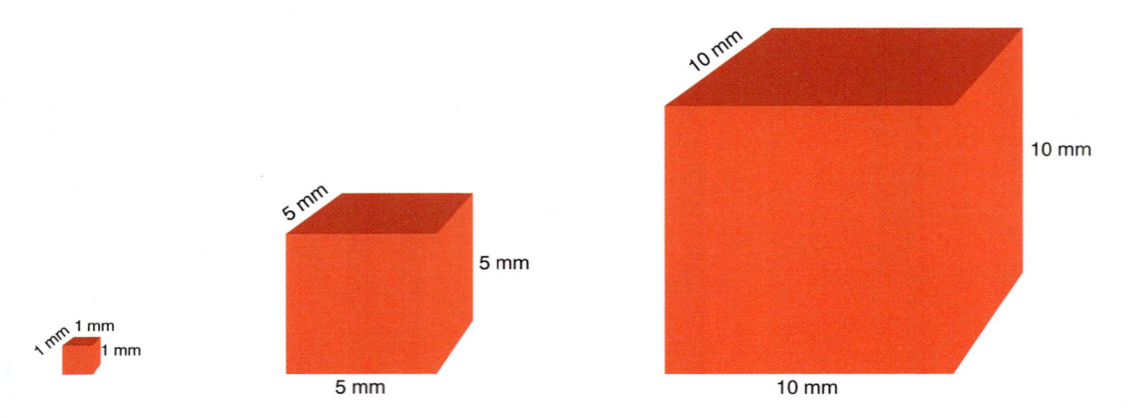

a) What is the name of the process that causes the alkali to penetrate the agar jelly? (1)

Diffusion ✓ ①

b) The student used cubes of three dimensions:

Dimensions of cube, mm
1 × 1 × 1
5 × 5 × 5
10 × 10 × 10

For each cube, calculate its:

  i) surface area

  ii) volume

  iii) surface area : volume ratio. (9)

Dimensions of cube/mm	i) Surface area of cube/mm²	ii) Volume of cube/mm³	iii) Surface area : volume ratio
1 × 1 × 1	1 ✗	1 ✓ ①	1 : 1 ✗
5 × 5 × 5	150 ✓ ①	125 ✓ ①	150 : 125 ✓ ①
10 × 10 × 10	600 ✓ ①	1000 ✓ ①	600 : 1000 ✓ ①

c) Explain the relationship between surface area and volume as the cube increases in size. (1)

As the volume of the cube increases, so does the surface area, but not to the same extent. ✓ ①

d) During the biology lesson, the alkali penetrated the two smaller cubes, but by the end of the lesson, the 10 mm × 10 mm × 10 mm cube had still not turned completely blue.

   i) Explain these results. (3)

The alkali diffuses into the cubes over their surface. ✓ ①

In the small and medium cubes, there is sufficient surface area for the alkali to penetrate quickly. ✓ ①

For the 10 mm × 10 mm × 10 mm cube, there is insufficient surface area, in relation to the cube's volume, for the alkali to penetrate quickly. ✓ ①

   ii) Explain what implications this has for organisms of increasing size. (3)

As organisms increase in size, and therefore volume, diffusion of useful substances into their bodies and useful substances out ✓ ①

Will not occur quickly enough to meet their needs. ✓ ①

14/17 **(Total 17 marks)**

TEACHER'S COMMENTS

1. a) Correct

   b) The student has:
   - produced a table, which is a good way of organising the answer. The correct units for surface area and volume have been used.
   - appreciated that a cube has six faces, so the total area is six times the surface area of one face.

   However, they have made a simple error for the 1mm^3 cube. The correct version of the table is:

Dimensions of cube/mm	Surface area of cube/mm^2	Volume of cube/mm^3	Surface area: volume ratio
1 × 1 × 1	6	1	6:1
5 × 5 × 5	150	125	150: 125
10 × 10 × 10	600	1000	600: 1000

   c) Correct. A better way of expressing this, however, would be to say that as the volume of the cubes increases, so does the surface area, but not at the same rate relative to the volume.

   d) i) This is a good answer.

   The student has given a full explanation for IGCSE level. They have begun the answer well by explaining that the alkali diffuses into the cubes over their surface. The student has appreciated that in the small and medium cubes, there is sufficient surface area, proportionally, for the alkali to penetrate quickly and completely.

   They have also appreciated that for the 10 mm × 10 mm × 10 mm cube, there is insufficient surface area, in relation to the cube's volume, for the alkali to penetrate quickly. In fact, the alkali may not penetrate fully, and at some point before this has happened, the agar may begin to diffuse outwards (particularly if the laboratory is warm).

   ii) The answer is correct, but the student has missed one of the marking points.

   Organisms solve the problem of increasing size by providing additional surface area, that is additional absorbing surfaces.

   You can see this in plants, for instance in different leaf shapes, and the internal structure of the leaf. Many animals increase their absorbing surface with long digestive systems, with folded walls with tiny projections, and with the huge surface area of their lungs. You can use examples such as this to illustrate and explain your answer.

Exam-style questions continued

Question 2

Most living organisms are made up of cells.

a) Copy and complete the sentences by writing the most appropriate word in each space.

The _____ holds the cell together and controls substances entering and leaving the cell. The _____ is the jelly-like substance contained within the cell. It is where many different chemical processes occur.

The _____ is the control centre of the cell and contains genetic material as _____. These control how a cell grows and works.

Plant cells also have features that are not found in animal cells. These include the _____, made of _____, which gives the cell extra support and defines its shape.

Many plant cells have a large central, permanent _____ that contains cell sap. It is used for storage of some chemicals, and to support the shape of the cell.

Plant cells exposed to the light contain _____. These contain the green pigment _____, which absorbs the light energy that plants use for the process of _____.

(10)

b) The levels of organisation in multicellular organisms include cells, tissues, organs and systems.

State whether each of the following structures is a cell, tissue or organ.

Structure	Cell	Tissue	Organ
Blood			
Brain			
Liver			
Muscle			
Neurone			
Ovum			
Skin			
Sperm			

(8)

(Total 18 marks)

Exam-style questions continued

Question 3

The table below lists some molecules that are important biologically.

a) Give the units that each one of the following biological molecules is made up of.

Biological molecule	Units that make up the molecule	
Glycogen		(1)
Lipids		(2)
Proteins		(1)
Starch		(1)

b) Describe a test that can be carried out in the laboratory for the following carbohydrates.

   i) glucose (5)

   ii) starch (4)

c) In humans, proteases are produced by the stomach, pancreas and small intestine. Copy the diagram and show the location of these organs. (3)

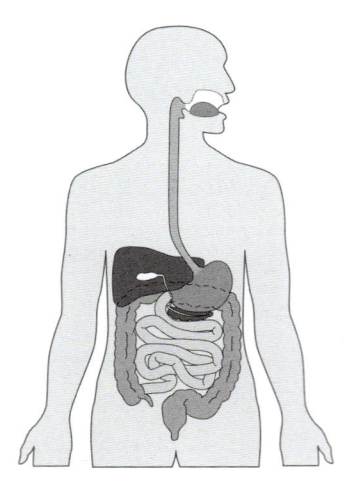

d) The graphs show the effect of pH on the activity of two proteases that break down proteins.

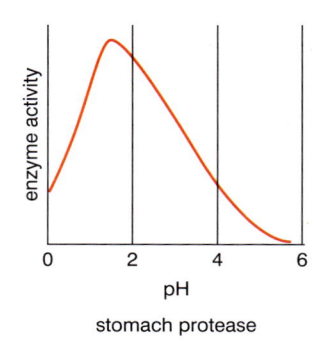

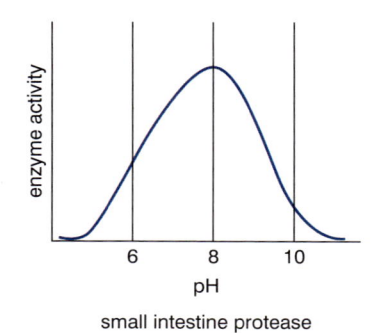

stomach protease

small intestine protease

i) What are the optimum pHs of stomach and small intestine proteases?

Stomach protease (1)

Small intestine protease (1)

ii) Describe an experiment to investigate the effect of temperature on the breakdown of a named food molecule. (5)

(Total 24 marks)

Exam-style questions continued

Question 4

A student cut a number of cylinders from a potato and weighed them. These were placed in sucrose solutions of different concentrations.

After one hour, the cylinders were removed, blotted dry and reweighed. The student calculated the percentage change in mass for each cylinder. The results are shown below.

Concentration of sucrose/g per cm^3	Percentage change in mass of potato cylinders				Average percentage change in mass
	Experiment 1	Experiment 2	Experiment 3	Experiment 4	
0.0	+31.4	+33.7	+31.2	+32.5	
0.2	+20.9	+22.2	+22.8	+21.3	
0.4	−2.7	−1.8	−1.9	−2.4	
0.6	−13.9	−12.8	−13.7	−13.6	
0.8	−20.2	−19.7	−19.3	−20.4	
1.0	−19.9	−20.3	−21.1	−20.3	

a) Calculate the average percentage changes in mass for each of the sucrose concentrations. (6)

b) i) Draw a graph of these results. Join the points with a line of best fit. (4)

   ii) At what concentration of sucrose was there no net movement of water? (2)

   iii) Describe the changes in mass over the range of sucrose concentrations. (3)

   iv) State the process involved in these changes in the potato cylinders. (1)

(Total 16 marks)

Question 5

The diagram shows the urinary system of humans.

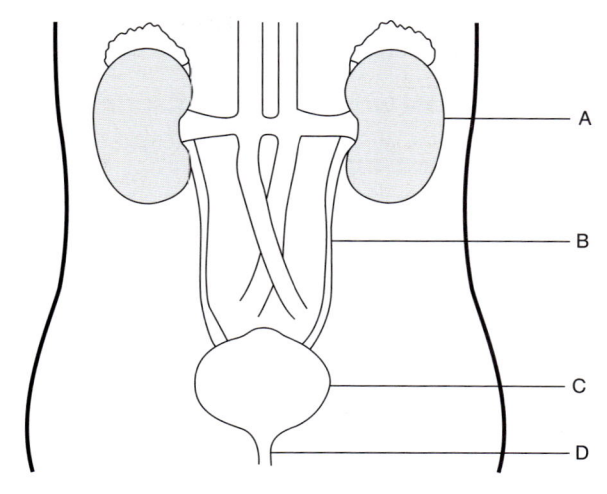

a) Identify the parts of the urinary system, A, B, C and D. (4)

b) Name another excretory organ of humans and one chemical that it excretes. (2)

c) The table below shows the composition of urine.

Substance	Concentration in urine
Water	96.0 cm³ per 100 cm³
Glucose	0.0 g per 100 cm³
Salts	1.8 g per 100 cm³
Urea	2.0 g per 100 cm³

Under normal conditions, on average, a person excretes 1500 cm³ of urine every day.

  i) Calculate:

    the volume of water excreted in a day (1)

    the mass of urea excreted in a day. (2)

  ii) Why and where is urea produced in the body? (2)

  iii) State **one** factor that affects the volume and concentration of urine produced, and explain how these are affected. (3)

(Total 14 marks)

Exam-style questions continued

Question 6

This question is about plants' responses towards stimuli.

Copy and complete the sentences by writing the most appropriate word in each space.

Growth in response to the direction of light is called _____.
If the growth is towards light, it is called _____,
as shown by plant _____.

Growth in response to gravity is called _____. Plant roots are _____. This response helps the plant to grow _____, so the plant can obtain the _____ it needs.

(Total 8 marks)

Question 7

The diagram shows a section through a leaf.

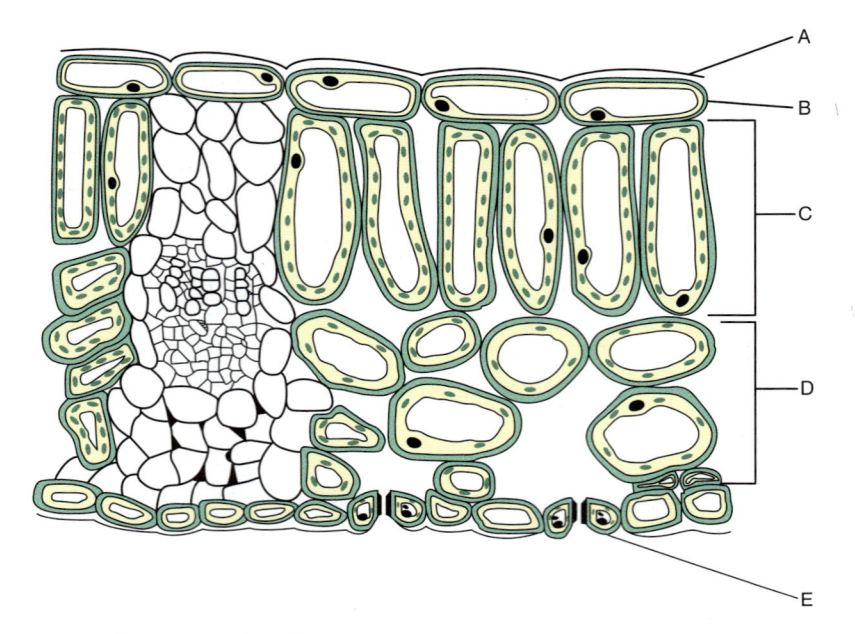

a) Name parts A–E shown in the diagram. (5)

b) The leaf is the main organ of photosynthesis. Write a word equation for photosynthesis. (3)

c) Define the term *transpiration*. (3)

d) Describe a technique used to investigate the effect of temperature on transpiration rate. (7)

(Total 18 marks)

Question 8

The circulatory system has several functions, including the transport of substances, temperature regulation and defence.

a) The diagram shows the structure of the heart.

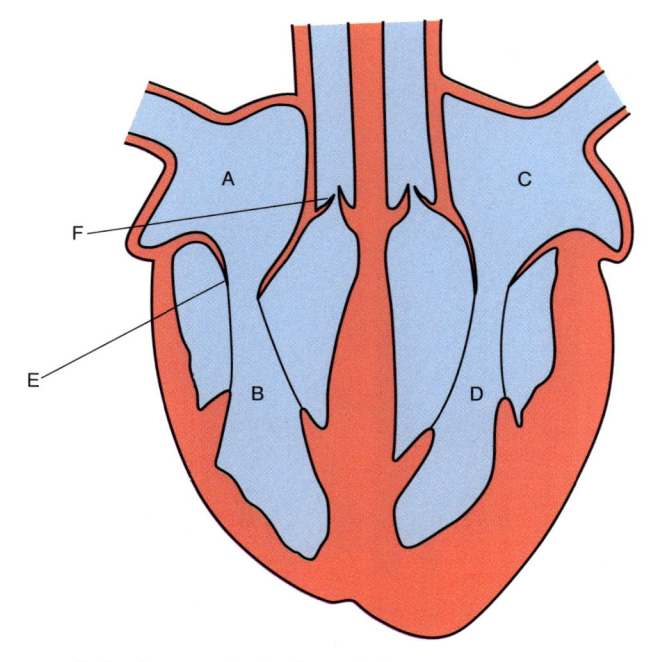

a) Name the chambers of the heart, A, B, C and D. (4)

b) Copy and complete the table for each of the different components of the blood.

Component of blood	Function
Red blood cells	
White blood cells	
Platelets	
Plasma	

(8)

(Total 12 marks)

Exam-style questions continued

Question 9

The diagram shows the structure of an industrial fermenter used to culture a microorganism that produces a medicinal drug.

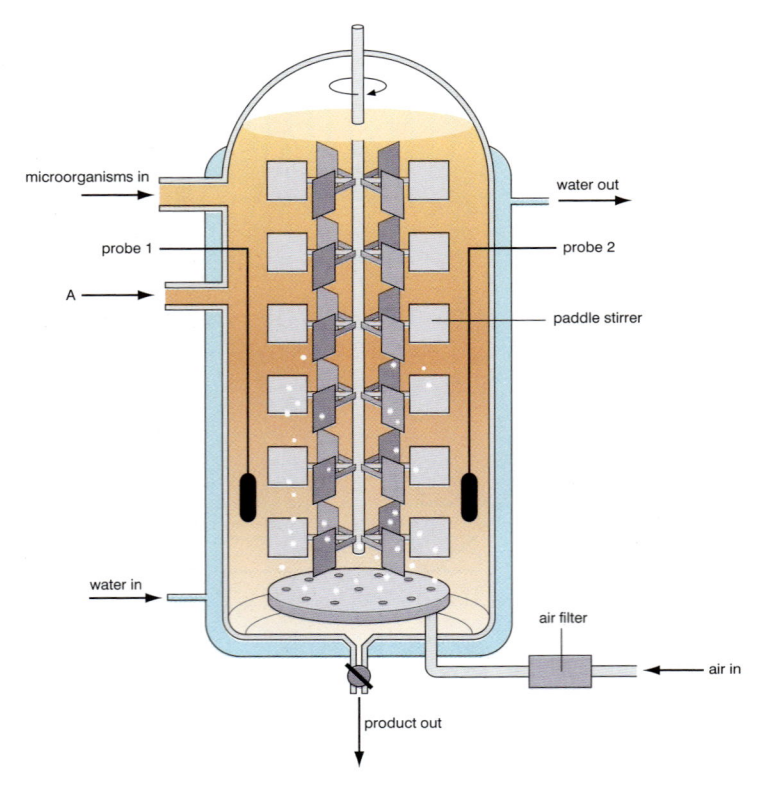

a) State why many drugs are produced with microorganisms rather than manufactured from chemical raw materials. (2)

b) i) Suggest what may be pumped into the fermenter at point A. (1)

   ii) Suggest two factors that probes 1 and 2 may be designed to measure. (2)

c) In the fermenter, explain fully the use of:

   i) the paddle stirrer (2)

   ii) water circulating around the fermenter. (3)

d) Give one way in which the manufacturer ensures that this microorganism grows rather than others. (1)

e) What type of respiration is the microorganism carrying out in the fermenter? Explain your answer. (2)

(Total 13 marks)

Question 10

a) Yeast, a fungus, is used in the production of biofuel.

   i) Give the word equation for the production of ethanol by yeast. (2)

   ii) Describe how yeast cells are involved in the process. (1)

b) Compare the energy produced by yeast undergoing anaerobic respiration with yeast undergoing aerobic respiration. (1)

(Total 4 marks)

Question 11

Fruit juices are often produced using the enzyme pectinase.

Describe an experiment to show that fruit juice can be extracted from fruit more effectively by the addition of pectinase. (8)

(Total 8 marks)

EXTENDED Question 12

The light micrograph shows a cell from the liver of a human.

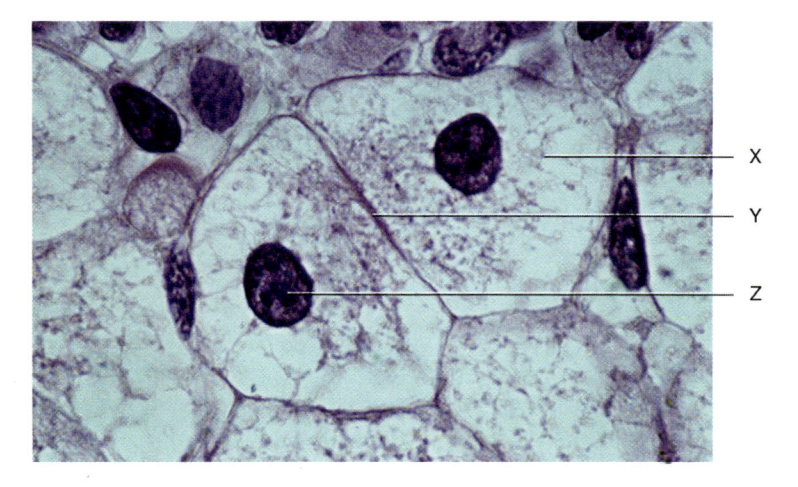

a) For structures X, Y and Z, state their names and functions. (6)

b) Calculate the width, in micrometres, of the longer dimension of cell A, as shown by the line. Show your working. (2)

c) How many liver cells, laid side-to side, would there be in a piece of liver tissue 1 cm across? Show your working. (2)

(Total 10 marks)

Exam-style questions continued

EXTENDED Question 13

A company that produces enzymes publishes information sheets of their performance. The graphs below show the performance of an enzyme, pectinase, at different pHs and temperatures.

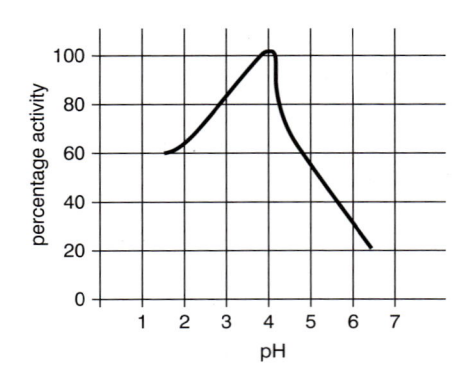

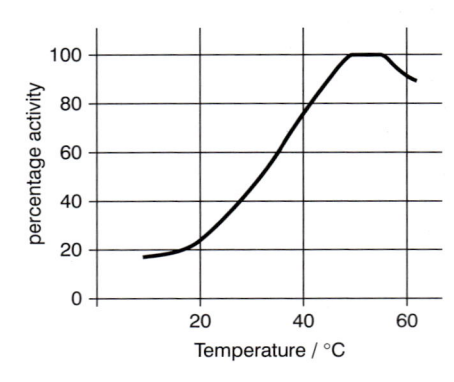

a) Describe and explain the effect of:

  i) pH (6)

  ii) temperature, on the activity of the enzyme pectinase (5)

b) A student finds an information sheet on the effect of pH on the activity of a protease called papain, from the papaya plant.

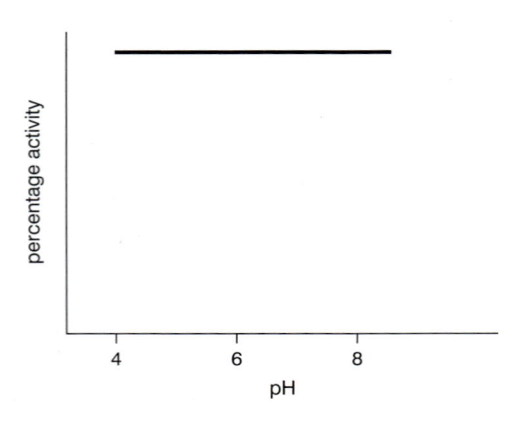

  i) State two proteases produced by the human gut. (2)

  ii) A student says that the graph shows that papain is unaffected by pH. Is the student correct? Explain your answer. (2)

(Total 15 marks)

EXTENDED Question 14

A student cut a number of discs cut from beetroot and weighed them. These were placed in sucrose solutions of different concentrations.

After one hour, the cylinders were removed, blotted dry and reweighed.

a) Describe and explain fully what would have happened to the beetroot discs over the one-hour period. **(7)**

b) The student examined the beetroot discs and found that the lower the sucrose concentration used, the firmer the discs.

Explain the reasons for this, and why this is important in plants. **(3)**

c) The diagram below shows a section of a plant root surrounded by soil particles.

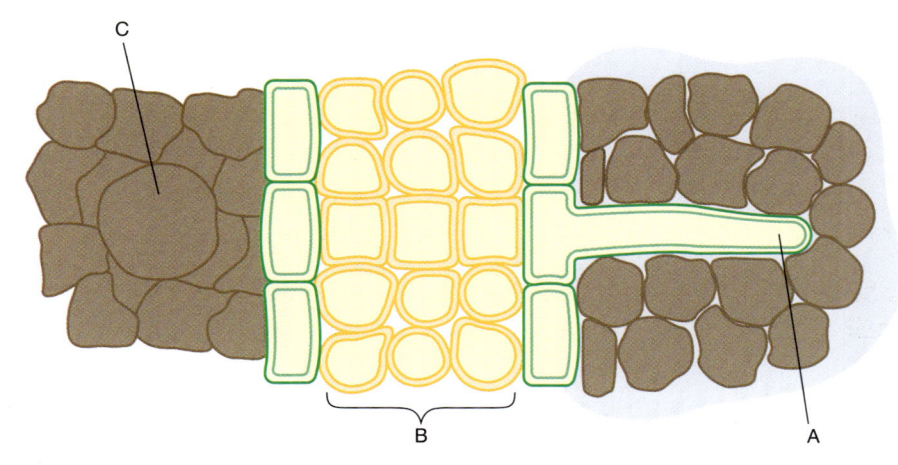

   i) Identify the parts of the plant root, A, B and C. **(3)**

   ii) Explain the importance of water potential in the uptake and transport of water by plant roots. **(7)**

d) Explain how mineral ions are taken up by plant roots. **(3)**

e) The effect of osmosis on animal cells is different from its effect on plant cells.

   i) Suggest and explain what will happen to red blood cells if placed in distilled water. **(3)**

   ii) Explain how osmosis is involved in the body's response to a cholera infection. **(5)**

(Total 31 marks)

Exam-style questions continued

EXTENDED Question 15

The diagram shows part of a kidney.

a) Name parts A–H shown in the diagram. (8)

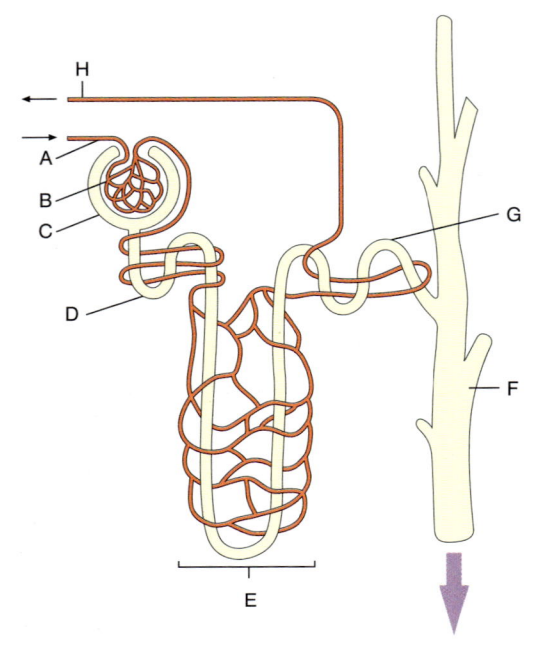

b) The table shows the concentration of certain substances in the plasma, glomerular filtrate and urine of a healthy person.

Substance	Concentration in plasma/g per 100 cm³	Concentration in glomerular filtrate/g per 100 cm³	Concentration in urine/g per 100 cm³
Amino acids	0.05	0.05	None
Glucose	0.10	0.10	None
Salts	0.90	0.90	<0.90 – 3.60
Protein	8.00	None	None
Urea	0.02	0.02	2.0

i) Referring to the structures in **a)**, explain the changes in concentrations of each of the four types of substance. (4)

ii) Where and why is urea produced in the body? (2)

c) The diagram shows what happens when someone is using a dialysis machine.

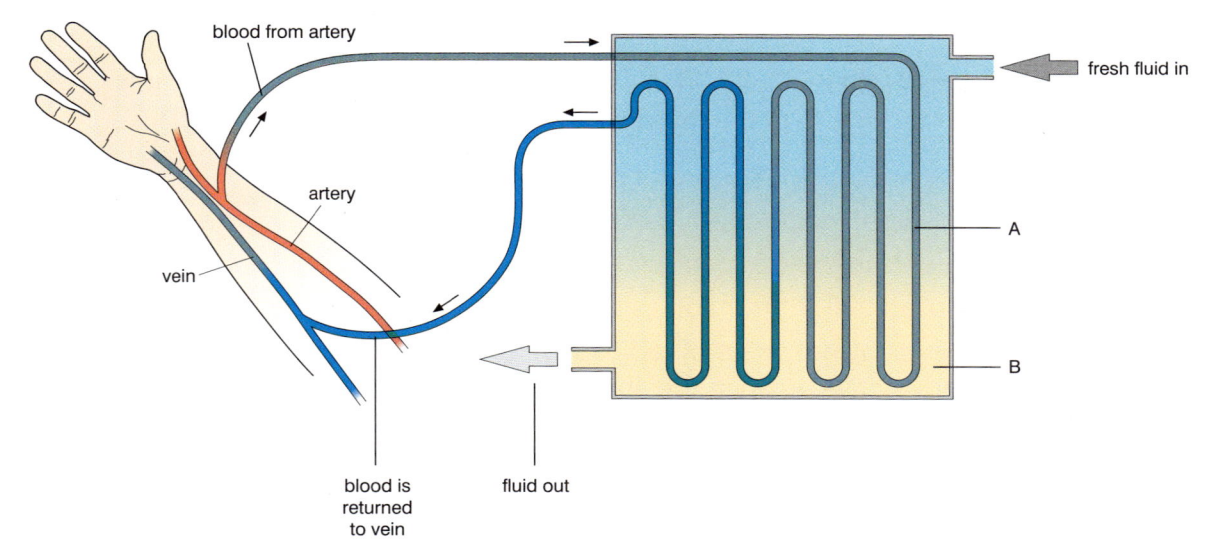

i) Explain how, when a patient is receiving dialysis, blood cells and blood proteins are prevented from leaving the blood. (1)

ii) Suggest how the treatment is able to remove urea from the blood, but keep the concentration of glucose in the blood at a normal level. (3)

d) Discuss two advantages and two disadvantages of the use of kidney transplants compared with dialysis. (4)

(Total 22 marks)

EXTENDED Question 16

Plants respond to the light available.

a) The graph below shows the effect of light intensity on photosynthesis in a single-celled plant.

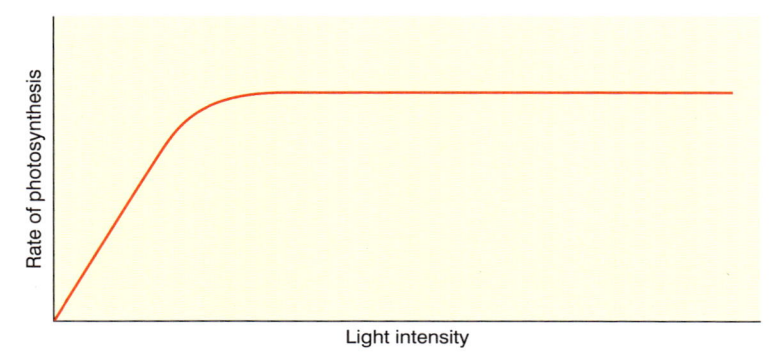

i) Describe and explain the effect of light intensity on the plant. (4)

Exam-style questions continued

The investigation was also carried out in a high concentration of carbon dioxide.

    ii) Sketch a graph of what you would expect so as to compare it with the graph above. Explain the shape of the graph you have drawn. (3)

b) A scientist investigating the response of plants to light placed:
- one group in the light, given even illumination
- one group of plants in the dark, and
- one group exposed to light from one side.

The plants were in an atmosphere of radioactive carbon dioxide, and after five hours, the amount of radioactive auxin in the area below the shoot tip was measured. The scientist's results are shown below.

	Plants in the light	Plants in the dark	Plants exposed to light from one side	
			Dark side	Lighted side
Total radioactive auxin/counts per minute	2985	3004	2173	878

Explain fully what these results show about the effect of light on auxin in the plants. (4)

c) Explain how the plant hormone, 2,4-D makes an effective herbicide. (2)

(Total 13 marks)

EXTENDED Question 17

The leaf is the main organ of photosynthesis.

a) Write a word and symbol equation for photosynthesis. (5)

b) Explain how the leaf is adapted to exchanging gases required for photosynthesis. (5)

c) Chemical substances in a plant are transported in the xylem and phloem.

Copy and complete the table below.

	Phloem	Xylem
Substances transported	(2)	(2)
Substances are transported:		
from	(1)	(1)
to	(1)	(1)

(8)

(Total 18 marks)

EXTENDED Question 18

The nervous system is involved in the body's response to stimuli.

a) When a person puts her hand on a hot object, she removes it quickly using a reflex action.

  i) Draw a diagram to show the reflex arc involved. (6)

  ii) Shortly after she removes her hand, she realises that she has touched a hot object. Explain how this occurs. (3)

b) What is the name of the other system involved in the body's coordination? How does the response of this system differ from the nervous system? (4)

(Total 13 marks)

EXTENDED Question 19

The circulatory system has several functions, including the transport of substances, temperature regulation and defence.

a) Describe how the back flow of blood in the heart is prevented. (4)

b) Explain why the wall of the left ventricle is four times as thick as the wall of the right ventricle. (4)

c) The blood system is involved in the body's immunity.

  i) Describe how the body can develop passive immunity. (1)

  ii) Explain how a vaccination can make a person immune to a particular disease. (5)

(Total 14 marks)

Scientists believe that there has been life on Earth for over 3500 million years. Nobody knows yet what triggered non-living molecules to become organised into living things that can reproduce themselves, but the earliest traces of life are in the form of bacteria-like structures from very ancient rocks. Reproduction led to different combinations of characteristics, and mutation produced new characteristics.

The environment determined which of these combinations of characteristics were the most successful, and so which individuals survived and which went extinct. Those that survived were able to pass on their genes to their offspring, through reproduction. This led to the evolution of new species and eventually to the millions of species that are alive on Earth today.

STARTING POINTS

1. How can you produce more plants without using flowers?
2. Why do flowers have to be pollinated before they make seed?
3. Why are some flowers large and brightly coloured and others small and pale?
4. How do the structures of the male and female human reproductive system support their function?
5. What controls the menstrual cycle every month?
6. How do growth and development differ?
7. Why are there two types of cell division?
8. How can we predict the inheritance of a characteristic controlled by a gene?
9. What is natural selection and how can it bring about evolution?
10. How can we change the characteristics of an organism?

SECTION CONTENTS

a) Reproduction
b) Growth and development
c) Inheritance
d) Exam-style questions

3
Development of the organism and the continuity of life

△ Reproduction can occur in a variety of ways. Here, a bee is involved in pollination, one of the ways in which flowering plants can reproduce.

Reproduction

INTRODUCTION

Most multicellular organisms reproduce sexually, which requires the transfer of gametes from the male to the female for fertilisation. Some flowering plants and a few animals can reproduce asexually. This is where there is no transfer of gametes and females do not need fertilisation to produce new individuals. Until recently, scientists believed that asexual reproduction in animals only happened in addition to sexual reproduction. However, we now know that some species of rotifers (microscopic aquatic animals) have not reproduced sexually for over 40 million years. Males of these species simply don't exist.

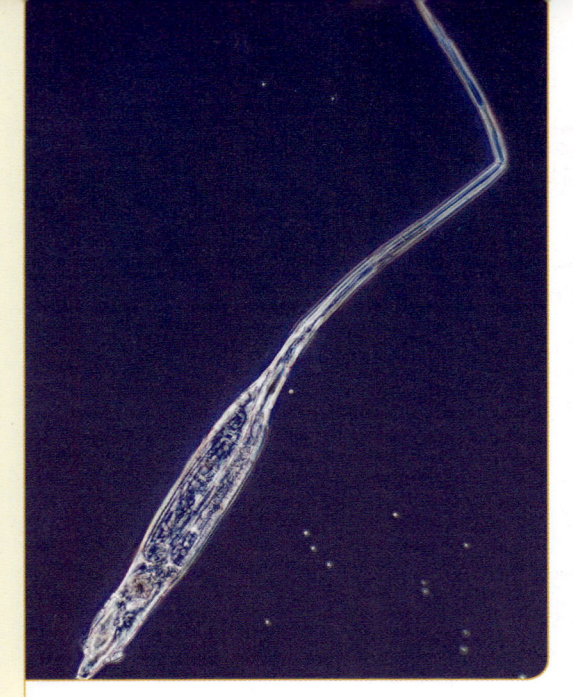

△ Fig. 3.1 Scientists think that this species of rotifer has not produced sexually for over 40 million years.

KNOWLEDGE CHECK

✓ Know that the flower is the reproductive structure in flowering plants.
✓ Know that the human reproductive system consists of organs, tissues and cells that are specially adapted for their role in reproduction.
✓ Know that sexual reproduction is the production of new individuals as a result of fertilisation; asexual reproduction is the production of new individuals without fertilisation.

LEARNING OBJECTIVES

✓ Be able to define *asexual reproduction* as the process resulting in the production of genetically identical offspring from one parent.
✓ Be able to describe asexual reproduction in bacteria, spore production in fungi and tuber formation in potatoes.
✓ Be able to define *sexual reproduction* as the process involving the fusion of haploid nuclei to form a diploid zygote and the production of genetically dissimilar offspring.
✓ Be able to identify and draw the sepals, petals, stamens, anthers, carpels, ovaries and stigmas of an insect-pollinated flower, and examine the pollen grains under a light microscope or in photomicrographs.
✓ Be able the state the functions of the sepals, petals, anthers, stigmas and ovaries.
✓ Be able to identify and describe the anthers and stigmas of a wind-pollinated flower, and examine the pollen grains under a light microscope or in photomicrographs.
✓ Be able to define *pollination* as the transfer of pollen grains from the male part of the plant (anther of stamen) to the female part of the plant (stigma).
✓ Be able to name the agents of pollination.

- ✓ Be able to compare the different structural adaptations of insect-pollinated and wind-pollinated flowers.
- ✓ Be able to describe the growth of the pollen tube and its entry into the ovule followed by fertilisation.
- ✓ Be able to investigate and describe the structure of a non-endospermic seed in terms of the embryo (radicle, plumule and cotyledons) and testa, protected by the fruit.
- ✓ Be able to outline the formation of a seed and fruit.
- ✓ Be able to state that seed and fruit dispersal by wind and by animals provides a means of colonising new areas.
- ✓ Be able to describe seed and fruit dispersal by wind and by animals.
- ✓ Be able to identify on diagrams of the human male reproductive system the scrotum, sperm ducts, prostate gland, urethra and penis, and the functions of these parts.
- ✓ Be able to identify on diagrams of the human female reproductive system the ovaries, oviducts, uterus, cervix and vagina, and the functions of these parts.
- ✓ Be able to outline sexual intercourse and describe fertilisation in terms of the joining of the nuclei of male gamete (sperm) and female gamete (egg).
- ✓ Be able to outline early development of the zygote in terms of the formation of a ball of cells that becomes implanted in the wall of the uterus.
- ✓ Be able to outline the development of the fetus.
- ✓ Be able to describe the function of the placenta and umbilical cord in relation to exchange of dissolved nutrients, gases and excretory products.
- ✓ Be able to describe the antenatal care of pregnant women, including special dietary needs and maintaining good health.
- ✓ Be able to outline the processes involved in labour and birth.
- ✓ Be able to describe the roles of testosterone and oestrogen in the development and regulation of secondary sexual characteristics at puberty.
- ✓ Be able to outline the following methods of birth control: natural, chemical, mechanical, surgical.
- ✓ Be able to describe the symptoms, signs effects and treatment of gonorrhoea.
- ✓ Be able to describe the methods of transmission of human immunodeficiency virus (HIV), and the ways in which HIV/AIDS can be prevented from spreading.
- ✓ **EXTENDED** Be able to discuss the advantages and disadvantages of asexual reproduction to a species.
- ✓ **EXTENDED** Be able to discuss the advantages and disadvantages of sexual reproduction to a species.
- ✓ **EXTENDED** Be able to distinguish between self-pollination and cross-pollination.
- ✓ **EXTENDED** Be able to discuss the implications of self-pollination and cross-pollination to a species.
- ✓ **EXTENDED** Be able to compare male and female gametes in terms of size, numbers and mobility.
- ✓ Be able to describe the menstrual cycle in terms of changes in the uterus and ovaries.
- ✓ **EXTENDED** Be able to explain the role of hormones in controlling the menstrual cycle, including FSH, LH, progesterone and oestrogen.
- ✓ **EXTENDED** Be able to indicate the functions of the amniotic sac and amniotic fluid.
- ✓ **EXTENDED** Be able to describe the advantages and disadvantages of breast-feeding of a baby compared with bottle-feeding using formula milk.
- ✓ **EXTENDED** Be able to describe the sites of production and the roles of oestrogen and progesterone in the menstrual cycle and in pregnancy.

- ✓ **EXTENDED** Be able to outline artificial insemination and the use of hormones in fertility drugs, and their social implications.
- ✓ **EXTENDED** Be able to outline how HIV affects the immune system in a person with HIV/AIDS.

ASEXUAL REPRODUCTION

Some organisms increase in number by **asexual reproduction**. For this type of reproduction it is not necessary to have two parents. During asexual reproduction, cells from the body of the mother divide to produce the offspring. This means that offspring produced by asexual reproduction are genetically identical to their parent and to each other.

Asexual reproduction is used by many different species. Bacteria reproduce asexually using binary fission. When a bacterial cell is large enough, the genetic material is copied exactly and then the cell splits in half. The new cells start to grow and the process begins all over again. This cycle can occur in as little as 20 minutes in ideal conditions. It produces large numbers of identical bacteria very quickly.

△ Fig. 3.2 The spores of Mucor on this strawberry are released from the black tips of hyphae high above the mycelium so that they can blow away to new sources of food.

Almost all fungi can reproduce asexually. Different types of fungi use different means of asexual reproduction, but the most important method is spore formation. This can be seen in *Mucor*, the common pin mould that often grows on bread and fruit. When this fungus has a plentiful supply of nutrients, an aerial hypha grows vertically. The tip of the cell swells with cytoplasm containing many nuclei. From this tip many spores are produced and released into the air. Spores are not like the seeds of flowering plants as they do not contain an embryo.

Another form of asexual reproduction is seen in plants such as potatoes that produce tubers. Tubers form from the ends of stems that grow into the soil. The stems

△ Fig. 3.3 Each of the potato tubers formed by this plant could produce a new plant in the next growing season.

swell into organs that store starch produced by the growing plant. When the plant dies back at the end of the growing season, the tubers remain in the soil until the next growing season.

The buds on the side of each tuber then produce new shoots and form a new plant. Each plant produces many potatoes, so each plant can produce many new plants in the next growing season.

EXTENDED

Asexual reproduction has advantages and disadvantages.

Advantages

- Only one parent is required. There is no need for a parent animal to find a mate or for pollination in plants.
- Large numbers of organisms can often be produced in a short time, particularly when conditions suddenly become favourable.
- All the offspring produced are identical, so they should survive well in the conditions in which the parent grows.

Disadvantages

- The lack of variation in the offspring means that all individuals are equally affected by any change in the environment that reduces their chance of survival.
- Because the offspring do not vary, they are not suited to moving away and living in environments with different conditions.

END OF EXTENDED

REMEMBER

For the highest marks, you must be able to describe as many differences between asexual and sexual reproduction as you can.

QUESTIONS

1. Define the term *asexual reproduction*.
2. Give one example of asexual reproduction in each of the following groups. In each case, explain why it is not an example of sexual reproduction.

    a) bacteria

    b) fungi

    c) flowering plants

SEXUAL REPRODUCTION

Sexual reproduction is the most common method of reproduction for the majority of larger organisms, including almost all animals and plants. It involves **haploid** cells, which are cells that have just one set of genetic information. It produces **diploid** cells that have two sets of genetic information (see Chromosomes, DNA and genes)

To produce a new organism the nuclei of two haploid gametes (sex cells) fuse. This is known as **fertilisation** and the resulting diploid cell is called a **zygote**. The zygote that is formed contains some genes from each of the parents, so it is not genetically identical to either parent. Some plants produce male and female gametes in the same flowers, so they are able to reproduce by self-fertilisation. But this is still sexual reproduction because it involves the fusion of a male and female gamete.

EXTENDED

Sexual reproduction has advantages and disadvantages.

Advantages

- Fusion of gametes combines genetic information from two parents, which results in variety in the offspring.
- Individuals may be better adapted to different conditions than the parents and each other. This gives the species a greater chance of survival in changing conditions.

Disadvantages

- Sexual reproduction usually requires a second parent for fertilisation.
- Finding a mate can require the individual to use up energy.
- Finding a mate takes time, so sexual reproduction takes longer to produce offspring than asexual reproduction.

END OF EXTENDED

QUESTIONS

1. Draw up a table to summarise the differences between sexual and asexual reproduction.
2. Give an example of
   a) a haploid cell
   b) a diploid cell.
3. **EXTENDED** Give one example when offspring from sexual reproduction would have a survival advantage over those produced by asexual reproduction.
4. **EXTENDED** Aphids are common pests of crop plants. Explain the advantages for female aphids to reproduce asexually over the summer months in the UK.

SEXUAL REPRODUCTION IN PLANTS

Flowering plants are the most successful group of plants. These are the only plants that have true flowers and produce seeds with a tough protective coat. During sexual reproduction in flowering plants:

- male and female haploid gametes are produced – some species may produce male and female gametes in the same flowers; other species may have male-only flowers and female-only flowers on the same plant; and in other species male flowers and female flowers are produced on different plants
- male pollen is transferred to the female part of the flower so that **pollination** can take place
- the male gamete and female gamete fuse during fertilisation to form a zygote
- the zygote develops to form an **embryo** within a seed, which protects the embryo and provides food during germination of the seed
- seeds are dispersed, so that they germinate away from the parent.

FLOWER STRUCTURE

Most flowers have a similar basic arrangement. Their structures are stacked one on top of each other along a short stem, arranged either in a spiral or in separate rings.

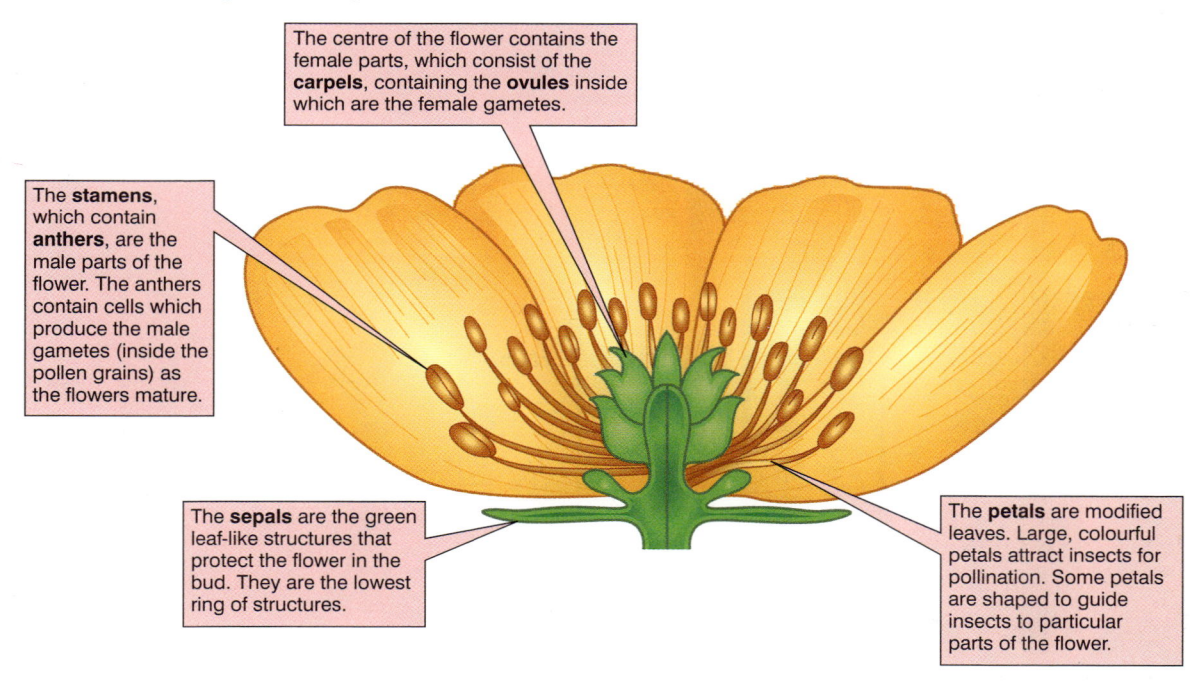

The centre of the flower contains the female parts, which consist of the **carpels**, containing the **ovules** inside which are the female gametes.

The **stamens**, which contain **anthers**, are the male parts of the flower. The anthers contain cells which produce the male gametes (inside the pollen grains) as the flowers mature.

The **sepals** are the green leaf-like structures that protect the flower in the bud. They are the lowest ring of structures.

The **petals** are modified leaves. Large, colourful petals attract insects for pollination. Some petals are shaped to guide insects to particular parts of the flower.

△ Fig. 3.4. Structure of an insect-pollinated flower.

The male parts of the flower

The male part of a flower is the ring of **stamens**. There may be just a few stamens or as many as 100. Each stamen consists of two parts: the **anther** at the top and a stalk called the filament.

The pollen grains contain the male gametes in flowering plants. Pollen develops in the pollen sacs of the anthers. Cells lining the inside of the pollen sacs divide by **meiosis** (see Cell division) to give four cells. Each of these cells develops into a pollen grain. As a grain matures, it develops a thick outer wall to protect the delicate male gamete inside. When all the pollen grains in the anther are mature, the anther splits open to release them.

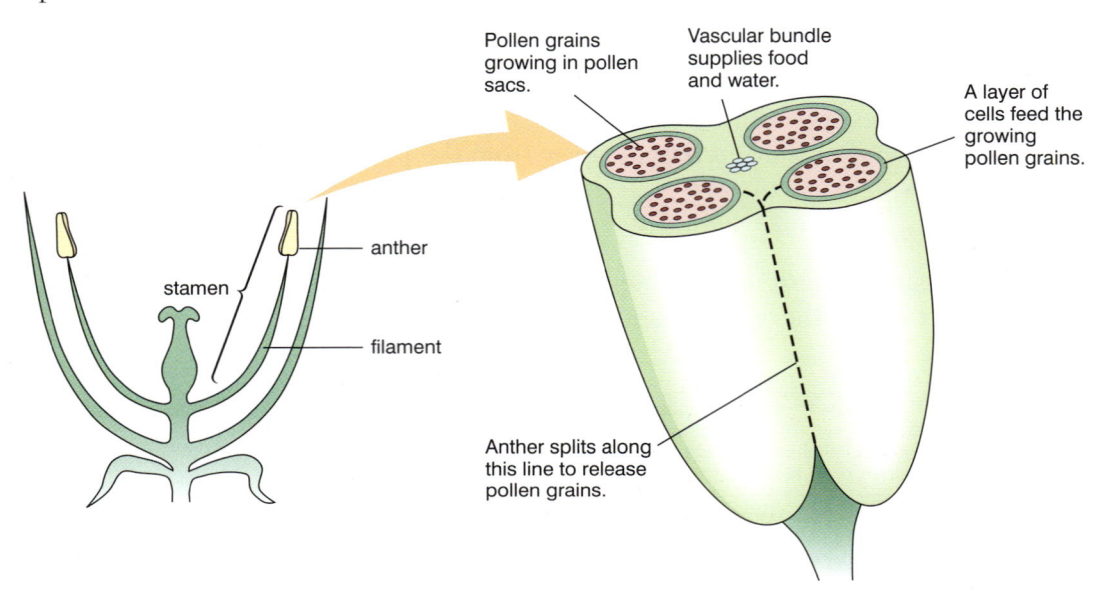

△ Fig. 3.5 The structure of a flower.

△ Fig. 3.6 Alder catkins shedding pollen.

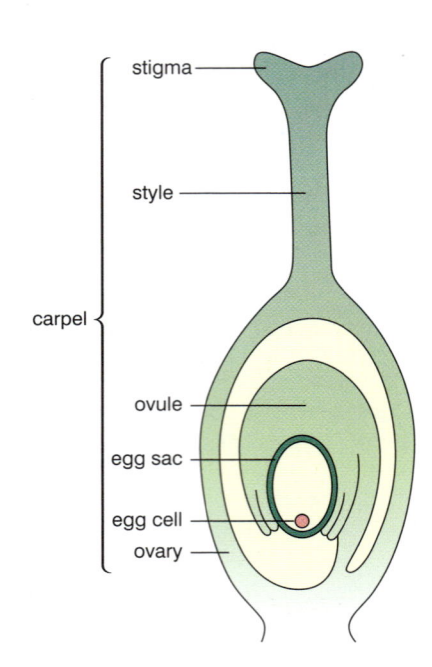

△ Fig. 3.7 The carpel.

The female parts of the flower

The female part of the flower is the **carpel**. A flower may have more than one carpel, each with its own **style** and stigma. The **stigma** is the part of the carpel where the pollen lands during pollination.

The ovary at the base of the carpel protects the female gamete from the dry air outside. The ovary contains one or more ovules, and each ovule contains an egg sac that surrounds the egg cell (female gamete).

△ Fig. 3.8 Even complex flowers like daisies, which contain thousands of male and female parts, have carpels surrounded by stamens.

QUESTIONS

1. What is the function of the stamen in a flower?
2. Where are the male gametes of a plant found?
3. Name the main parts of a carpel and explain their role in a flower.

POLLINATION

Before fertilisation can take place, the male gametes have to reach the female gametes. This involves transferring the pollen to the stigma, a process known as **pollination**. In many plants this means transferring the pollen from one flower to another. Some plants use the wind to transfer their pollen between flowers. Others use animals, especially insects, to carry the pollen. Flowers have different features depending on whether they are pollinated by wind or by insects.

Wind-pollinated plants	Insect-pollinated plants
Small petals, which do not get in the way when wind blows the pollen	Large petals for insects to land on
Green or tiny petals	Brightly coloured petals to attract insects
No scent	Often scented to attract insects
No nectaries	Nectaries present at the base of the flower produce a sugary liquid to attract insects, such as bees and butterflies
Many anthers which are often large and hang outside the flower so that pollen is easily dispersed	A few small anthers, usually held inside the flower
Pollen grains have smooth outer walls	Pollen grains have sticky or spiky outer walls
Stigmas are large and feathery, often hanging outside the flower to trap pollen	Stigmas are small and held inside the flower
Produce large amounts of pollen	Produce smaller amounts of pollen
Pollen is lightweight	Pollen is heavier

△ Table 3.1 Differences between wind-pollinated and insect-pollinated plants.

△ Fig. 3.9 In insect-pollinated plants, nectaries secrete a sugary liquid to attract insects.

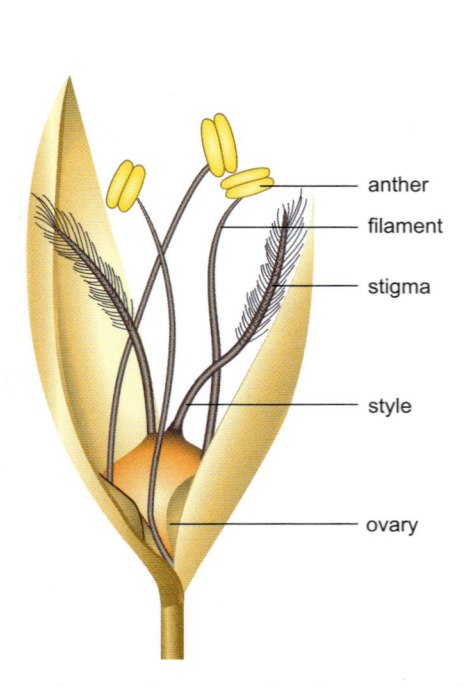

- anther
- filament
- stigma
- style
- ovary

△ Fig. 3.10 These grass plants have anthers that hang outside the flowers and release large amounts of pollen to the wind. The stigmas also hang outside the flower to collect pollen from other grass plants.

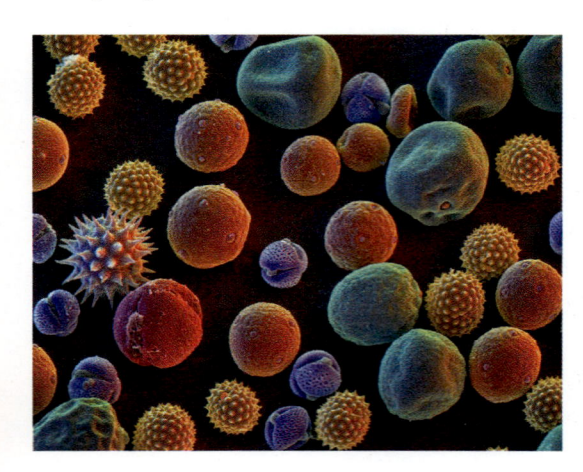

◁ Fig.3.11 Different plant species can be identified by differences in the shape of their pollen.

> **SCIENCE IN CONTEXT**
>
> ## POLLINATOR RELATIONSHIPS
>
> Different pollinators are attracted to different features of animal-pollinated flowers. Tube-shaped flowers attract insects with long tubular mouth-parts, such as butterflies, or birds with long beaks such as hummingbirds. Blue and violet-coloured flowers are more attractive to bees. Butterflies often prefer red. Plants pollinated by moths or bats tend to open at night and may not be brightly coloured but instead produce a strong sweet scent. Plants, like the titan arum, that rely on flies to pollinate them, often smell like rotting meat.
>
> The most bizarre partnerships between flowers and insects occur where a particular species of orchid produces flowers that mimic the female of a particular species of wasp. Male wasps are attracted to the flowers to mate with what they think are female wasps. During the 'mating' the flowers deposit pollen on the insect, which is then carried to the next flower that the insect is attracted to.
>
> △ Fig. 3.12 A male wasp receiving pollen while 'mating' with an orchid flower.

REMEMBER

Be very careful not to confuse *pollination* with *fertilisation*.

EXTENDED

Self-pollination and cross-pollination

Cross-pollination occurs when the pollen from one plant transfers to the stigma of a different plant. However, plants can produce both male and female gametes and it is possible for the pollen from one plant to pollinate its own stigma. This is called **self-pollination**. Most flowering plants try to prevent self-pollination by using self-incompatibility systems.

Some plants produce only flowers of a single sex, which makes self-pollination impossible. For example, pistachio trees are single sex.

Some plants have flowers that contain male and female parts, but the different parts mature at different times. For example, on a hemp plant the male flowers mature before the female. The female flowers mature before the male in the titan arum or bunga bangkai 'corpse flower' (*Amorphophallus titanum*).

Even if pollen from one flower manages to reach its own stigma, the pollen grain will tend to grow more slowly than a foreign grain.

This means that nuclei from foreign pollen grains are more likely to fertilise the egg cell in the ovule.

In just a few species, a flower will self-pollinate if cross-pollination doesn't occur. This means that seed will be produced so that new plants can grow.

END OF EXTENDED

QUESTIONS

1. Describe the main features of a wind-pollinated flower.
2. Explain how wind-pollinated flowers are adapted to distribute their pollen.
3. Describe the main features of an insect-pollinated flower.
4. Explain how insect-pollinated flowers are adapted to distribute their pollen.
5. **EXTENDED** Explain how plants try to avoid self-pollination.

FERTILISATION AND SEED FORMATION

Fertilisation occurs when the male nucleus from the pollen grain fuses with the nucleus of the female egg cell. To get from the tip of the stigma to the ovule, the pollen grain produces a thin tube called the **pollen tube**. The diagram shows how the pollen tube grows down through the style, into the ovule and delivers the male gamete to the egg. Once inside the egg, the male gamete fuses with the female egg cell to produce a zygote.

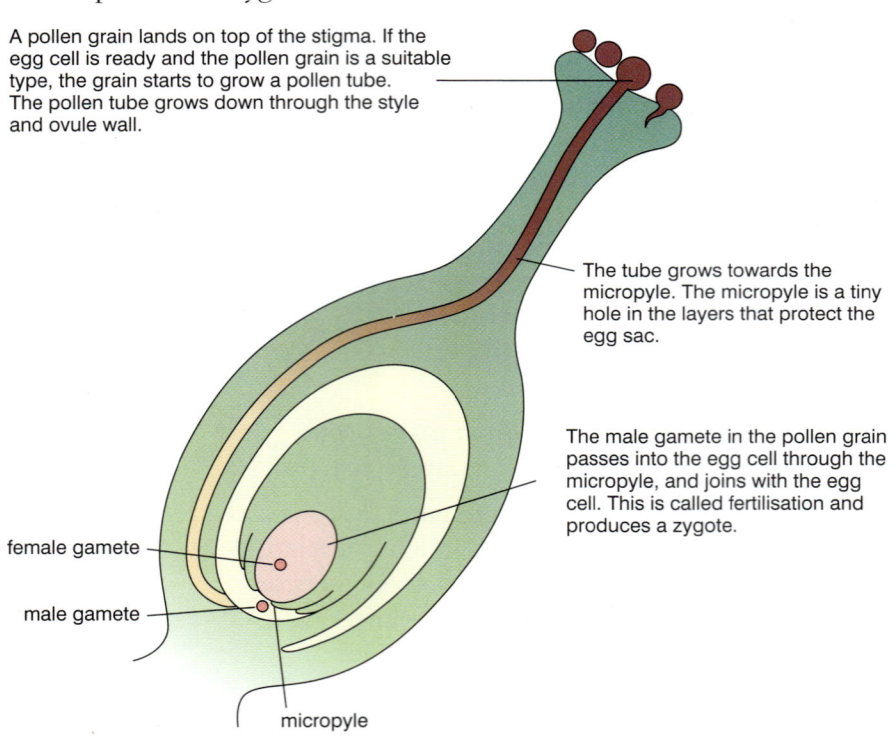

A pollen grain lands on top of the stigma. If the egg cell is ready and the pollen grain is a suitable type, the grain starts to grow a pollen tube. The pollen tube grows down through the style and ovule wall.

The tube grows towards the micropyle. The micropyle is a tiny hole in the layers that protect the egg sac.

The male gamete in the pollen grain passes into the egg cell through the micropyle, and joins with the egg cell. This is called fertilisation and produces a zygote.

female gamete
male gamete
micropyle

△ Fig. 3.13 Pollination leads to fertilisation.

The zygote develops quickly to form an embryo plant, which needs to be protected from drying out. This embryo plant contains differentiated tissues when it starts to grow or germinate:

- the **radicle** will become roots
- the **plumule** will become stem and leaves.

Surrounding the zygote, other tissue grows to form food stores of carbohydrate, lipids and proteins. These food stores are called **cotyledons** – some species of plants have two cotyledons, for example beans. Other plants, such as maize and grasses, have only one cotyledon. The ovule wall hardens as the zygote grows, to form a tough or hard protective casing around the seed called the **seed coat**. The ovary tissue surrounding the ovule often develops into a fruit. If the ovary contained several ovules, the fruit will contain several seeds, such as in apples and grapes.

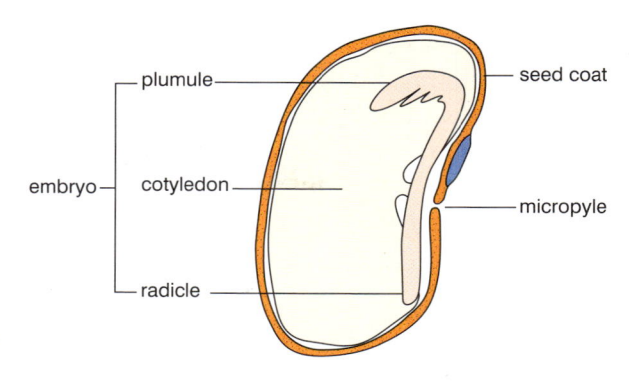

△ Fig. 3.14 The structure inside a bean seed.

Fruits often have soft and sweet flesh to attract animals to eat them. The hard coat of the seed protects the embryo as the seed passes through the animal's gut and is egested. The seed is deposited usually at a distance from the parent plant with a useful supply of nutrients in the animal's faeces.

△ Fig. 3.15 In a plum the ovule turns into a hard seed.

QUESTIONS

1. Distinguish between *pollination* and *fertilisation* in a plant.
2. Describe the process of fertilisation in a flowering plant, from the point of pollination of the flower.
3. Identify what the following structures in the carpel will develop into when fertilisation has taken place: female gamete (egg cell), ovule wall, ovary.

Seed dispersal

Seeds that fall from a plant onto the ground below could germinate to produce new individuals. These would compete with the parent plant and possibly reduce the success of the parent and the new young plants. So plants tend to **disperse** (spread) their seeds into new areas.

Plants use a variety of different methods to disperse their seeds including dispersal by the wind and by animals.

Seeds carried by the wind tend to be very light. The seeds of the orchid are like dust and are easily blown about. Other plants have special adaptations to their seeds to help them travel further like the 'parachutes' of *Taraxacum* (dandelion) species. The seeds of the sycamore tree are much bigger than orchid seeds but have special 'wings' that allow them to be carried by the wind.

Animals can act to disperse seeds either by eating them and passing the seeds out undamaged or by transporting the seeds away. Seeds may transported deliberately to be eaten elsewhere or may stick to the feet and fur of animals and be carried to another area.

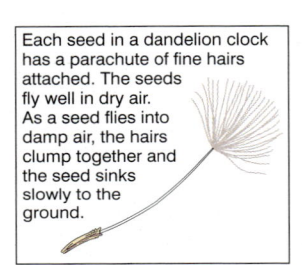

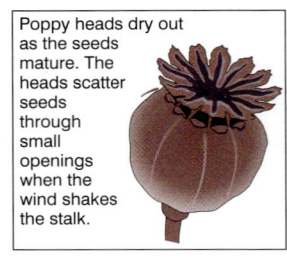

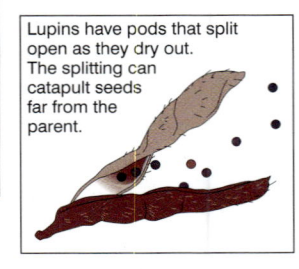

Each seed in a dandelion clock has a parachute of fine hairs attached. The seeds fly well in dry air. As a seed flies into damp air, the hairs clump together and the seed sinks slowly to the ground.

The sticky seeds of the cleavers attach to animals and hitch a ride to new areas.

Poppy heads dry out as the seeds mature. The heads scatter seeds through small openings when the wind shakes the stalk.

Lupins have pods that split open as they dry out. The splitting can catapult seeds far from the parent.

Seed dispersal

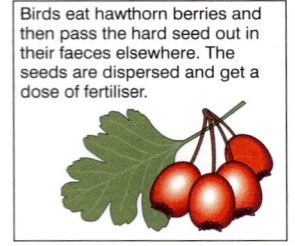

Sycamore seeds have wings which make them spin as they fall. This keeps them in the air longer and so lets them spread further from the parent.

Birds eat hawthorn berries and then pass the hard seed out in their faeces elsewhere. The seeds are dispersed and get a dose of fertiliser.

Squirrels help to distribute nuts by collecting them and burying them in the ground. Often the buried seeds germinate before the squirrel finds them again.

The fibrous outer husk of a coconut allows it to float in sea currents to new areas. Many tropical islands have a line of coconut palms along the top of the beach.

△ Fig. 3.16 Seed dispersal.

QUESTIONS

1. Explain why it is usually an advantage to plants to disperse their seed.
2. Give two examples of seeds that are dispersed by animals.
3. Give two examples of seeds that are dispersed by wind.

SEXUAL REPRODUCTION IN HUMANS
Male reproductive system

In humans, the male has two **testes**. These are the organs in which **sperm** are produced. The testes are supported outside the body in the **scrotum**. This keeps them at a lower temperature than the core of the body, as warm temperatures inhibit the production of sperm.

EXTENDED

Sperm are among the smallest cells in the human body, at about 45 μm long. Most of the cell is formed of the tail which propels the sperm through the female uterus to the egg for fertilisation. Over 100 million sperm cells are produced in the testes each day.

END OF EXTENDED

Sperm ducts carry the sperm from the testes to the penis, through the prostate gland and seminal vesicles. The prostate gland and seminal vesicles together produce the liquid in which the sperm are able to swim. The liquid also contains sugars that sperm can use to supply energy for swimming, and other chemicals that protect the sperm as they swim to the egg. Semen is the mixture of sperm cells and liquid.

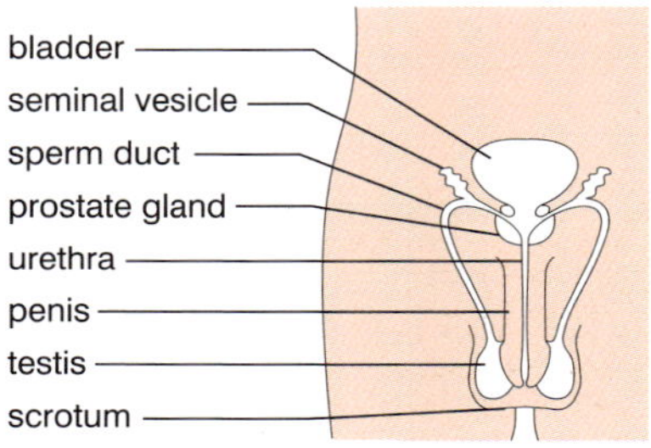

Semen passes along the sperm duct to the urethra to outside the body. The urethra also carries urine from the bladder to outside of the body. When the man is sexually excited, large spaces in the penis fill with blood. This causes the penis to become larger and stiffer in what is called an erection. At the same time a muscle ring (sphincter) at the top of the urethra contracts, preventing urine entering the urethra from the bladder.

△ Fig. 3.17 The male reproductive system. (Note that the bladder is not part of the reproductive system.)

QUESTIONS

1. Sketch a diagram of the human male reproductive system.
2. Add labels to your sketch to name the main parts of the system.
3. Describe the role of each of the main parts of the system in human reproduction.

Female reproductive system

The two **ovaries** are the organs in human females that produce the eggs. They are positioned within the abdominal cavity on either side of the uterus and joined to it by the **oviducts**.

Every month from puberty until menopause, when a woman is around 50, one ovary releases one egg which travels down the oviduct to the **uterus** (womb). If it is not fertilised, the egg will be flushed from the uterus during the monthly period (bleed).

EXTENDED

The egg cell is one of the largest human cells, at about 0.2 mm in diameter. It cannot move on its own, but is pushed gently along the oviduct by cilia on the inside of the tube. An ovary may contain thousands of egg cells, but usually only one is released from one ovary at **ovulation** each month.

END OF EXTENDED

The lower end of the uterus, at the **cervix**, leads into the **vagina**, an elastic, muscular tube. The vagina opens to the outside of the body at the vulva, which is formed by the meeting of folds of skin called the labia.

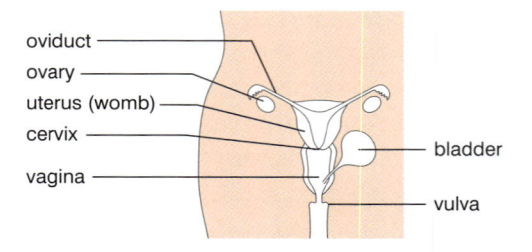

△ Fig. 3.18 The female reproductive system.

QUESTIONS

1. Sketch a diagram of the human female reproductive system.
2. Add labels to your sketch to name the main parts of the system.
3. Describe the role of each of the main parts of the system in human reproduction
4. **EXTENDED** Draw up a table to compare the size, numbers and mobility of human egg and sperm cells.

SEX HORMONES

Testosterone (the male sex hormone) is secreted from the testes. At puberty, the increased secretion of testosterone causes the development of the following **secondary sexual characteristics** in boys:

- an increase in rate of growth until adult size
- hair growth on face and body including pubic hair
- penis, testes and scrotum growth and development
- deepening of voice
- increased muscle development
- sperm production.

The female sex hormones are **progesterone** and **oestrogen**, which are both produced in the ovaries. At puberty, increased secretion of oestrogen causes the development of the following secondary sexual characteristics in girls:

- an increase in rate of growth until adult size
- breast development
- vagina, oviducts and uterus development
- start of menstrual cycle (periods)
- hips widening
- pubic hair and under-arm hair growth.

These hormones also control the changes that occur during the **menstrual cycle**.

The menstrual cycle

The menstrual cycle is a sequence of changes that occur in a woman's body that occur every month. The average cycle is 28 days long, but it is normal for it to vary in different women.

The cycle begins with the monthly period, or bleeding, which is produced from the breakdown of the thickened lining of the uterus. After this, the uterus lining starts to thicken again. Ovulation occurs about halfway through the cycle, when an egg is released from one of the ovaries. The egg travels along the oviduct to the uterus.

If the egg is fertilised during this time, it will implant in the uterus lining and the lining will continue to develop for pregnancy. If the egg is not fertilised, the cell and the uterus lining are shed during the monthly period at the start of the next cycle.

EXTENDED

The release of the egg from an ovary and the development of the uterus lining are controlled by the hormones oestrogen and progesterone from the ovaries, and by **LH** (luteinising hormone) and **FSH** (follicle-stimulating hormone) produced by the pituitary gland in the brain.

- The first day of the cycle is taken to be the point when the lining of the uterus prepared for the previous egg starts to break down.
- At this point FSH is released from the pituitary gland and stimulates the development of an egg cell in the ovary.
- Cells surrounding the developing egg secrete increasing amounts of oestrogen.
- The oestrogen stimulates the lining of the uterus to repair and thicken.
- High levels of oestrogen in the body cause the pituitary gland to release more LH.
- About two weeks through the cycle (around day 14) the high levels of LH cause an egg to be released from the ovary into the oviduct. This is called **ovulation**. When this happens, cells in the ovary start

to secrete progesterone, which reduces the amounts of LH and FSH released from the pituitary gland.
- Progesterone stimulates the uterus lining to thicken even more in preparation to receive a fertilised egg.
- If the egg is not fertilised, the concentrations of oestrogen and progesterone start to fall.
- The fall in hormone concentration causes the uterus lining to break down—it is lost from the body during menstruation (a period).
- When the hormone concentrations are low enough, the pituitary gland starts to release more FSH and another egg starts to develop in one of the ovaries. The cycle begins again.

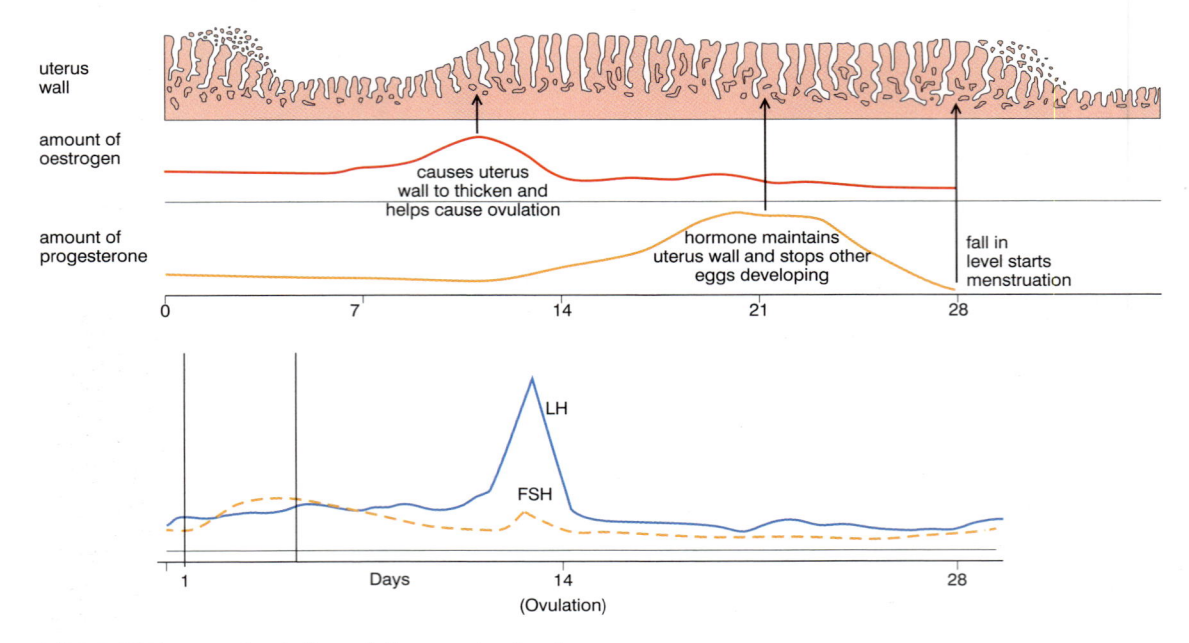

△ Fig. 3.19 Hormone levels through the menstrual cycle.

If the egg is fertilised, then progesterone continues to be released from the ovary. This maintains the uterus lining during pregnancy and prevents further ovulation.

END OF EXTENDED

QUESTIONS

1. Name the male sex hormone.
2. Name two female sex hormones produced in the ovary.
3. Describe the role of secondary sexual characteristics in humans.

4. Draw the menstrual cycle as a circle of 28 days. On your diagram label:

   a) ovulation

   b) menstruation

   c) increases and decreases in oestrogen secretion

   d) increases and decreases in progesterone secretion.

5. a) **EXTENDED** Where are the hormones LH and FSH produced?

   b) Describe the role of LH and FSH in the control of the menstrual cycle.

FERTILISATION AND DEVELOPMENT OF THE FETUS

Stiffening of the penis makes it possible for a man to insert his penis into the vagina of a woman to deliver the sperm during sexual intercourse. Rapid contractions of muscles in the penis during ejaculation send the sperm shooting out into the vagina, on their way to the egg cell.

Sperm deposited near the cervix swim up into the uterus, and then along the oviduct to the egg. Many sperm fail to make the journey but some will reach the oviducts at the top end of the uterus. The egg will have been travelling along the oviduct while the sperm have been swimming up from the uterus. Fertilisation takes place in the oviduct. The nucleus of one sperm cell fuses with the nucleus of the egg cell, forming a fertilised egg or **zygote**.

After fertilisation in the oviduct, the fertilised egg (zygote) travels on towards the uterus. This journey takes about three days, during which time the zygote will divide several times to form a ball of 64 cells, which is now called an **embryo**.

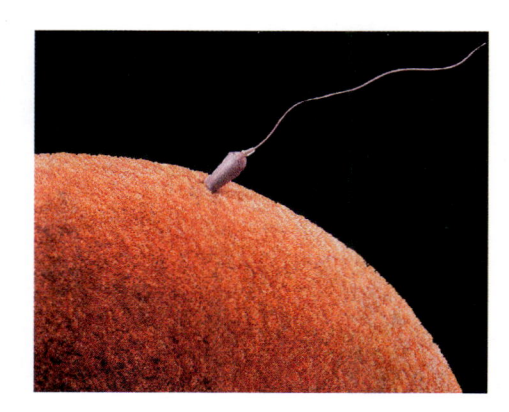

△ Fig. 3.20 The moment of fertilisation as a sperm fuses with an egg.

In the uterus, the embryo embeds (implantation) in the thickened lining, and cell division and growth continue. For the first three months, the embryo gets nutrients from the mother by diffusion through the uterus lining.

By the end of three months, the placenta has developed, and the embryo has developed into a **fetus** in which all the main organs of the body can be identified.

Over the next 28 weeks the fetus will increase its mass roughly 8 million times. At no other point in an individual's lifetime will it grow at such a high rate.

This period of development in the uterus is known as **gestation**, and lasts about 40 weeks in humans, measured from the time of the woman's last period.

The rapid growth during gestation depends on a good supply of food and oxygen, provided by the mother.

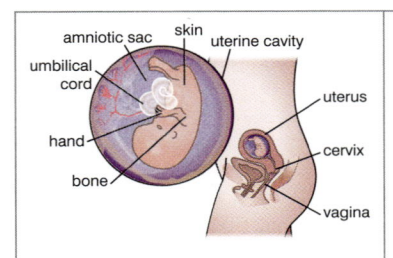

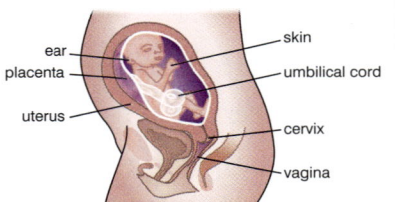

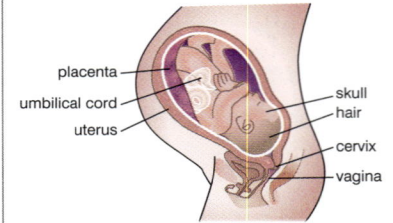

11 weeks:
- fetus about 4 cm long
- most of the main body structures have formed

23 weeks:
- fetus about 29 cm long and weighs about 500 g
- fetus hears sound from outside, and moves

40 weeks:
- fetus is about 51 cm long and weighs about 3.4 kg
- all organs fully developed ready for birth

△ Fig. 3.21 The fetus in the uterus at three times during gestation.

SCIENCE IN CONTEXT: ULTRASOUND SCANS

Ultrasound is very high frequency sound, far above the frequency that can be heard. It is used in medical imaging for showing soft tissues inside the body. It is particularly useful for looking at the developing fetus in the uterus, because it does not harm either the fetus or the mother.

An ultrasound scan is commonly done before about halfway through gestation, to make sure that the fetus is developing normally. At about this stage, if the fetus is lying at the right angle, it may even be possible to tell if it is a male fetus because the testes can be distinguished at this age.

Ultrasound scans may be done at other times during gestation if there is any concern about the development of the fetus.

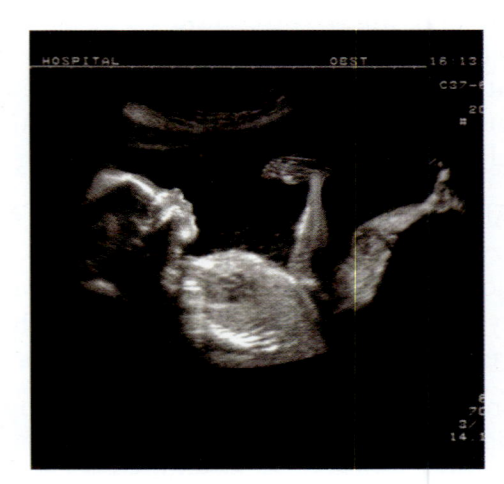

△ Fig. 3.22 This ultrasound scan was taken in the 20th week of gestation and shows that the fetus is developing normally.

EXTENDED

The fetus develops inside a bag of fluid called **amniotic fluid.** This fluid is produced from the amniotic membrane that forms the outer layer of the bag (amnion). The fluid protects the fetus from mechanical damage, for instance if the mother moves suddenly. It also reduces the effect of large temperature variations which would affect the fetus' rate of development. One of the signs that birth will occur soon is when this bag bursts during labour.

END OF EXTENDED

The placenta

The **placenta** is an organ that allows a constant exchange of materials between the mother and fetus. It develops from fetal tissues. The placenta and the uterus wall have a large number of blood vessels that run very close to each other but do not touch. Maternal and fetal blood do not mix. If they did, the higher blood pressure in the mother could damage the fetus. It also helps to prevent pathogens and some chemicals getting into the blood of the fetus.

The fetus is joined to the placenta by the umbilical cord, which carries the blood vessels of the fetus. Dissolved food molecules, oxygen and other nutrients that the fetus needs for growth diffuse from the mother's blood into the blood of the fetus. Waste products from metabolism in the fetus diffuse across into the mother's blood.

From the formation of the placenta until birth, this is the only way that the developing fetus exchanges materials with the outside world. Birth occurs when all the organs of the fetus are fully developed and ready to carry out the life processes on their own.

QUESTIONS

1. Define the following terms in your own words:
    a) *zygote*
    b) *embryo*
    c) *fetus.*
2. Where in the human body does fertilisation of the egg cell occur?
3. Briefly describe how the fetus develops up to the point of birth.
4. Describe the role of the placenta during the development of a fetus.

Antenatal care

Antenatal care is the care that pregnant women get before their baby is born. During this time, they will receive checks from a doctor or midwife to make sure that they are healthy and to make sure the fetus is developing well. Health checks include:

- encouraging the woman to give up smoking during pregnancy
- advice on how little alcohol should be drunk, and which drugs to avoid
- advice on food preparation to avoid food-borne infections such as from cheese
- checks on blood for iron content, with prescription for iron if this is low
- possible screening tests for some conditions in the fetus
- advice on a healthy diet, including taking folic acid to avoid development problems (such as spina bifida) in the fetus
- advice on what exercise is suitable and how to remain fit during pregnancy.

In addition, the mother-to-be will get information on what to expect during labour and birth, and breast-feeding after birth.

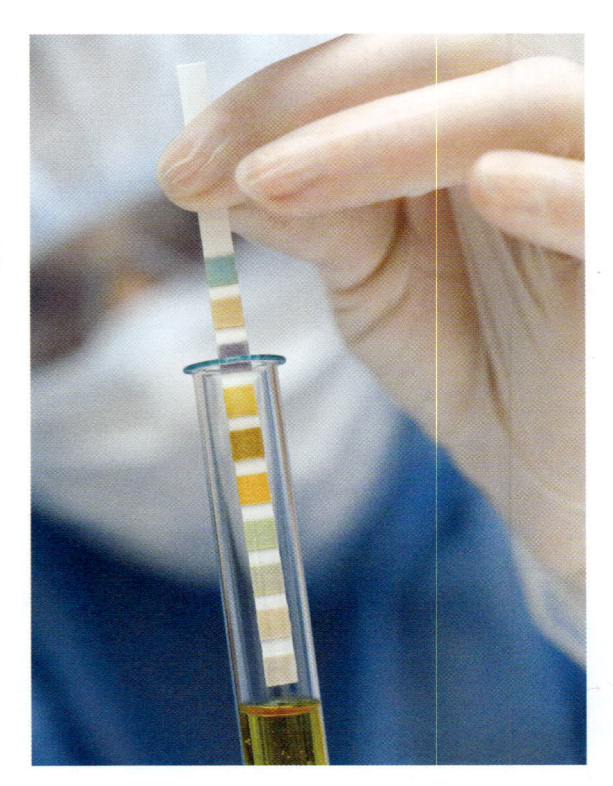

△ Fig. 3.23 Urine and blood tests of pregnant women can show if the fetus is developing properly.

LABOUR AND BIRTH

The **birth** of a baby is the final part of a process called **labour** which usually takes hours, sometimes days. Labour starts with contractions of the uterus wall, which becomes hard during these contractions. As labour proceeds, the contractions become longer, more frequent and more painful. Another sign of labour starting may be a show of blood-tinged mucus, which is the plug at the entrance to the uterus. A flow of water from the vagina, often called the 'breaking of the waters', shows labour is near, as does unexpected diarrhoea or sickness.

At the hospital the mother to be will be checked to make sure all is progressing well. She may be linked to monitors that can follow her blood pressure and heart rate, as well as the heart rate of the fetus.

The next stage of labour can last some hours. During this stage the muscles of the uterus are opening up (dilation) the cervix so that the baby can pass through. Periodically a doctor or midwife will check to see how far the cervix has dilated. Contractions at this point can become quite painful and doctors can prescribe different kinds of pain relief.

When the cervix has fully dilated, the mother will feel an urge to push. The final stage of labour is shorter than the earlier part and within minutes the baby is born. In most cases the baby is born head first.

A breech birth happens when the baby is born feet first. Breech births need to be very carefully managed by the midwife to prevent damage to the mother or baby due to tangling or damage to the umbilical cord.

The baby is followed after 10 minutes or so by the afterbirth. This is the placenta. The midwife or doctor will check to see the complete placenta has been delivered to prevent any infection developing in the uterus.

Sometimes the baby is too large to pass through the vagina or needs to be helped with forceps, and will tear the outer tissues slightly. Often a few stitches help to keep the torn sides of the wound together after birth to help with healing.

EXTENDED

Feeding after birth

After birth, the first milk produced by the mother does not look like milk at all. This is the watery **colostrum** which contains antibodies that protect the baby from infection in the first few days and months of life. By about the third day after birth the normal milk begins to come through.

Human breast milk contains a mixture of chemicals that are different from the milk of any other animal. The levels of these chemicals can be measured and used to produce an artificial 'human' milk by modifying milk from other animals. These are called formula milks. However, it is impossible to produce a formula milk exactly like human milk. Most doctors believe that breast milk is the best milk for growing babies if it is available and the mother is able to feed her baby.

	Breastfeeding	**Formula (bottle) milk feeding**
Advantages	Perfectly matched to baby's needs Contains antibodies that help to protect the baby from infection Always sterile and safe for baby	Does not require a mother to provide it Can be more convenient in some circumstances
Disadvantages	Nipples can become painful Some women find it difficult to produce enough milk at first	Does not provide the antibodies of breast milk so child is at greater risk of infection Sterile water and clean conditions are required to make up a feed, which can be difficult in certain areas

△ Table 3.2. Advantages and disadvantages of breastfeeding and bottle feeding.

END OF EXTENDED

QUESTIONS

1. Explain what is meant by *antenatal care*.
2. Explain why a midwife will advise on what a woman should eat and drink during pregnancy, giving examples.
3. State two signs that labour is beginning.
4. **EXTENDED** a) Why is breastfeeding a baby usually better for its health?
   b) Suggest one situation where a baby might need to be bottle-fed.

METHODS OF BIRTH CONTROL

Sometimes people want to have sexual intercourse but not produce children. This is particularly important for couples who want to plan how many children they have, and when they have them. Techniques for preventing pregnancy are called **contraception**. The following discussion covers the most common methods of contraception.

Natural methods

Avoiding sexual intercourse completely is called **abstinence**. Without sexual intercourse, fertilisation and pregnancy cannot occur. However, many couples find abstinence difficult to maintain.

Another method is the **rhythm method**, which relies on the fact that an egg is only likely to be fertilised for a few days after ovulation. Avoiding intercourse during this time, or withdrawing the penis before ejaculation, may reduce the chance of pregnancy, but this is the least reliable method of birth control. Sperm can survive for up to 48 hours in the uterus, so if intercourse takes place one night and the temperature spike occurs the next morning, pregnancy may result. A variation on this method involves measuring the woman's body temperature to monitor the moment of ovulation, but this only helps to predict ovulation more accurately.

Chemical methods

Use of a cream or foam containing a **spermicide** (chemical that kills sperm) can kill sperm in the vagina. This method is not considered very effective by itself and is usually used with other methods such as condoms or diaphragms. Some women are allergic to the chemicals in the spermicides.

The **contraceptive pill** (commonly called 'the pill') may contain just progesterone or a mixture of progesterone and oestrogen. Because these hormones control ovulation in the menstrual cycle, they can be used to stop the cycle and prevent ovulation occurring. The pill is very effective when taken regularly. However, if a pill is missed, other forms of contraception are needed until the woman's next period occurs to prevent pregnancy. These hormones can also be delivered from a small skin implant, which lasts for several months and so increases the effectiveness of preventing fertilisation.

Mechanical methods

Mechanical methods work by preventing the sperm from reaching the egg. A **condom** is a very thin sheath of rubber that is placed over the man's erect penis prior to intercourse. This prevents sperm from being released into the vagina. A **diaphragm** or cap can be used by the woman in a similar way. This fits over the entrance to the cervix and prevents sperm in the vagina from entering the uterus. There is also a female version, called a **femidom**, which the woman can insert in her vagina before intercourse to capture the sperm. A diaphragm or cap can be used by the woman in a similar way. This fits over the entrance to the cervix. Both condoms and diaphragms are used with spermicides that help to kill sperm in case of any leakage. Another advantage of condoms is that they offer some protection against sexually transmitted diseases including HIV. These methods are fairly successful in preventing pregnancy unless they break or are not positioned correctly.

An **IUD** (intrauterine device) is a small coil-shaped object that can be fitted inside the uterus by a doctor or nurse. IUDs help to prevent sperm from reaching the egg, They also prevent a fertilised egg from implanting in the uterus wall, so it cannot develop. An IUD can be left in place for several years, during which time the contraceptive effect is 100%.

All of the methods above are temporary. When a couple decides that they want children, they can stop using them and pregnancy should follow as normal.

Surgical methods

To prevent fertilisation permanently, surgery can be used to cut the tubes along which the sex cells travel. In men this means cutting the sperm ducts, which is called **vasectomy**. In women, **female sterilisation** involves cutting the oviducts. These operations are very difficult if not impossible to reverse, so they are only recommended for couples who are certain that they want no more children.

> EXTENDED

SOLVING FERTILITY PROBLEMS

Not all people who wish to have children find it easy to conceive. In the past there was little that could be done about this, and couples who had difficulty conceiving had to accept a childless life. Nowadays there are different forms of fertility treatment.

One method is **artificial insemination**. Here semen is introduced into the female reproductive tract by artificial means instead of by intercourse. This might be done if the man is making only a few sperm or sperm that are not very strong. The sperm can be prepared to make them more effective before placing them in the woman. If the man makes no sperm at all, the semen may come from a sperm donor. In some instances the woman may be unable to carry a baby to full term. In such a case, artificial insemination of a surrogate mother using the father's sperm can be successful.

If the woman is not releasing eggs properly, **fertility drugs** may be used to stimulate her ovaries. With these drugs there is a risk of releasing several eggs at once, resulting in several fetuses developing at the same time. A multiple birth (such as twins or triplets) can cause problems for the mother. It can also cause problems for the babies because they are more likely to be born early when they are not fully ready to cope with life outside the uterus.

Fertility treatment has helped large numbers of people to have a family and thus improved the quality of their lives. However, many of the procedures are very expensive and the multiple births that sometimes occur with fertility drugs can be a burden on family resources. Some women have used treatment to conceive at a later time in life than would be normal for childbearing, and some people question whether this is a good idea.

More recently, there have been concerns about the legal rights of sperm donors and surrogate mothers, and about children's right to know who their 'natural' parents are.

END OF EXTENDED

QUESTIONS

1. Suggest why some couples may wish to use birth control methods.
2. Which methods of birth control are most effective if used correctly?
3. Explain why most natural methods of birth control are the least successful in preventing pregnancy.
4. Give one advantage of using mechanical methods of birth control apart from preventing pregnancy.
5. Explain why surgical methods of birth control prevent pregnancy.
6. **EXTENDED** Explain how artificial insemination may assist a couple to have a child.
7. **EXTENDED** Describe one advantage and disadvantage of using fertility drugs with a childless couple.

SEXUALLY TRANSMITTED DISEASES

Unfortunately, sexual intercourse may spread infection, because of the exchange of body fluids which may contain pathogens. There are many STDs (sexually transmitted diseases), including gonorrhoea and AIDS (acquired immune deficiency disease).

Gonorrhoea

Gonorrhoea is a bacterial infection, also known as 'clap'. About 50 per cent of infected women and 10 per cent of infected men can have the infection without showing any symptoms.

Symptoms that are seen in women include:
- vaginal discharge
- pain on passing urine
- irritation or discharge from the anus
- low abdominal pain or pelvic tenderness.

Symptoms in men include:
- a discharge from the tip of the penis
- inflammation of the testicles and prostate gland
- pain on urinating
- irritation or discharge from the anus.

The highest rates of gonorrhoea are seen in young men and women in their teens or early twenties. Without treatment, gonorrhoea can lead to long-term health problems and infertility in both women and men. In women it may cause pelvic inflammatory disease, which can lead to long-term pelvic pain, blocked oviducts and ectopic pregnancy (a pregnancy outside the uterus).

Fortunately, gonorrhoea responds well to a single dose of an appropriate antibiotic. Anyone undergoing treatment should be checked for infection after a month. If the infection has not been cleared, a second treatment may be required. Any sexual partners should be treated at the same time.

AIDS

AIDS (acquired immune deficiency syndrome) is a disease of the immune system caused by the HIV virus (human immunodeficiency virus). The virus is present in sexual fluids, and so can be transmitted during sexual intercourse. It may also be passed to another person in blood, as through a scratch or bite, or through the sharing of needles for intravenous injection of drugs such as heroin. Infection can also pass from a mother to her fetus, through the placenta, or to her baby through breastfeeding after birth. A person may have no obvious symptoms early in infection.

There is no cure for AIDS, so prevention of infection is essential. This is most easily done by abstinence, or by limiting sexual partners to those who do not carry the virus. Barrier methods such as the condom or femidom are most effective in reducing the risk of infection during intercourse.

△ Fig. 3.24 The use of condoms during sexual intercourse can prevent the spread of the HIV virus.

EXTENDED

Infection with HIV damages the immune system that protects the body against infection. Normally white blood cells seek out and destroy invading bacteria and viruses. HIV avoids being recognised and destroyed by the immune system by repeatedly changing its outer 'coat'. It multiplies within a special type of white blood cell that helps other immune cells to attack and destroy disease-causing organisms. This decreases the body's ability to fight off infection, eventually leading to Acquired Immune Deficiency Syndrome (AIDS). People with AIDS are often infected with other pathogens that cause diseases such as tuberculosis. Many AIDS patients die from these other infections.

END OF EXTENDED

QUESTIONS

1. Explain what we mean by a *sexually transmitted disease*.
2. Why do many people not know they have gonorrhoea, and why is this a problem?
3. Describe the methods of transmission of HIV.
4. EXTENDED Describe the effect of HIV on the body.
5. EXTENDED Describe two ways to prevent the transmission of HIV during sexual intercourse.

End of topic checklist

Key terms

abstinence, AIDS (acquired immune deficiency syndrome), amniotic fluid, amniotic sac, antenatal care, anther, artificial insemination, asexual reproduction, carpel, cervix, condom, contraception, contraceptive pill, cotyledon, cutting, diaphragm, diploid, disperse, embryo, female sterilisation, femidom, fertilisation, fertility drug, fetus, fruit, FSH (follicle-stimulating hormone), gamete, gestation, gonorrhoea, haploid, HIV (human immunodeficiency virus), IUD (intrauterine device), labour, LH (luteinising hormone), meiosis, menstrual cycle, oestrogen, ovary, oviduct, ovulation, ovule, placenta, plumule, pollen grain, pollen tube, pollination, progesterone, radicle, rhythm method, scrotum, secondary sexual characteristic, seed coat, sexual reproduction, sperm, sperm ducts, spermicide, stamen, stigma, style, testa, testis, uterus, vagina, vasectomy, zygote

During your study of this topic you should have learned:

○ The definition of *asexual reproduction* as the process resulting in the production of genetically identical offspring from one parent.

○ How to describe asexual reproduction in bacteria, spore production in fungi and tuber formation in potatoes.

○ **EXTENDED** How to discuss the advantages and disadvantages of asexual reproduction to a species.

○ The definition of *sexual reproduction* as the process involving the fusion of haploid nuclei to form a diploid zygote and the production of genetically dissimilar offspring.

○ **EXTENDED** How to discuss the advantages and disadvantages of sexual reproduction to a species.

○ How to identify and draw the sepals, petals, stamens, anthers, carpels, ovaries and stigmas of an insect-pollinated flower, and examine the pollen grains under a light microscope or in photomicrographs.

○ About the functions of the sepals, petals, anthers, stigmas and ovaries.

○ How to identify and describe the anthers and stigmas of a wind-pollinated flower, and examine the pollen grains under a light microscope or in photomicrographs.

○ The definition of *pollination* as the transfer of pollen grains from the male part of the plant (anther of stamen) to the female part of the plant (stigma).

End of topic checklist continued

- About the agents of pollination.
- How to compare the different structural adaptations of insect-pollinated and wind-pollinated flowers.
- How to describe the growth of the pollen tube and its entry into the ovule followed by fertilisation.
- How to investigate and describe the structure of a non-endospermic seed in terms of the embryo (radicle, plumule and cotyledons) and testa, protected by the fruit.
- About the formation of a seed and fruit.
- That seed and fruit dispersal by wind and by animals provides a means of colonising new areas.
- How to describe seed and fruit dispersal by wind and by animals.
- **EXTENDED** How to distinguish between self-pollination and cross-pollination.
- **EXTENDED** How to discuss the implications of self-pollination and cross-pollination to a species.
- How to identify on diagrams of the human male reproductive system the scrotum, sperm ducts, prostate gland, urethra and penis, and the functions of these parts.
- How to identify on diagrams of the human female reproductive system the ovaries, oviducts, uterus, cervix and vagina, and the functions of these parts.
- **EXTENDED** How to compare male and female gametes in terms of size, numbers and mobility.
- How to describe the menstrual cycle in terms of changes in the uterus and ovaries.
- **EXTENDED** How to explain the role of hormones in controlling the menstrual cycle, including FSH, LH, progesterone and oestrogen.
- About sexual intercourse and how to describe fertilisation in terms of the joining of the nuclei of male gamete (sperm) and female gamete (egg).
- About early development of the zygote in terms of the formation of a ball of cells that becomes implanted in the wall of the uterus.
- About the development of the fetus.
- **EXTENDED** About the functions of the amniotic sac and amniotic fluid.

- How to describe the function of the placenta and umbilical cord in relation to exchange of dissolved nutrients, gases and excretory products.
- How to describe the antenatal care of pregnant women, including special dietary needs and maintaining good health.
- About the processes involved in labour and birth.
- **EXTENDED** How to describe the advantages and disadvantages of breast-feeding of a baby compared with bottle-feeding using formula milk.
- How to describe the roles of testosterone and oestrogen in the development and regulation of secondary sexual characteristics at puberty.
- **EXTENDED** How to describe the sites of production and the roles of oestrogen and progesterone in the menstrual cycle and in pregnancy.
- About the following methods of birth control:
  - natural (abstinence, rhythm method)
  - chemical (contraceptive pill, spermicide)
  - mechanical (condom, diaphragm, femidom, IUD)
  - surgical (vasectomy, female sterilisation).
- **EXTENDED** About artificial insemination and the use of hormones in fertility drugs, and their social implications.
- How to describe the symptoms, signs, effects and treatment of gonorrhoea.
- How to describe the methods of transmission of human immunodeficiency virus (HIV), and the ways in which HIV/AIDS can be prevented from spreading.
- **EXTENDED** How HIV affects the immune system in a person with HIV/AIDS.

End of topic questions

Note: The marks awarded for these questions indicate the level of detail required in the answers. In the examination, the number of marks awarded to questions like these may be different.

1. The photograph shows a catkin on a goat willow tree. A catkin is formed from a group of flowers.

    a) What is the purpose of the flowers on a goat willow tree? **(1 mark)**

    b) Name the yellow parts of the flowers shown in this photograph. **(1 mark)**

    c) Describe their purpose in a flower. **(1 mark)**

    d) Are goat willow flowers pollinated by the wind or by insects? Explain your answer using clues from the photograph. **(2 marks)**

2. a) Explain the advantage to a flower of having adaptations for attracting insects rather than relying on wind for pollination. **(1 mark)**

    b) Describe one disadvantage for an insect-pollinated plant that relies on one or just a small number of insect species for pollination. **(1 mark)**

3. **EXTENDED** Describe one advantage and one disadvantage of self-pollination in flowers. **(2 marks)**

4. Draw up a table to show the different ways in which seeds may be dispersed, and how the seeds in each group are adapted to make dispersal effective. **(5 marks)**

5. a) Where are sperm cells made in the human body? **(1 mark)**

    b) Where are egg cells made in the human body? **(1 mark)**

    c) Where is an egg cell fertilised by a sperm cell? **(1 mark)**

    d) Starting from the point of their formation, explain how a sperm cell reaches the egg cell at fertilisation. **(4 marks)**

6. Describe fully how the hormones oestrogen and progesterone control the menstrual cycle. **(4 marks)**

7. a) What is the placenta? **(1 mark)**

    b) What role does the placenta play in supporting the fetus? **(1 mark)**

    c) How are substances exchanged across the placenta? **(2 marks)**

    d) What is the advantage of keeping the mother's blood separated from the blood of the fetus? **(1 mark)**

8. Explain as fully as you can why, during antenatal care, pregnant women are advised to give up smoking. **(5 marks)**

9. Draw up a table to show the advantages and disadvantages of different methods of birth control. **(14 marks)**

10. a) Explain why gonorrhoea can be treated with antibiotics but AIDS cannot. **(2 marks)**

  b) Explain why all sexual partners of someone being treated for gonorrhoea should be treated at the same time. **(2 marks)**

Growth and development

INTRODUCTION

One of the fastest growing organisms on Earth is the giant kelp, a form of seaweed (algae) that grows off the coast of North America, from Alaska to California. The strands may grow at a rate of over 60 cm every day, and eventually grow to over 45 m long. They form large kelp forests in which many other species live. Their rapid rate of growth helps them to outcompete other species of algae that grow in similar conditions.

△ Fig. 3.25 A giant kelp forest under the sea.

KNOWLEDGE CHECK

✓ Know that growth is one of the key characteristics of living organisms.
✓ Know that nutrition supplies the substances needed to build new tissue, and respiration provides the energy for growth.
✓ Know that mitosis is the division of body cells to make more body cells.

LEARNING OBJECTIVES

✓ Be able to define *growth* as the permanent increase in size and dry mass by an increase in cell number or cell size or both.
✓ Be able to define *development* as an increase in complexity.
✓ Be able to investigate and state the environmental conditions that affect seed germination.

GROWTH

Over time, living organisms tend to increase in size and change in structure. We say that they grow and develop. Growth and development are closely connected.

Cells swell and increase in mass when they take in water, but this is not true growth. It is temporary and can be reversed. True growth is a permanent increase in mass, through the building of new tissue. This may result in new cells, or larger cells, or both.

Growth can be measured in terms of an increase in **dry mass**, that is the mass of the organism with all water extracted.

> ### SCIENCE IN CONTEXT: OVERNIGHT WATER LOSS
>
> If you weigh yourself just before going to bed and as soon as you get up, you may find that you have lost 1–2 kg overnight. This is not true weight loss. It is mostly from the loss of water through evaporation from your skin as you sleep. By the end of the day, you will have easily replaced this water, and have increased in mass again.

DEVELOPMENT

As a multicellular organism grows, its cells differentiate into tissues and organs. This differentiation is called **development**. It results in an increasing complexity of cells, tissues, organs and systems in an organism.

QUESTIONS

1. Define the following terms:
    a) growth
    b) dry mass
    c) development.
2. Explain why an increase in mass may not be an increase in growth.

GERMINATION

The plant embryo in a seed is made from a relatively small number of cells, which show only a little differentiation into plumule and radicle. After **germination**, the cells of the embryo start to divide and differentiate, resulting in growth and the development of all the tissues in the new plant.

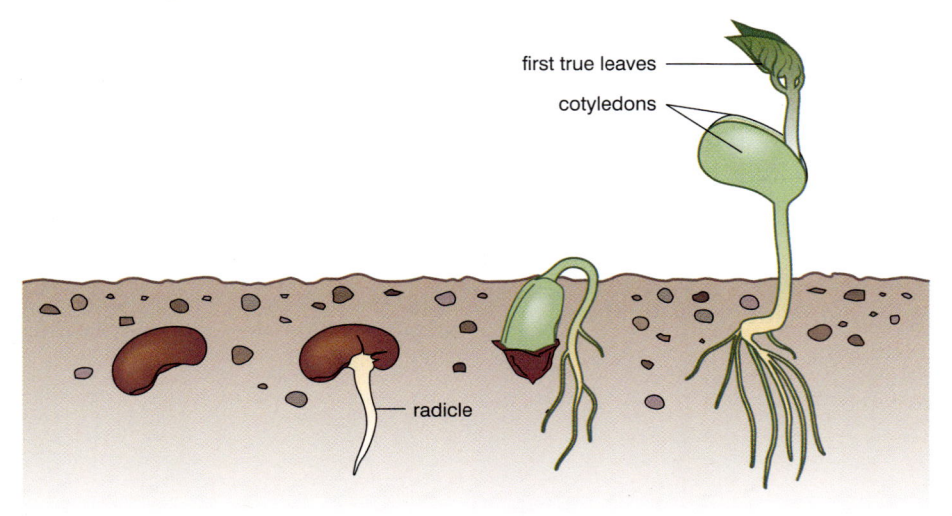

◁ Fig. 3.26 Germination of a bean seed.

There are three environmental conditions that seeds need for germination:

- temperature
- moisture
- oxygen.

The presence of light is not usually needed for germination. This is because most seeds germinate below ground, so they cannot get their food from photosynthesis.

Temperature

Many seeds lie dormant for long a time during cold periods, such as winter, and start to grow as the earth warms. A seed will not start to germinate until the conditions around it reach a suitable temperature. This is the temperature when the enzymes in the seed can work effectively. However, if the temperature becomes too hot the enzymes may denature and the seed may die. This is why it is very important to store seeds in the correct conditions and to control the temperature in glasshouses carefully, using ventilation and shading.

Moisture

Water is required to swell the seed and burst the seed coat. All seeds contain some moisture but during germination metabolic reactions are being carried out rapidly. More water is needed for:

- activation of hormones and enzymes
- hydrolysis of storage compounds, such as by conversion of starch to glucose
- transport of materials to be used for respiration and growth
- metabolic reactions and enzyme actions which occur in solution.

Oxygen

Active living cells respire and the most useful form of respiration, aerobic respiration, requires oxygen. Seeds can use anaerobic respiration for a short while but the rate at which energy is released is very slow (not useful in an actively growing organism) and the by-product ethanol is toxic. That is why most seeds will only germinate successfully if there is plenty of oxygen in the soil.

△ Fig. 3.28 Waterlogged soil excludes oxygen, making it difficult for these seeds to germinate and grow.

Developing investigative skills

You can investigate the particular conditions for germination.

Using and organising techniques, apparatus and materials

❶ Using the apparatus shown, write a plan to investigate the effect of (a) light, (b) water and (c) temperature on the germination of seeds. Think carefully about what controls to use in each case.

Observing, measuring and recording

❷ An investigation was carried out using two petri dishes containing 20 seeds of the same species. Both dishes received the same amount of light and moisture, but they were kept at different temperatures. The table shows the number of seeds that germinated over a period of 8 days.

△ Fig. 3.29 A simple set-up for germinating seeds.

Day	Total number germinated seedlings	
	Cool/10 °C	Warm/20 °C
1	0	0
2	0	0
3	0	5
4	1	11
5	6	15
6	16	17
7	18	17
8	18	17

❸ Display the results of this investigation in a suitable way.

Handling experimental observations and data

❹ Describe the patterns shown by these results.

❺ Draw a conclusion from these results.

❻ Explain the results using your scientific knowledge.

SCIENCE IN CONTEXT

CONDITIONS FOR GERMINATION

Different plants are suited to different climates. Those that are adapted to colder climates germinate at lower temperatures. They may also need a very cold period followed by an increase in temperature before they will germinate.

Other seeds do not germinate until they have been exposed to very high temperatures, such as the heat from a forest fire. The extreme heat weakens the seed coat so that water can enter the seed and germination can begin.

Germinating after a fire means there is likely to be less competition with other species that usually cover the ground. Also, the ash left from the burning acts as a natural fertiliser for the new plants.

◁ Fig. 3.27 Fire clears the ground of competing plants, and stimulates these seeds to germinate in ideal conditions.

QUESTIONS

1. What is germination?
2. What effect do the following conditions have on germination of seed?
   a) oxygen
   b) moisture
   c) warmth
3. Explain why seeds need these conditions for germination.

End of topic checklist

Key terms

development, dry mass, germination

During your study of this topic you should have learned:

○ The definition of *growth* as a permanent increase in size and dry mass as a result of an increase in cell number or cell size or both.

○ The definition of *development* as an increase in complexity.

○ How to investigate and state the environmental conditions that affect germination of seeds: water, oxygen and a suitable temperature.

End of topic questions

Note: The marks awarded for these questions indicate the level of detail required in the answers. In the examination, the number of marks awarded to questions like these may be different.

1. A student has two plants of the same species. The larger plant has been watered more frequently than the smaller plant. How could the student show that the difference in size was not just due to the larger plant containing more water in its cells? **(2 marks)**

2. Explain why multicellular organisms show development as they grow. **(3 marks)**

3. Explain why seeds will not germinate successfully in waterlogged soil. **(2 marks)**

4. A gardener has some packets of seeds for planting. The packets explain how to plant the seeds to get the best germination.

    a) What is meant by *germination*? **(1 mark)**

    b) All the packets say that the seeds need to be planted in moist compost and kept warm. Explain why the seeds need these conditions. **(2 marks)**

    c) The larger seeds need to be planted deeper in the compost, and the tiniest seeds need to be scattered on the surface of the compost. Explain why different seeds need to be planted at different depths. (Hint: think about food reserves.) **(3 marks)**

    d) Some seeds that come from plants in high-latitude regions (such as Canada or Russia) need to be placed in the freezer for a few weeks before they will germinate. This makes them respond as if they had been through a cold winter. Explain the survival advantage of this adaptation. **(2 marks)**

Inheritance

INTRODUCTION

Unless a zygote (fertilised egg) divides completely into two separate cells on its first division, and develops as two identical twins, the baby that develops from that zygote is genetically unique. Each cell in a zygote contains genetic information, half of it from the father and half from the mother. During gamete formation, some of that genetic information will have changed. So the baby will have about 100 variations in its genes that neither of its parents has. Interestingly, although the baby is genetically unique, virtually all of the cells in its body are genetically identical.

△ Fig. 3.30 This baby will have inherited some characteristics from her mother and some from her father.

KNOWLEDGE CHECK

- ✓ Know that organisms show variation in their features.
- ✓ Know that variation can be inherited or caused by the environment.
- ✓ Know that genes are small parts of the genetic information (DNA) found in the nucleus of a cell.

LEARNING OBJECTIVES

- ✓ Be able to define *inheritance* as the transmission of genetic information from generation to generation.
- ✓ Be able to define the terms *chromosome*, *gene*, *allele*, *haploid nucleus*, *diploid nucleus*.
- ✓ Be able to describe the inheritance of sex in humans (XX and XY chromosomes).
- ✓ Be able to define *mitosis* as nuclear division giving rise to genetically identical cells in which the chromosome number is maintained by the exact duplication of chromosomes.
- ✓ Be able to state the role of mitosis in growth, repair of damaged tissues, replacement of worn out cells and asexual reproduction.
- ✓ Be able to define *meiosis* as reduction division in which the chromosome number is halved from diploid to haploid.
- ✓ Be able to state that gametes are the result of meiosis.
- ✓ Be able to state that meiosis results in genetic variation so the cells produced are not all genetically identical.
- ✓ Be able to define the terms *genotype*, *phenotype*, *homozygous*, *heterozygous*, *dominant*, *recessive*.
- ✓ Be able to calculate and predict the results of monohybrid crosses involving 1 : 1 and 3 : 1 ratios.

- ✓ Be able to state that continuous variation is influenced by genes and environment, resulting in a range of phenotypes between two extremes, e.g. height in humans.
- ✓ Be able to state that discontinuous variation is caused by genes alone and results in a limited number of distinct phenotypes with no intermediates, e.g. A, B, AB and O blood groups in humans.
- ✓ Be able to define *mutation* as a change in a gene or chromosome.
- ✓ Be able to describe mutation as a source of variation, as in Down's syndrome.
- ✓ Be able to outline the effects of ionising radiation and chemicals on the rate of mutation.
- ✓ Be able to describe the role of artificial selection in the production of varieties of animals and plants with increased economic importance.
- ✓ Be able to define *natural selection* as the greater chance of passing on of genes by the best-adapted organisms.
- ✓ Be able to define *genetic engineering* as taking a gene from one species and placing it in another species.
- ✓ **EXTENDED** Be able to explain codominance by reference to the inheritance of ABO blood groups: A, B, AB and O blood groups and genotypes I^A, I^B and I^o.
- ✓ **EXTENDED** Be able to describe sickle cell anaemia, and explain its incidence in relation to that of malaria.
- ✓ **EXTENDED** Be able to describe variation and state that competition leads to differential survival of, and reproduction by, those organisms best fitted to the environment.
- ✓ **EXTENDED** Be able to assess the importance of natural selection as a possible mechanism for evolution.
- ✓ **EXTENDED** Be able to describe the development of strains of antibiotic-resistant bacteria as an example of natural selection.
- ✓ **EXTENDED** Be able to explain why, and outline how, human insulin genes were put into bacteria using genetic engineering.

CHROMOSOMES, DNA AND GENES

Inheritance is the passing on of characteristics from one generation to the next, from parents to offspring. Since characteristics are coded for by genetic information in the cell nucleus, inheritance is also the passing on of genes from the haploid nuclei of the parent's gametes to the diploid zygote, which will divide and grow to form all the cells in the offspring.

Inside almost every cell in the body is a nucleus, which contains long threads called **chromosomes**. These threads are usually stretched out and fill the nucleus, but when the chromosomes condense (coil up) just before cell division, they can be seen through a microscope. The distinctive X-shape is formed by the replication of the chromosome before cell division. The chromosomes are made of a chemical called **deoxyribonucleic acid (DNA)**.

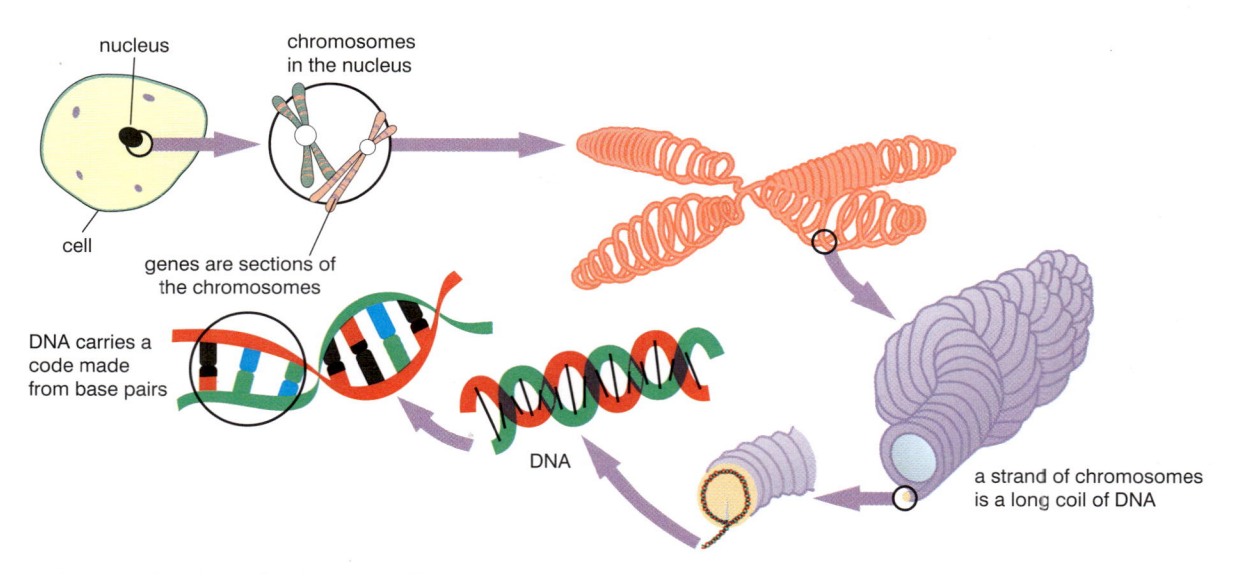

△ Fig. 3.31 The relationship between cell, nucleus, chromosome, DNA and genes.

Small sections of the DNA code for particular proteins or characteristics. These small sections are called **genes**. During cell division the sequence of genes on each chromosome is copied and passed on to the new cells.

Enzymes are proteins, so many of the effects of genes are the result of the way the enzymes that are produced control reactions inside the cell.

Genes code for particular characteristics, such as eye colour. However, variations in a characteristic or protein are caused by different forms of a gene, called **alleles**. For example, some people have alleles that code for brown eye colour; others have alleles that code for hazel eyes, or blue eyes or grey eyes.

If you take all the chromosomes from a body cell, you can arrange them into pairs. This is because you inherit one of each chromosome pair from your father and one from your mother. In humans there are 46 chromosomes in a body cell, which make up 23 pairs. Of these pairs, 22 are always identical, but the final pair may be identical or different. This pair are the **sex chromosomes**, that determine whether you are a male or female. Women have two X chromosomes and are XX. Men have one X and one Y chromosome, and so are XY. (The inheritance of these chromosomes is covered later in Sex determination and inheritance.)

In the identical pairs of chromosomes, both chromosomes have the same genes in the same order, but the genes on each chromosome may be in the form of the same allele or different alleles. This causes differences in inheritance.

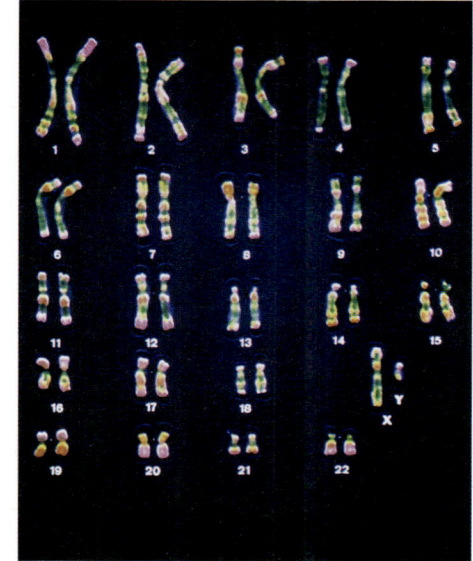

△ Fig. 3.32 The chromosomes from a man's body cell arranged in their pairs. The chromosomes in each pair look very similar except in the sex chromosome pair (marked X and Y).

QUESTIONS

1. Put the following in order of size, starting with the smallest: cell, chromosome, gene, nucleus.

2. Using an example, explain the difference between *gene* and *allele*.

3. Give the combination of sex chromosomes in men and women.

4. Define the term *inheritance* as fully as you can.

INHERITING CHARACTERISTICS

Some characteristics, such as the colour of your eyes, are passed down (**inherited**) from your parents, but other characteristics may not be passed down. Sometimes characteristics appear to miss a generation: for instance, you and your grandmother might both have dry earwax but both of your parents may have wet earwax.

Leopards occasionally have a cub that has completely black fur instead of the usual spotted pattern. It is known as a black panther but is still the same species as the ordinary leopard.

Just as in humans, leopard chromosomes occur in pairs. One pair carries a gene for fur colour. There are two copies of the gene in a normal body cell (one on each chromosome). The version of the gene (allele) may be identical but sometimes they are different, one being for a spotted coat and the other for a black coat.

△ Fig. 3.33 Two spotted leopard parents may produce offspring with spotted coats or black coats.

Leopard cubs receive half their genes from each parent. Eggs and sperm cells contain only half the number of chromosomes of normal body cells. This means that egg and sperm cells contain only one of each pair of alleles. When an egg and sperm join together at fertilisation, forming a zygote that will develop into the new individual, it now has two alleles of each gene.

Different combinations of alleles will produce different fur colour:

spotted coat allele	+	spotted coat allele	=	spotted coat
spotted coat allele	+	black coat allele	=	spotted coat
black coat allele	+	black coat allele	=	black coat

The black coat only appears when *both* of the alleles for the black coat are present. As long as there is at least one allele for a spotted coat, the coat will be spotted because the allele for a spotted coat overrides the allele for a black coat. It is the **dominant allele**. Alleles like the one for the black coat are described as **recessive**.

- An individual with two identical alleles for a gene is said to be **homozygous** ('homo' means 'the same') for that gene.
- An individual with two different alleles for a gene is said to be **heterozygous** for that gene ('hetero' means 'different').

A leopard with a spotted coat may be homozygous for the spotted allele, or heterozygous. A leopard with a black coat can only be homozygous for the black allele.

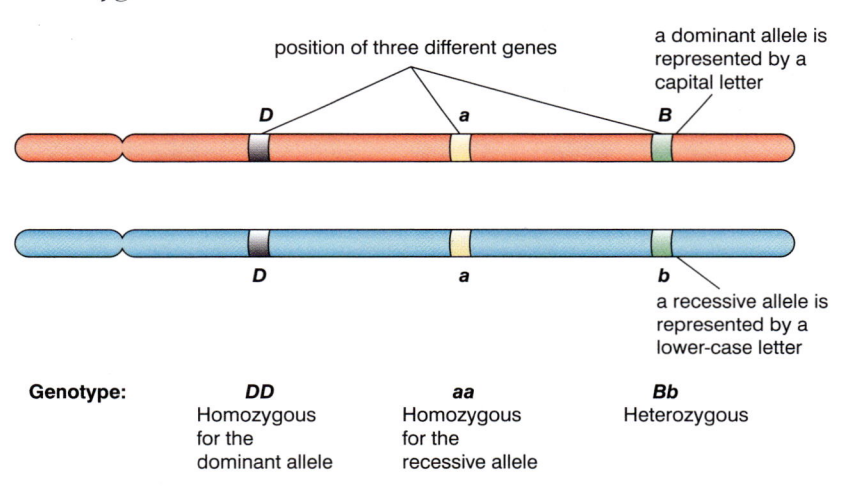

△ Fig. 3.34 Definitions in genetics.

If you are female, you have two copies of every gene on all your chromosomes. In males, a few genes are present on the X sex chromosome but not on the Y chromosome. These genes can produce sex-linked characteristics, where men are more likely to be affected by a recessive allele than women.

QUESTIONS

1. Define the following terms in your own words:
   a) dominant
   b) recessive
   c) homozygous
   d) heterozygous.

2. How many alleles for a particular gene would be found in:

   a) a body cell

   b) a gamete

   c) a zygote?

MONOHYBRID CROSSES

An individual's combination of genes is his or her **genotype**. An individual's combination of physical features is his or her **phenotype**. Your genotype influences your phenotype.

We can show the influence of the genotype in a **genetic diagram**. This uses a capital letter for the dominant allele and a lower case letter for the recessive allele.

REMEMBER

To avoid the risk of confusion when drawing genetic diagrams, we choose a letter that is easily distinguished in capital and lower case. For example, S or C for the coat gene might seem reasonable to use for the leopard example, but this gives alleles S/s or C/c which is not ideal.

Using the example of the leopards, **B** stands for the dominant allele for a spotted coat and letter **b** stands for the recessive allele for the black coat. Two spotted parents who have a black cub must each be carrying a **B** and a **b**. The genetic diagram below shows the possible genotypes and phenotypes of the offspring.

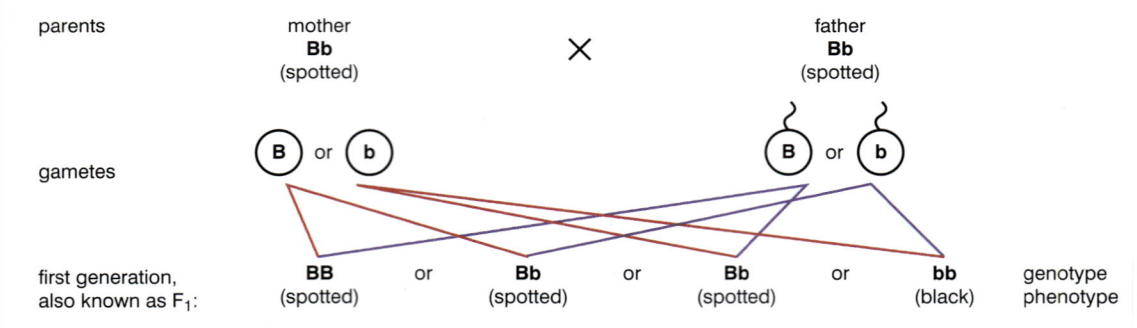

| parents | mother **Bb** (spotted) | × | father **Bb** (spotted) |

gametes: B or b   B or b

first generation, also known as F$_1$: **BB** (spotted) or **Bb** (spotted) or **Bb** (spotted) or **bb** (black) — genotype phenotype

△ Fig. 3.35 Three different genotypes are possible from the cross in this diagram. The probability of each genotype is 1BB : 2Bb : 1bb.

Because we are looking at a characteristic (fur colour) controlled by one gene, this is an example of a **monohybrid cross**. 'Mono' means one, and a 'hybrid' is produced when two different types breed or cross.

Another type of genetic diagram is known as a **Punnett square**. The example above can be shown in a Punnett square. The four boxes at the bottom right show the possible combinations in the offspring.

			male Bb spotted	
			gametes	
			B	b
female Bb spotted	gametes	B	BB spotted	Bb spotted
		b	Bb spotted	bb black

Probabilities and predictions

In a monohybrid cross, when two heterozygous parents are crossed, the phenotype of the offspring with the dominant allele and the offspring with the recessive allele appears in the ratio of 3:1. The 3:1 ratio refers to the probabilities of particular combinations of alleles, so the chance of having an offspring with the phenotype of the dominant allele (B) is three times the chance of having an offspring with the phenotype of the recessive allele (b).

In the example of the leopards, there is a 1 in 4 or 25% chance of a leopard cub being black. This is because there must be two recessive alleles (homozygous) in order for the phenotype (b) to be expressed (visible).

With a large number of offspring in an *actual* cross of two heterozygous leopard parents, you would expect something near the 3:1 ratio of spotted to black cubs. However, because it is a matter of chance which sperm cell fertilises which egg cell, you should not be too surprised if a small litter – for example, of four cubs – contained two black cubs, or none.

Using the example of leopard coat colour and a Punnett square, we can also look at what happens if we cross homozygous recessive and heterozygous individuals:

			male Bb spotted	
			gametes	
			B	b
female bb black	gametes	b	Bb spotted	bb black
		b	Bb spotted	bb black

The predicted outcome from this cross is a 1 : 1 ratio of spotted to black colouring. This gives a 1 in 2 or 50% chance of a cub from these parents being spotted or being black.

> **REMEMBER**
>
> In order to gain higher marks, you will need to be able to predict probabilities of outcomes from any monohybrid cross. Practise drawing genetic diagrams to make sure you are confident with them.

Can you predict what will happen if a homozygous spotted coat leopard is crossed with a heterozygous spotted coat leopard?

			male Bb spotted	
			gametes	
			B	b
female BB spotted	gametes	B	BB spotted	Bb spotted
		B	BB spotted	Bb spotted

In this case, although some of the cubs born are likely to be homozygous and some heterozygous, they will all have spotted coats. They will have the same phenotype but not the same genotype.

QUESTIONS

1. Define *monohybrid inheritance* in your own words.

2. Rabbits have a gene for coat colour – the allele for brown coat is dominant over the allele for black coat colour. Using the letter B for the dominant allele and b for the recessive allele, write down all the possible genotypes and phenotypes for this gene. Explain your answers.

3. Using your answers from Question 2:

   a) Construct a genetic diagram to show the possible offspring from a cross between a male rabbit and a female rabbit that are both heterozygous for this gene.

   b) What is the probability of producing a black baby rabbit from this cross?

Developing investigative skills

In an investigation into the inheritance of a characteristic, students used red beads to represent dominant alleles and blue beads to represent recessive alleles.

Because a homozygous dominant individual produces gametes that only contain the dominant allele, all the red beads were placed into a beaker to represent the gametes for this individual. Since a homozygous recessive individual produces gametes that only contain the recessive allele, all the blue beads were placed into another beaker to represent the gametes for this individual.

To model what would happen in a cross between these two individuals, they took one gamete (bead) from one pot and paired it with one gamete (bead) from the other pot and wrote down the genotype and phenotype for that 'offspring'. This showed that all the offspring from these parents would be heterozygous (one red, one blue bead).

Using and organising techniques, apparatus and materials

❶ Describe and explain how you would adapt this method to represent a cross between two heterozygotes. (Hint: Make sure you use enough beads to get a reasonable approximation of the actual result to the expected result.)

Observing, measuring and recording

Some students carried out an investigation like this that started with two 'parents' heterozygous for a characteristic.

❷ Each pot started with 40 beads. How many red beads and how many blue beads were in each of the two pots? Explain your answer.

Only 20 selections were made from the two beakers, to produce the 'offspring'. The results are shown in this table.

Number of red/red pairs in 'offspring'	5
Number of red/blue pairs in 'offspring'	12
Number of blue/blue pairs in 'offspring'	3

❸ Draw a genetic diagram for this cross, to show the predicted probabilities of genotypes and phenotypes. (Hint: remember to choose letters for the alleles and explain which allele is modelled by the red beads and which by the blue beads.)

❹ Describe how the actual results differ from the expected results.

Handling experimental observations and data

❺ Comment on the difference between the expected and actual results.

❻ Explain how you would adjust the method to help improve the results.

SCIENCE IN CONTEXT

TEST CROSSES

Breeders can use a cross with a homozygous recessive individual to test whether an individual with the dominant phenotype is homozygous or heterozygous for the dominant allele. For example, a plant breeder may have a tall pea plant. The allele for tallness (T) is dominant over the allele for dwarfness (t). How can the plant breeder find out if the plant is TT or Tt?

- A cross between a homozygous dominant (TT) and homozygous recessive (tt) will produce all heterozygous (Tt) offspring, and so all offspring will have the phenotype of the dominant allele.

- A cross between a heterozygous dominant (Tt) and a homozygous (tt) recessive will have a 50% chance of producing offspring that have the phenotype of the dominant allele (tall) and a 50% chance of producing offspring with the phenotype of the recessive allele (dwarf).

(Try drawing genetic diagrams to confirm this for yourself.)

This kind of cross is known as a test cross.

EXTENDED

CODOMINANCE

In the leopard example just discussed, one allele of the gene pair for coat colour was dominant and the other was recessive. When both alleles of a gene pair in a heterozygote are expressed in the phenotype, with neither dominant or recessive to the other, this is called **codominance**.

Human blood types are determined by three different alleles of the same gene: I^A, I^B and I^o (note that I represents the gene and the superscript letter shows the allele). I^A results in the production of the A antigen in blood, I^B results in the production of B antigen, and I^o produces no antigen. The I^o allele is recessive but the I^A and I^B alleles are codominant. These three possible alleles can give us the following allele pairs:

$I^A I^A$ $I^B I^B$ $I^o I^o$ $I^A I^B$ $I^A I^o$ $I^B I^o$

These six different genotypes give us four different phenotypes: the four different human blood groups A, B, AB and O.

- The phenotype of blood group A can have the genotype of I^AI^A or I^AI^o because I^A is dominant over recessive I^o.
- The phenotype of blood group B can have the genotype of I^BI^B or I^BI^o allele pairs because I^B is dominant over recessive I^o.
- The phenotype of blood group AB has the genotype of the two codominant alleles, I^AI^B.
- The phenotype of blood group O can only have the genotype I^oI^o, the recessive allele pair.

You can use genetic diagrams to predict the outcomes of crosses that involve codominant alleles.

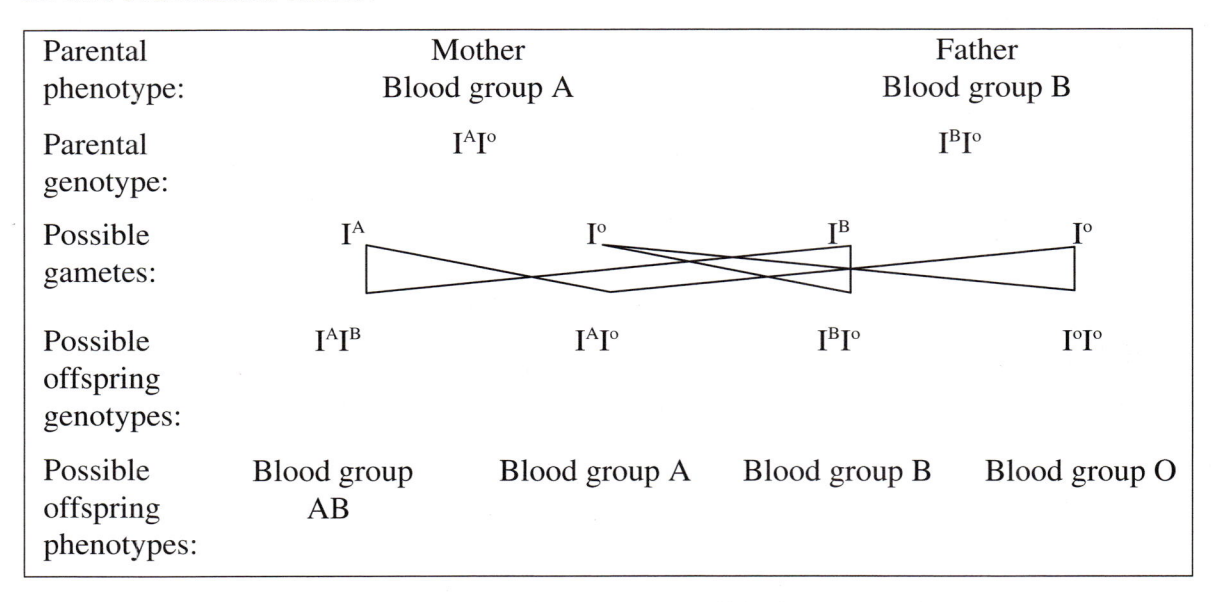

Parental phenotype:	Mother Blood group A		Father Blood group B	
Parental genotype:	I^AI^o		I^BI^o	
Possible gametes:	I^A	I^o	I^B	I^o
Possible offspring genotypes:	I^AI^B	I^AI^o	I^BI^o	I^oI^o
Possible offspring phenotypes:	Blood group AB	Blood group A	Blood group B	Blood group O

△ Fig. 3.36 A genetic diagram predicting the crosses between parents of blood groups A and B.

A cross between a parent who is heterozygous for blood group A ($I^A I^o$) and a parent who is heterozygous for blood group B ($I^B I^o$) produces a predicted ratio of 1 : 1 : 1 : 1 for children with each of the four blood groups, giving a 25% or 1 in 4 chance that a child will inherit any one of the four blood groups.

Gregor Mendel (1822–1884) was the first person to study genetic inheritance thoroughly and scientifically. He chose characteristics in peas to study because he could see clear differences in characteristics and patterns in their inheritance. He started by crossing plants with the same characteristics many times, until he was certain that they were pure-breeding (that is, they would only produce offspring with that characteristic).

He then made hundreds of crosses of the same kind. He started by removing the anthers of each flower. Then he brushed pollen from a plant he had chosen for one parent on to the stigma of the other 'parent' and covered the flower to prevent other pollen getting in.

From his results, Mendel was able to show that alleles generally do not mix effects in the phenotype, but that a dominant allele in a heterozygote prevents the recessive allele being expressed.

1. Why was it important that the parent plants were pure-breeding?
2. Why did Mendel need to carry out hundreds of crosses before drawing a conclusion?
3. Pea flowers are pollinated by insects. How could Mendel be certain that no chance fertilisations took place?
4. At the time Mendel carried out his work, people couldn't understand how a characteristic could be present in one generation, 'disappear' in the next generation and then reappear in the next. Using genetic diagrams, and a characteristic of your choice, show how this happens when starting with pure-breeding parents for the dominant and recessive characteristics.
5. Explain the importance of a thorough and scientific method for drawing reliable conclusions.

△ Fig. 3.37 The results of one of Mendel's crosses for pea form and colour. (Note each pea is the result of a separate cross between a pollen grain and an egg cell.)

QUESTIONS

1. **EXTENDED** Define *codominance*.
2. **EXTENDED** Draw up a table to show the genotypes and phenotypes of the four major blood groups in humans.
3. **EXTENDED** Draw a genetic diagram to show the inheritance of blood group from a mother who is blood group O and a father who is AB.

END OF EXTENDED

SEX DETERMINATION AND INHERITANCE

In the nucleus of a human cell there are 46 chromosomes that form 23 pairs. In all but one of these pairs, the two chromosomes of the pair always look identical. We call these 22 pairs the autosomal chromosomes. The chromosomes of the other pair are identical in women, but differ in men. These are called the **sex chromosomes**. The sex chromosomes of women are called XX and in men are called XY.

When the gametes are produced, they each receive one of the sex chromosomes. So egg cells all contain an X chromosome, but sperm cells may contain an X chromosome or a Y chromosome. As a result of the way the sperm cells are produced about 50% are X and 50% are Y.

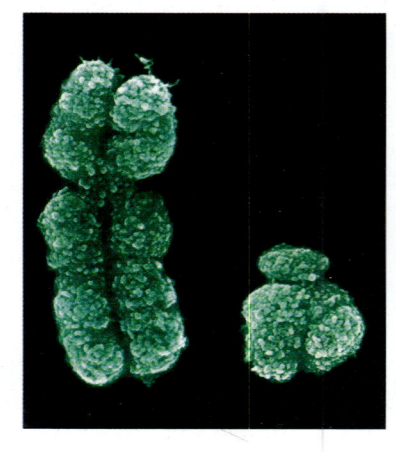

△ Fig. 3.38 The human X and Y chromosomes are different shapes.

During fertilisation, one sperm cell fuses with one egg cell. We can use a Punnett square to show the possible combinations outcomes of sex chromosomes in the offspring.

		father's gametes	
		X	Y
mother's gametes	X	XX (female)	XY (male)
	X	XX (female)	XY (male)

This shows that there is a 50% or 1 in 2 probability of any child being a boy or a girl. The ratio of boys to girls born in a family is often not 1 : 1, but over the whole human population about equal numbers of baby boys and baby girls are born.

QUESTIONS

1. Which sex chromosomes would be found in the cells of an adult woman?
2. Which sex chromosomes would be found in the cells of a baby boy?
3. A couple have three boys. What is the chance of their next child being a girl? Explain your answer.

CELL DIVISION

Mitosis

Organisms grow by the division of cells, when the body cells split in two. This kind of division is used in normal growth and repair. It is also the way that single-celled organisms reproduce and is the only type of cell division involved during asexual reproduction (reproduction that does not involve sex cells), as you saw in Asexual reproduction.

Before a cell splits, its chromosomes duplicate themselves. The new cells formed, sometimes called the **daughter cells**, contain chromosomes identical with the original cell. This type of cell division, in which the new cells are genetically identical to the original, is known as **mitosis**.

Mitosis takes place in all normal body cells.

Meiosis

During sexual reproduction, a male gamete fuses with a female gamete. If each gamete had the same number of chromosomes as a normal body cell, the zygote would end up with twice as many chromosomes as normal. Instead, gametes are produced by a different form of cell division called **meiosis**. Cells produced by meiosis have one chromosome of each pair – half the normal number of chromosomes. These cells are called **haploid** cells. When the gametes fuse during fertilisation, they restore the normal number of chromosomes, creating a **diploid** cell with pairs of chromosomes again.

Cells produced by meiosis are not identical. This means that, during fertilisation where there is a random chance that any one male gamete will fuse with the female gamete, the offspring produced will be different from each other. We say they show variation.

QUESTIONS

1. The diagram shows the life cycle of a human.

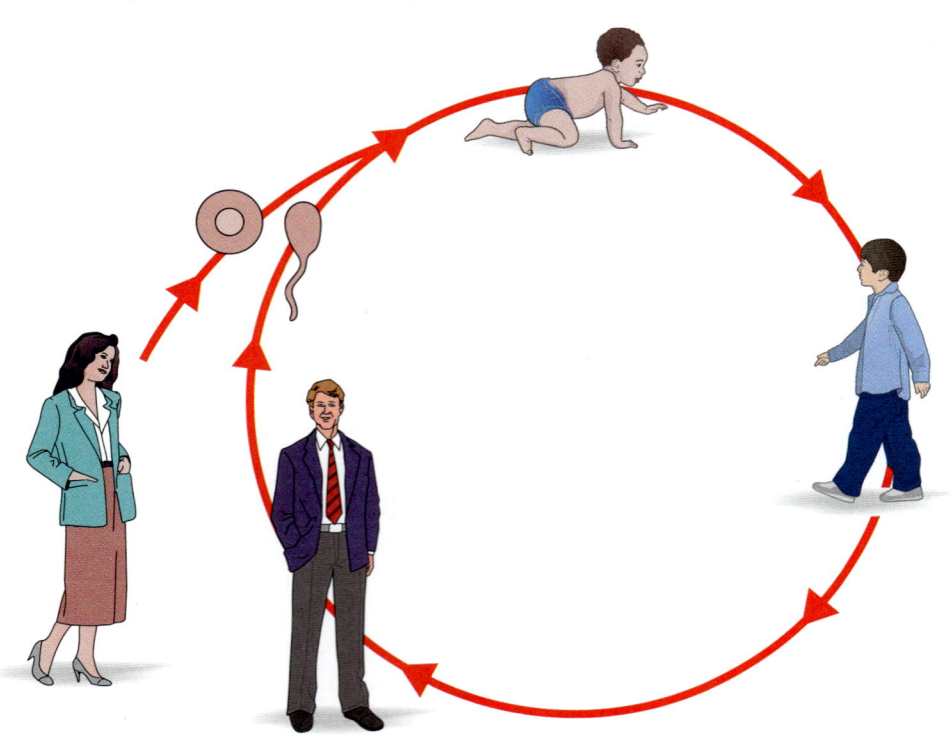

   Copy the diagram and annotate it to show:

   when meiosis and mitosis occur

   which cells are diploid and which are haploid.

2. a) Which form of cell division produces new body cells?

   b) Explain why this is important to the organism.

3. Name the form of cell division that is found in organisms that reproduce by: a) sexual reproduction, b) asexual reproduction.

4. Explain the importance of meiosis in producing variation in offspring.

VARIATION

No two people are the same. Similarly, no two trees (even if they are of the same species) are exactly the same in every way. They have different heights, different trunk widths and different numbers of leaves.

The variation in appearance of characteristics in an individual may have various causes.

- Environmental causes such as your diet, the climate you live in, accidents, your surroundings, the way you have been brought up and your lifestyle can influence your characteristics.
- Genetic causes are characteristics that are controlled by your genes, such as eye colour and gender.
- Many characteristics are influenced by both *environment* and *genes*. For example, people in your family might tend to be tall, but unless you eat correctly when you are growing you will not become tall, even though genetically you have the tendency to be tall. Other examples are more controversial, such as human intelligence, where it is unclear how far environment or genes contribute to variation.

Variation can be divided into two different groups.

- **Discontinuous variation** (sometimes called discrete variation) is where a characteristic can have one of a certain number of specific alternatives: for example, gender, where you are either male or female, and blood groups, where you are either A, B, AB or O. Discontinuous variation is caused by genetic variation alone.
- **Continuous variation** is where a characteristic can have any value in a range, for example body weight and length of hair. This is usually a combination of variation caused by genes and the environment.

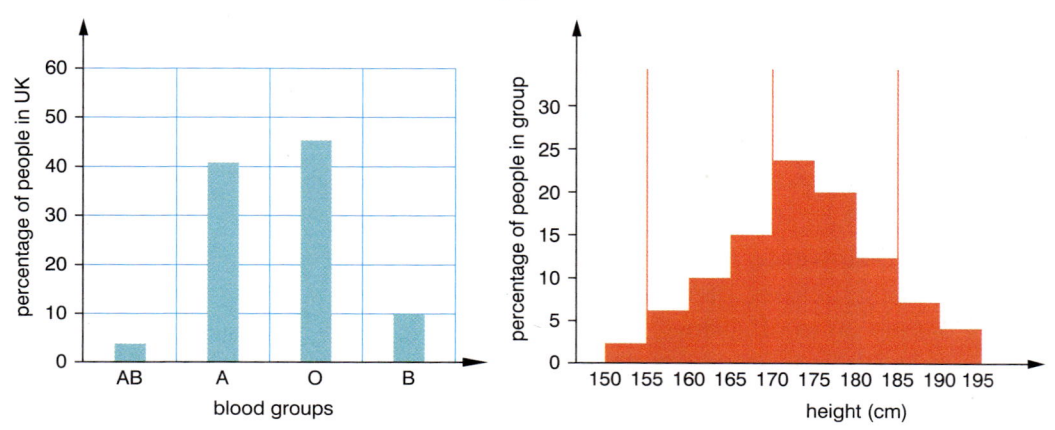

△ Fig. 3.39 Charts showing discontinuous variation (left) and continuous variation (right).

Mutation

A mutation is a change in the DNA of an individual. Occasionally it is caused by the chromosomes not separating correctly when gametes are formed, so that a gamete ends up with two chromosomes of the pair, not the usual one. The zygote formed from that gamete will then have three copies of a chromosome, not the usual pair. People who have

Down's syndrome have three copies of chromosome 21. The third copy can change how some of the genes on the other two chromosomes work, and can produce characteristic features, such as a change in shape of the eyes, a difference in intelligence and a permanently happy nature.

A gene mutation is a change in a gene that can happen when the DNA of a cell is copied, such as in cell division. This is rather like mis-spelling a word when you are copying text, where you use one wrong letter and end up with a different word (such as *bold* instead of *bond*). The different word can completely change the meaning of the sentence. Similarly, the error in the DNA can produce a different form of the gene and a different form of the protein it produces.

△ Fig. 3.40 Down's syndrome child with her mother.

A gene mutation may produce a new version that is:
- beneficial for the organism, giving it an advantage over other individuals of the species; for example, fair skin is a mutation that has happened in human populations several times as they moved into areas of northern Europe and Asia because it allows the individual to make more vitamin D from sunlight than those with darker skins.
- neutral, which means it has no obvious effect; many mutations fall into this category and can only be identified by looking in detail at the genetic code.
- harmful to the organism, either causing the early death of the embryo, or making the individual less able to survive than other individuals; many inherited diseases are caused by mutations of genes that produce proteins that are important in key processes.

Mutations are random changes, so they can occur anywhere in the DNA. They are also rare, happening, on average, once in every 10 to 100 million DNA bases that are copied. If the mutation occurs in a body cell, then only the cells produced by mitosis from that cell will carry the mutation. However, if the mutation occurs in a cell that produces gametes during meiosis, there is a possibility that the mutation may be passed on to offspring.

EXTENDED

Sickle cell anaemia

A mutation in the gene that produces haemoglobin can result in sickle cell haemoglobin. This form of haemoglobin changes the shape of red blood cells when oxygen concentration is low. In people who are homozygous for the sickle cell allele, this can cause many health problems when red blood cells get stuck in capillaries. It can also lead to a shortage of red blood cells in the blood as the liver removes the sickle-shaped cells. This leads to **sickle cell anaemia**.

Sickle cell anaemia (from the homozygous condition) is harmful and can shorten life. However, the heterozygous condition seems to be advantageous in warm humid areas where **malaria** occurs. Malaria is caused by a blood parasite, injected into the blood by the bite of an *Anopheles* mosquito. It causes around 1 million deaths each year, mainly in young children. People who have one sickle cell allele are more likely to survive infection with malaria than people who have no sickle cell alleles. So the occurrence of the sickle cell allele is more common in areas where malaria is found (see Natural selection).

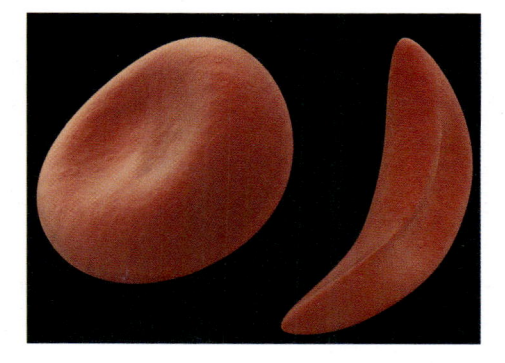

△ Fig. 3.41 Red blood cells from a person who is heterozygous for the sickle cell allele. This person has some red blood cells that are the normal disc shape and some that form the sickle shape when the concentration of oxygen in their blood is low.

END OF EXTENDED

Causes of mutation

The incidence of mutation can be increased beyond the natural rate as the result of:

- exposure to **ionising radiation**, such as gamma rays, X-rays and ultraviolet radiation
- chemical **mutagens**, such as some of the chemicals in tobacco.

These factors may change the genetic code directly or cause it to be mis-copied more frequently during cell division.

One of the most obvious effects of radiation or chemical mutagens is to damage the control mechanisms that instruct a cell to stop dividing at the right time. The continuing division of cells produces a lump of unspecialised tissue called a **cancer**. Cancers that take over the space of other tissues can eventually cause death. In Europe and North America, the different forms of cancer are the greatest cause of death from disease.

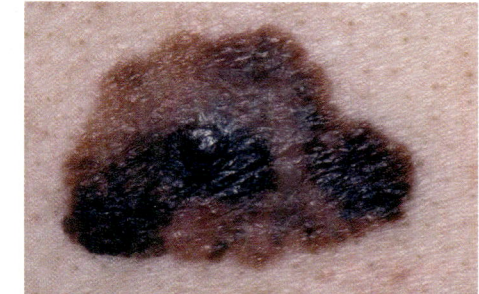

△ Fig. 3.42 This skin cancer was caused by too much ultraviolet radiation from sunlight.

QUESTIONS

1. **a)** Name two causes of variation of characteristics in a species, and give an example of each.

    b) Give one example of variation caused by a combination of these causes, and explain your answer.

2. Define the term *mutation* in your own words.

3. Explain how the condition of Down's syndrome is produced.

4. Name two different causes of gene mutation and give an example of each.

SELECTION

Artificial selection

Artificial selection is when a breeder chooses which parent organisms to breed from. The choice is usually for a particular characteristic that has economic importance, such as size of farm animals kept for meat, egg size in chickens, or colour of flowers in plants grown for horticulture. Since many characteristics are controlled by genes, by breeding together organisms that show the nearest form to desired characteristic the breeder is more likely to get offspring that also has the desired characteristic (such as large size). If the breeder continues to select individuals with the largest size, over time the average size of the organisms will increase.

Artificial selection may also be done to combine particular combinations of characteristics. For example, by breeding a wheat plant that has a long stalk and a large seedhead with a wheat plant that has a short stalk and small seedhead, there may be some offspring with the advantageous combination of short stalk and large seedhead. Selecting from the offspring that have the best combinations of these characteristics over many generations can produce a new variety with the perfect combination of these characteristics.

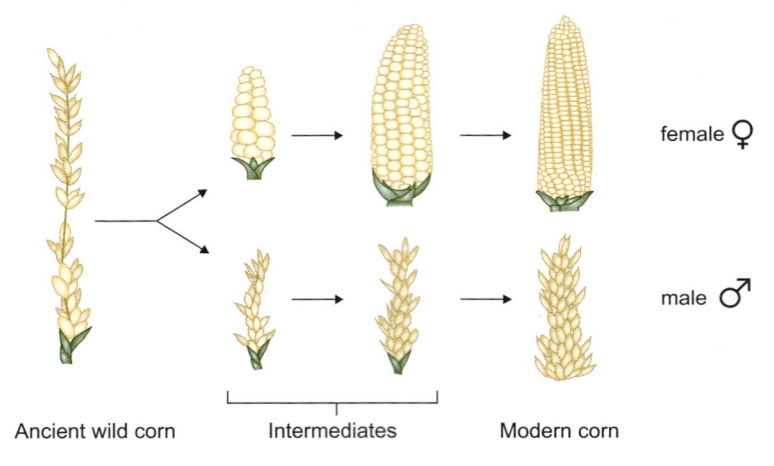

△ Fig. 3.43 Over centuries of selective breeding, ancient wild corn (maize) has developed into the modern breeds of corn that have the large seed cobs that we harvest.

Note that plants of the same species but with distinctively different characteristics are called **varieties**, while animals of the same species with distinct characteristics are called **breeds**. Since they are still of the same species, different plant varieties can interbreed and different animal breeds can interbreed.

> **REMEMBER**
>
> For the best marks, be prepared to explain the inheritance of characteristics through selective breeding in terms of the inheritance of genes and alleles you have already learned.

SCIENCE IN CONTEXT

SELECTIVE BREEDING OF TULIPS

Plants are also bred in horticulture, for gardens, for houseplants and cut flowers, to improve the colour, shape and form of the flowers and leaves. This is because people like new things.

For example, tulips were introduced to Europe in the 1500s from Turkey. They were so exotic that they became a luxury item that all wealthy people had to have. Plant breeders rapidly developed new varieties through selective breeding, such as flowers with different-coloured lines or specks on the petals.

At the peak of 'tulip mania' in the Netherlands in the 1630s, single tulip bulbs were being sold for more than 10 times the annual income of a skilled craftsman. Prices suddenly collapsed in 1637, mainly because people got bored with tulips.

△ Fig. 3.44 A completely black flower is almost impossible to breed, but that doesn't stop people wanting to try to produce it, because many people would pay a lot of money for something so rare.

QUESTIONS

1. Who carries out artificial selection and why?
2. Give two examples of artificial selection.
3. Describe the process of artificial selection.

EXTENDED

One of the problems with selective breeding is that when you breed from only a small number of individuals, you reduce not only the variation in the characteristics you are selecting for, but also the variation in other alleles. This means that you can lose other characteristics which might be useful in the future.

To protect against this, many wild varieties of rice, wheat, potatoes and other plants are collected and grown in case we need their characteristics in the future.

△ Fig. 3.45 There are many wild varieties of rice, but we eat only a few varieties selectively bred for particular characteristics such as larger grain size.

1. Using what you know about sexual reproduction, suggest why the amount of variation between selected individuals is smaller than in wild populations of a plant.
2. Why is it useful that selectively bred varieties have only limited genetic variation? Explain your answer as fully as you can.
3. Why could it be a problem in the future that selectively bred varieties have only limited genetic variation? Explain your answer as fully as you can.
4. Growing wild varieties of crop plants to keep them for the future takes a lot of space and time to look after them. This space and time could be used to grow varieties that produce more food. Do you think it is worth keeping wild varieties like this? Explain your answer as fully as you can.

END OF EXTENDED

Natural selection

Natural selection is like artificial selection, but caused by the environment.

- In any population there is variation between the individuals. For example, different individuals of a population of seed-eating birds have slightly different thickness of beak.
- Some variations are better adapted to the environment than others – so if drought conditions cause the trees to produce tougher seeds, the birds with the stronger, thicker beaks will get the most food.
- The individuals that get the most food will be able to produce the most offspring and feed them well until they are old enough to find food on their own.
- If the characteristic is inheritable, the best adapted variations will increase in the next generation – beak thickness is inheritable, thick-beaked birds produce more chicks than thin-beaked birds during drought years, so the next generation will contain more thick-beaked birds than the previous generation.

EXTENDED

Natural selection is increased by competition between individuals within a population for key resources, such as food, that affect survival and reproduction. Most organisms produce many more offspring than survive to adulthood. Many individuals die before they are old enough to reproduce. The reason why some die and others survive is sometimes due to chance, but is more often to do with how well adapted they are to the environment.

In any population of organisms, such as a species of flower in a field, there will be variation in characteristics as a result of the variation in their alleles. For our example, let's consider height.

△ Fig. 3.46 Taller plants are better adapted to receiving more sunlight on their leaves.

- Some individual plants of this species will grow taller than others, as a result of their genes.
- Taller plants grow their leaves higher up the stalk, capturing more sunlight for photosynthesis and shading shorter plants so that those get less sunlight.
- So taller plants are able to make more food, and therefore make more seed than shorter plants.

△ Fig. 3.47 Taller plants also have bigger seedheads.

- Embryos in the seeds from the tall plants will have inherited the 'tallness' genes from the parent plants.
- There will be more seed with 'tallness' genes, so when they germinate there are likely to be a greater proportion of tall plants in the next generation.
- Over more generations the height of this species of plant in this area will increase because taller plants are more likely to survive and reproduce than shorter plants. (In reality, this only happens up to a point where other factors limit the change – for example, a tall plant

△ Fig. 3.48 The taller plants are more likely to survive and reproduce.

also needs more energy to grow taller as well as make more seed, so there may come a point when growing even taller means making fewer seed.)

It is a factor of the environment (light intensity) that has caused this change in the population. We call this natural selection, because a natural factor appears to select individuals with some characteristics more than others, making it possible for them to pass on their genes to the next generation more successfully.

If the environment does not change, selection does not change. This will favour individuals with the same characteristics as the parents. If the environment changes, or a mutation produces a new allele, selection might now favour individuals with different characteristics or with the new allele. So the individuals that survive and reproduce will have a different set of alleles to pass on to their offspring. This will bring about a change in characteristics of the species: that is, it will cause **evolution**.

> ### REMEMBER
> Natural selection gives some individuals a survival or reproductive advantage over others with different characteristics. When describing the stages in evolution of a species by natural selection, make sure you identify why each new development gives the organism an advantage over those individuals that have not changed.

Evolution means change over time. In terms of organisms, it usually means how species change in their characteristics over time. Usually this takes hundreds, thousands or even millions of years, making it difficult to see evolution in action.

Evidence from the fossil record is helpful in talking about evolution. For example, over 100 million years ago some species of dinosaur evolved feathers to provide insulation of the skin over the whole body. Some longer feathers also possibly evolved for display, to attract a mate. Over time, some species of feathered dinosaur evolved larger forelimbs, with longer feathers, possibly to help them glide to escape predators. From these, the first birds evolved strong wing muscles that allowed them to fly.

△ Fig. 3.49 Reconstruction of a dinosaur with feathers, from fossil evidence.

Evolution of resistance to antibiotics

Antibiotics are chemicals that are used to kill bacteria when they cause infection. They were first used widely to treat injured soldiers during the Second World War and since then have saved millions of lives. However, more recently, the evolution of bacteria that are resistant to antibiotics has become a major problem, causing many human deaths from infections each year.

The evolution of antibiotic resistance is a good example of evolution through natural selection. It happens like this:

- A patient suffering from a bacterial infection is treated with an antibiotic, for example penicillin.
- The bacterial infection is caused by millions of bacteria of one species, and the individual bacteria within that population will show variation.
- Some bacteria, as a result of random mutation, may have a new characteristic that means the penicillin doesn't kill them as quickly as the other bacteria – we say these bacteria have developed **antibiotic resistance**.
- Unless the full course of antibiotics is taken (which kills all the bacteria), the few that are more resistant will survive and reproduce.
- The numbers of resistant bacteria in the patient will increase, making it more likely that the resistant bacteria will be passed to another person.
- The newly infected person, if they become ill as a result of the infection, cannot be treated with penicillin because it will not kill the resistant bacteria. So the doctor will have to use a different antibiotic to control the infection.

Over time, some bacteria have developed resistance to a larger range of antibiotics, and many species show multiple resistance – resistance to many kinds of antibiotics. Now there are very few new antibiotics to use on multiple-resistant types of bacteria, and doctors are concerned that there will be an uncontrollable increase in the numbers of deaths from bacterial infections that used to be treatable.

The rate of evolution of a new variation of a characteristic is related to how well it is favoured by natural selection. Antibiotics are what is called a *strong selecting factor* – only those bacteria which are most resistant to them will survive and the rest die. So evolution of antibiotic resistance happens quickly.

QUESTIONS

1. Define *natural selection* in your own words.
2. **EXTENDED** Explain the importance of the following factors in natural selection:

   a) variation in the population
   b) competition
   c) survival advantage.

3. **EXTENDED** Explain how natural selection can lead to evolution.
4. **EXTENDED** Draw diagrams, such as a cartoon strip, to explain how antibiotic resistance develops in a bacterial species.

END OF EXTENDED

GENETIC ENGINEERING

Genetic engineering (also called genetic modification) is the cutting out of a gene from one organism of one species and its insertion into the DNA of an organism of another species so that the gene is expressed and produces its characteristic. The DNA that is formed when the gene is inserted may be called the recombinant DNA, and the organism containing the recombinant DNA is called a **transgenic organism**, or sometimes a genetically modified organism (GMO).

There are many examples of genetically modified organisms that have been produced, including:

- female sheep that contain the gene for human growth hormone, which they produce in their milk
- crops of wheat and maize that contain a gene from a bacterium that produces a poison that kills insects, such as caterpillars that eat the plants
- many kinds of crops that are resistant to a herbicide (plant-killing chemical) which means the herbicide can be sprayed over the growing crop to kill weeds without harming the crop plants.

EXTENDED

An example of genetic engineering

One of the best examples of genetic engineering is the insertion of the gene that codes for human insulin into bacteria, causing the bacteria to produce human insulin. Insulin is a hormone needed by some people who are diabetic and unable to make their own insulin. They have to inject the hormone every day to control their blood glucose concentration.

- The human insulin gene is identified in human DNA and cut out. Many copies of the gene are then made.
- Bacterial plasmids (separate rings of DNA) are extracted and cut open.
- The plasmids and genes are mixed together so that the gene becomes joined to the plasmid.

- The plasmids are put back into bacterial cells which are then tested to check if they contain recombinant DNA. Any bacteria that do not have the inserted gene are discarded.
- Bacteria that contain the inserted gene can now produce human insulin and are cultured on a large scale in a fermenter.

This produces large quantities of human insulin quickly and relatively cheaply.

If new genes are introduced into the cells of an animal or plant embryo in an early stage of development then, as the cells divide, the new cells will also contain copies of the new gene.

Only if the gamete cells contain the new gene, can it be passed on to offspring.

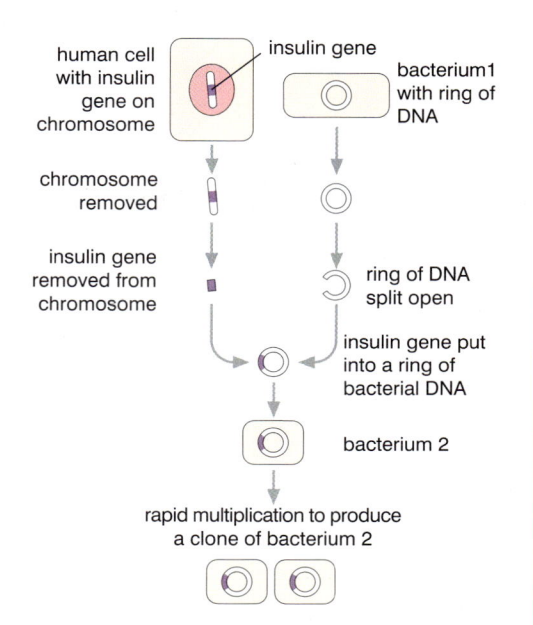

△ Fig.3.50 The process of genetic modification.

> **SCIENCE IN CONTEXT**
>
> **INSULIN PRODUCTION**
>
> Insulin is a hormone (see Control of blood glucose concentration). Some people are unable to produce insulin – they have Type 1 diabetes. They need to inject insulin regularly to keep their blood glucose concentration under control.
>
> Before the development of transgenic bacteria containing the human insulin gene, insulin was extracted from the pancreases of domesticated animals grown especially for this purpose. The process was slow and expensive process. Another problem was that the insulin produced by other species is slightly different from human insulin, so it could cause other problems in people.
>
> Since the gene expressed in transgenic bacteria is the human gene, they produce insulin that is identical to human insulin. Also, growing the bacteria in a fermenter produces much larger amounts of insulin, more rapidly and more cheaply than the old process.

END OF EXTENDED

QUESTIONS

1. Explain what is meant by *genetic engineering*.
2. Give two examples of organisms that have been genetically engineered.
3. **EXTENDED** Use the example of bacteria modified to produce human insulin to explain how genetic engineering is carried out.

End of topic checklist

Key terms

allele, antibiotic resistance, artificial selection, breed, chromosome, clone, codominance, continuous variation, daughter cell, diploid, discontinuous variation, dominant, Down's syndrome, DNA (deoxyribonucleic acid), evolution, genetic code, genetic diagram, genetic engineering, haploid, heterozygous, homozygous, inheritance, ionising radiation, malaria, meiosis, mitosis, monohybrid cross, mutagen, natural selection, phenotype, Punnett square, recessive, sex chromosome, sickle cell anaemia, variety

During your study of this topic you should have learned:

- The definition of *inheritance* as the transmission of genetic information from generation to generation.
- The definition of the terms:
    - *chromosome* – as a long thread of DNA, made up of a string of genes
    - *gene* – as a length of DNA that is the unit of heredity and codes for a specific protein
    - *allele* – as any of two or more alternative forms of a gene
    - *haploid nucleus* – as a nucleus containing a single set of unpaired chromosomes
    - *diploid nucleus* – as a nucleus containing two sets of chromosomes.
- How to describe the inheritance of sex in humans (XX and XY chromosomes).
- The definition of *mitosis* as nuclear division giving rise to genetically identical cells in which the chromosome number is maintained by the exact duplication of chromosomes.
- About the role of mitosis in growth, repair of damaged tissues, replacement of worn out cells and asexual reproduction.
- The definition of *meiosis* as reduction division in which the chromosome number is halved from diploid to haploid.
- That gametes are the result of meiosis.
- That meiosis results in genetic variation so the cells produced are not all genetically identical.
- The definition of the terms:
    - *genotype* – as the genetic makeup of an organism in terms of the alleles present
    - *phenotype* – as the physical or other features of an organism due to both its genotype and its environment

- *homozygous* – as having two identical alleles of a particular gene
- *heterozygous* – as having two different alleles of a particular gene
- *dominant* – as an allele that is expressed if it is present
- *recessive* – as an allele that is only expressed when there is no dominant allele of the gene present.

○ How to calculate and predict the results of monohybrid crosses involving 1 : 1 and 3 : 1 ratios.

○ **EXTENDED** How to explain codominance by reference to the inheritance of ABO blood groups: A, B, AB and O blood groups and genotypes I^A, I^B and I^O.

○ That continuous variation is influenced by genes and environment, resulting in a range of phenotypes between two extremes, e.g. height in humans.

○ That discontinuous variation is caused by genes alone and results in a limited number of distinct phenotypes with no intermediates, e.g. A, B, AB and O blood groups in humans.

○ The definition of *mutation* as a change in a gene or chromosome.

○ That mutation is a source of variation, as in Down's syndrome.

○ About the effects of ionising radiation and chemicals on the rate of mutation.

○ **EXTENDED** How to describe sickle cell anaemia, and explain its incidence in relation to that of malaria.

○ About the role of artificial selection in the production of varieties of animals and plants with increased economic importance.

○ The definition of *natural selection* as the greater chance of passing on of genes by the best-adapted organisms.

○ **EXTENDED** How to describe variation and state that competition leads to differential survival of, and reproduction by, those organisms best fitted to the environment.

○ **EXTENDED** How to assess the importance of natural selection as a possible mechanism for evolution.

○ **EXTENDED** How to describe the development of strains of antibiotic-resistant bacteria as an example of natural selection.

○ The definition of *genetic engineering* as taking a gene from one species and placing it in another species.

○ **EXTENDED** Why and how human insulin genes were put into bacteria using genetic engineering.

End of topic questions

Note: The marks awarded for these questions indicate the level of detail required in the answers. In the examination, the number of marks awarded to questions like these may be different.

1. Write two sentences that correctly link all the following words to explain how they are related: characteristic, chromosome, DNA, gene, nucleus, protein. **(2 marks)**

2. Explain why there are two sexes in humans, using the inheritance of sex to support your answer. **(4 marks)**

3. The form of earwax in humans is controlled by one gene. The dominant allele produces wet-type earwax, and the recessive allele produces dry earwax.

    a) Using appropriate symbols, draw a genetic diagram to show the inheritance of earwax between a man with dry earwax and woman who is heterozygous for the characteristic. **(3 marks)**

    b) Describe the predicted probability of genotypes and phenotypes in their children. **(2 marks)**

    c) Explain why it is possible that their three children all have dry earwax. **(2 marks)**

4. **EXTENDED** A plant breeder had two plants of the same species that she knew were homozygous. One had pure white flowers and one had pure red flowers. She transferred pollen from one plant to the stigmas of the other plant.

    All the seed produced grew into plants with flowers that were red with white splashes. Explain why the gene in this example shows codominance. **(3 marks)**

5. Draw up a table to summarise the similarities and differences between mitosis and meiosis in terms of: number of cells produced, whether daughter cells are genetically identical or different, whether daughter cells are haploid or diploid, and what its purpose is. **(8 marks)**

6. Explain why a life cycle needs a stage in which meiosis occurs before fertilisation. **(2 marks)**

7. Using an example of your own choice, explain how natural selection can result in some individuals of a population passing on more genes to the next generation than others. **(3 marks)**

8. **EXTENDED** Describe in words or pictures how the evolution of bacteria that are resistant to many types of antibiotics has evolved. **(4 marks)**

9. **EXTENDED** Skin cancer is the most common form of cancer in light-skinned people. The majority of skin cancers are not life-threatening if treated early.

    a) Cancer may be caused by a mutation to a gene in a cell. What does *mutation* mean? **(1 mark)**

b) Name two types of causes of mutation. (2 marks)

c) What is the most likely cause of skin cancer? Explain your answer. (2 marks)

d) The graph shows the number of new cases of skin cancer in Sweden between 1970 and 2005.

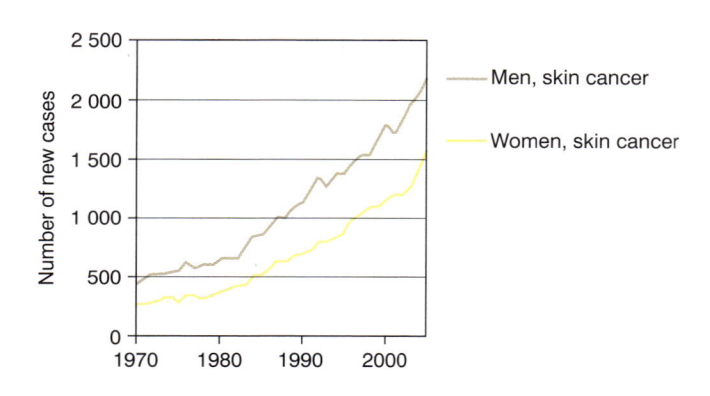

i) Describe the curves shown on the graph. (2 marks)

ii) Suggest one possible reason for the trend shown in both curves. Explain your answer. (2 marks)

10. EXTENDED Describe how an advantageous mutation can contribute to evolution. (4 marks)

11. Explain why genetically modified organisms are produced. (1 mark)

12. EXTENDED Describe the advantages of producing insulin from genetically modified bacteria. (3 marks)

Exam-style questions

Note: The questions, sample answers and marks in this section have been written by the authors as a guide only. The marks awarded for these questions indicate the level of detail required in the answers. In the examination, the number of marks awarded to questions like these may be different.

Sample student answer

Question 1

People are either able to roll their tongue into a U-shape, or unable to roll their tongue. Tongue rolling is controlled by a single pair of genes which has two alleles.

a) The diagram shows a pair of chromosomes:

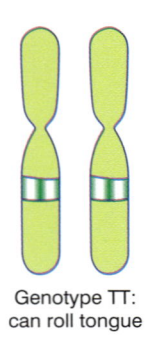

Genotype TT:
can roll tongue

i) Is the allele for tongue rolling dominant or recessive? Explain your answer. (2)

Dominant ✓ ①

The letters are capitals. ✗

ii) Write down the other possible genotypes related to tongue rolling, along with their phenotypes. (2)

Tt - The phenotype is a tongue roller. ✓ ①

tt - The phenotype is non tongue roller (cannot roll tongue). ✓ ①

TEACHER'S COMMENTS

a) i) Correct – dominant.

The rest of the answer is too vague. It should be more specific, such as 'The two letters making up the genotype are written in upper case.'

ii) Correct.

b) i) The answer is correct, but could be better worded. Rather than saying 'the couple', it is better to say that the man could be TT or Tt, and the woman could be TT or Tt.

ii) Both statements are correct, but only two marking points have been addressed. The student is correct, that unless both parents were Tt, all children would be tongue rollers, but the answer would benefit from two statements of explanation:

First of all, it should be made clear that as both the man and woman can roll their tongue, the must have at least one T allele.

There should then be a sentence of explanation to link the statements, such as:

'Without the presence of a t allele in both parents, all the children would be tongue rollers.'

iii) The diagram illustrates the cross correctly, but lacks detail.

The best way of illustrating the cross is to use a Punnett square, showing each stage of the cross:

the genotypes of the parents

the different alleles that could be passed on to the offspring from the mother and father (the alleles in the egg cells and sperm cells)

the possible combinations of alleles in the offspring (genotypes)

the possible phenotypes produced.

| | | Mother possible alleles in eggs ||
		T	t
Father possible alleles in sperm	T	TT Tongue roller	Tt Tongue roller
	t	Tt Tongue roller	tt Non tongue roller

A further point is the way in which the student has written the third possible genotype in their answer (tT). Although not incorrect, the convention is to write the dominant allele first, so it should be written Tt.

b) A couple who can both roll their tongues have children.

   i) Give the possible genotypes of the man and the woman. (2)

   The couple could be TT ✓ ①

   or Tt. ✓ ①

   ii) Their first child cannot roll his tongue; the second one can.

   What does this tell you about the genotypes of the couple? Explain your answer fully. (4)

   The genotype of both the man and women must be Tt. ✓ ①

   Because otherwise all the children would be tongue rollers. ✓ ①

   iii) Show the genetic cross involved. (4)

   Tt   ×   Tt ✓ ①
            ↓
   TT   Tt   tT   tt ✓ ①

   Total 14 marks

   ⑨⁄₁₄

Exam-style questions continued

Question 2

The diagram shows a cell and some of the structures in it.

Identify the structures 1, 2, 3, 4 and 5.

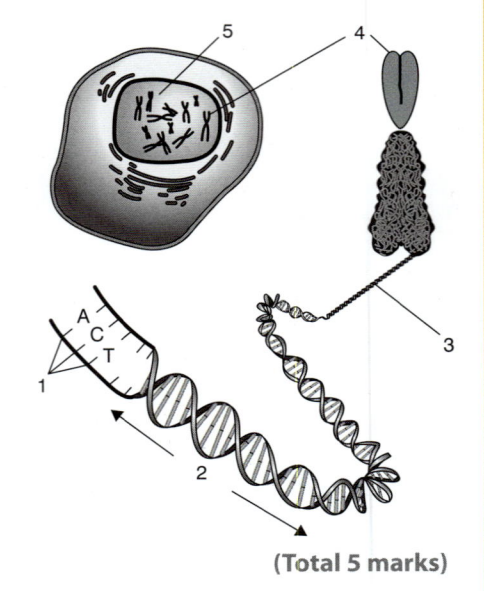

(Total 5 marks)

Question 3

A disease called cystic fibrosis is caused by a faulty allele of a recessive gene.

The family tree below shows the occurrence of cystic fibrosis in a family.

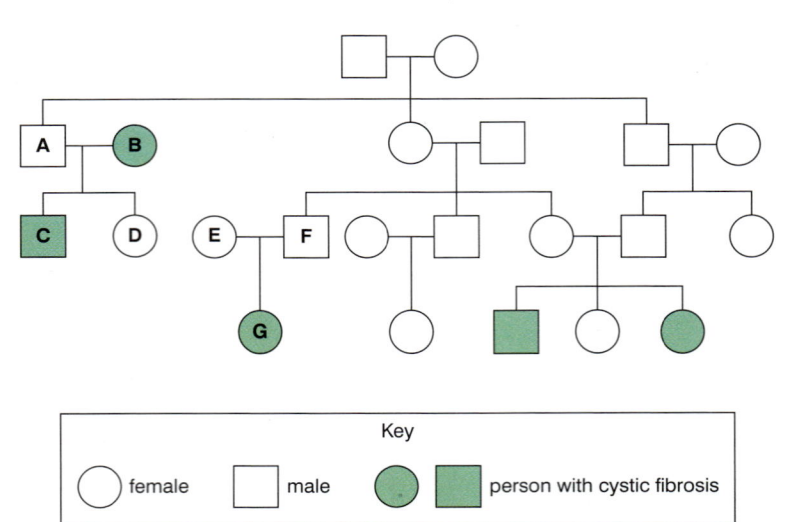

Key: ○ female □ male ● ■ person with cystic fibrosis

a) Give the genotypes of the following members of the family, explaining how you come to your decision:

   i) father A (2)

   ii) daughter D (2)

   iii) parents E and F. (3)

b) In the part of the family tree with the people A–G, identify the carriers of cystic fibrosis. (4)

(Total 11 marks)

Question 4

a) The passage below describes the process of sexual reproduction.

Use suitable words to complete the sentences in the passage.

Sexual reproduction is the most common method of reproduction for the majority of larger organisms, including almost all animals and plants. To produce a new organism, two .... fuse. This process is known as .... (2)

Usually, sexual reproduction involves .... parent organisms of the same species. The .... formed is genetically different from each of the parents. (2)

b) Give one advantage and one disadvantage of:

   i) asexual reproduction. (2)

   ii) sexual reproduction. (2)

c) **i)** Label the diagram of the human female reproductive system. (6)

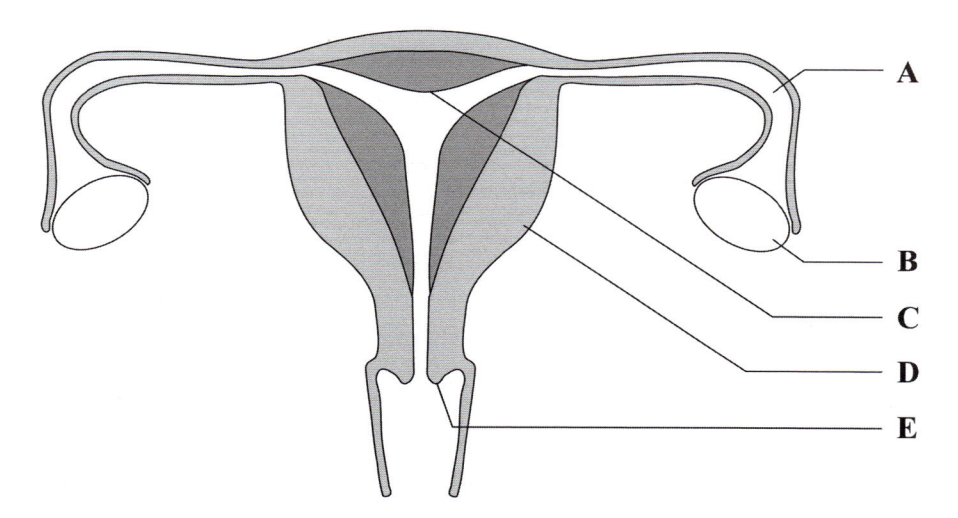

   ii) Draw a flow chart to illustrate the processes involved in labour and birth. (6)

d) State **one** method of each type of birth control:

   i) chemical (1)

   ii) barrier (1)

   iii) surgical. (1)

(Total 23 marks)

Exam-style questions continued

Question 5

Flowers are adapted to be pollinated by insects or by the wind.

a) Name the structures of an insect-pollinated flower shown in the diagram below. **(10)**

b) Explain how each of the structures is adapted for pollination in wind-pollinated flowers:

    i) petals (2)

    ii) stigma (2)

    iii) stamens (2)

    iv) pollen grains (2)

(Total 18 marks)

Question 6

Define fully the term *genetic engineering*. (4)

c) State one reason why improving the characteristics of an organism by genetic engineering might be preferable to improving characteristics by selective breeding. (1)

d) State one medical application of genetic engineering. (1)

(Total 6 marks)

EXTENDED Question 7

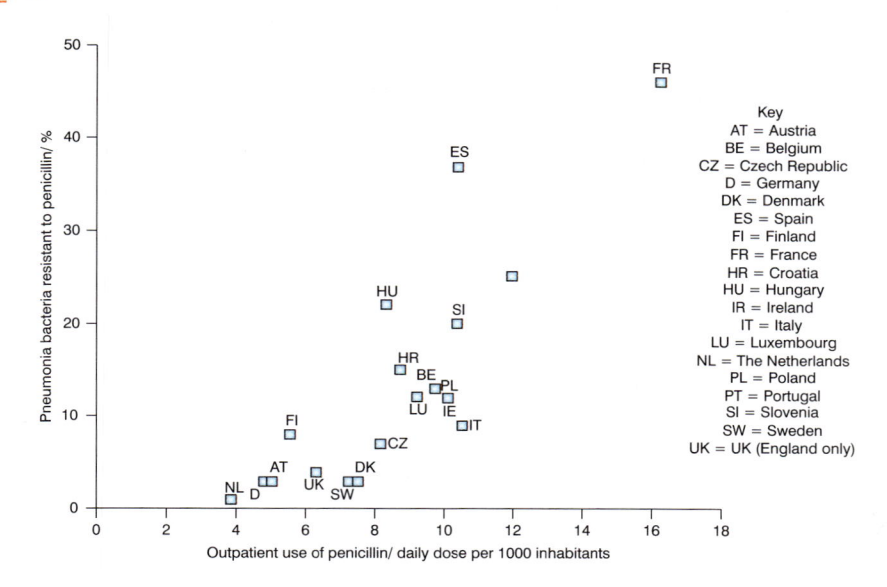

Antibiotics such as penicillin have been used routinely to control acne in patients across Europe. This scattergraph shows the relationship between antibiotic resistance and use of penicillin.

a) Describe the relationship between penicillin use and penicillin resistance in pneumonia bacteria. (1)

b) Explain how this resistance may have developed. (4)

(Total 5 marks)

EXTENDED Question 8

Codominance is seen in the genetic traits of a number of organisms, including humans.

a) The coat colour of some cattle shows incomplete dominance.

Some cattle have alleles that produce red hairs; others have alleles that produce white hairs. Cattle that have a combination of both alleles have red and white hairs. Their coat colour is described as *roan*.

  i) Explain the term *codominance*. (2)

  ii) Draw genetic diagrams to show the following crosses and predict the ratios of offspring produced.

Use C^R to represent the allele for red hair and C^W to represent the allele for white hair.

A red bull and a white cow. (6)

A roan bull and a white cow. (8)

Exam-style questions continued

b) In the inheritance of human blood groups, the alleles for blood group A and blood group B are dominant to the allele for blood group O. The alleles for blood group A and B are codominant.

Use I^A to represent the allele for blood group A, I^B to represent the allele for blood group B, and I^O to represent the allele for blood group O.

What is the probability of a father with blood group A and a mother with blood group B having:

  i) a child with blood group A (4)

  ii) a child with blood group O? (4)

(Total 24 marks)

EXTENDED Question 9

Haemophilia is a sex-linked condition in which blood fails to clot.

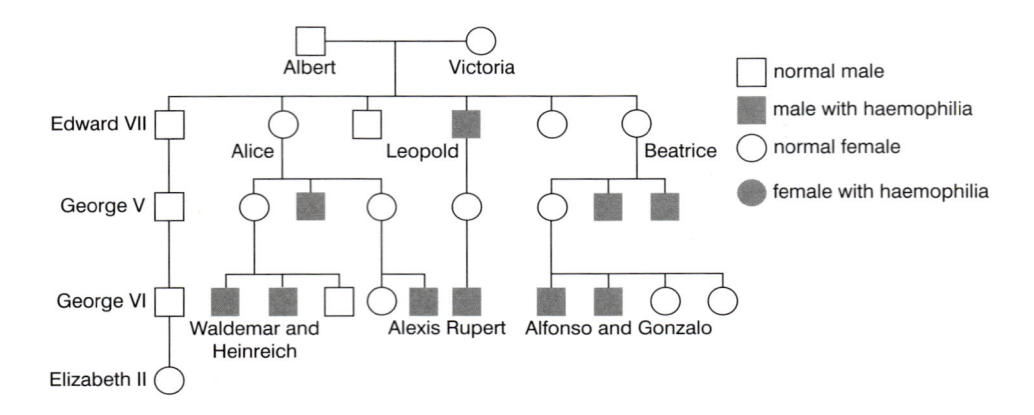

The simplified pedigree shows the inheritance of haemophilia through the royal families of Europe in the descendants of Great Britain's Queen Victoria and Prince Albert.

a) Using the letters H for normal, and h for haemophiliac, state the genotypes for Queen Victoria and Prince Albert. Explain your answer. (7)

b) Explain why it is mostly males who are affected by the condition. (4)

c) Explain why haemophilia has not occurred in the current British royal family (Edward VII through to Elizabeth II). (2)

(Total 13 marks)

EXTENDED Question 10

This question is about reproduction.

a) The diagram shows a human sperm.

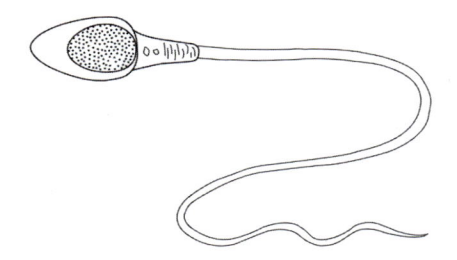

Explain how the sperm is adapted to fertilising a human egg. (4)

b) Explain the roles of hormones in controlling the menstrual cycle and preparing the uterus for a fertilised egg. (3)

c) Discuss the advantages of breast-feeding compared with bottle-feeding. (3)

(Total 10 marks)

EXTENDED Question 11

A horse has a chromosome number of 64.

a) State whether each statement refers to mitosis, meiosis, both or neither. (6)

   i) The chromosome number in each daughter cell is 64.
   ii) The daughter cells are haploid.
   iii) Two identical cells are produced.
   iv) The nuclear membrane breaks, disappears at the beginning, and is reformed at the end of the process.
   v) Some variability occurs in the alleles of parent and daughter chromosomes.
   vi) Occurs when new red blood cells are produced in the blood of the horse.

b) Explain why a horse's gametes are produced by meiosis. (5)

(Total 11 marks)

EXTENDED Question 12

a) Outline the technique used to modify bacteria so that they produce insulin. (6)

b) Explain why it is better to produce insulin using bacteria, rather than obtain it from livestock such as cattle. (4)

(Total 10 marks)

The Earth has a unique position in the Solar System. It is due to this, and our distance from the Sun, that there is such a vast abundance of living species on our planet. Earth is sufficiently far from the Sun that water has not all evaporated from the surface, and not too far that it is solid ice. Liquid water provides a solution for many organisms to live in, and the solution inside their cells in which many cell reactions take place. Millions of species share the Earth's surface, dependent on each other and on the air, land and water around them. Only the human species has developed technology, and this technology now risks damaging the environment for us as well as for all other living organisms.

STARTING POINTS

1. In order to study environments we need to know how abundant organisms are and how they are distributed. How can we do this?
2. There are many species in an ecosystem, but how can we show they are interdependent?
3. What are pyramids of biomass and energy and what do they show us?
4. The Earth will not run out of water, carbon and nitrogen, so how are these recycled through living organisms and the environment?
5. Human activity affects the environment, often resulting in pollution. What kinds of effects are produced by human activity and how can we reduce the damage we are causing?

SECTION CONTENTS

a) Energy flow, food chains and webs
b) Nutrient cycles
c) Population size
d) Human influences on the ecosystem
e) Exam-style questions

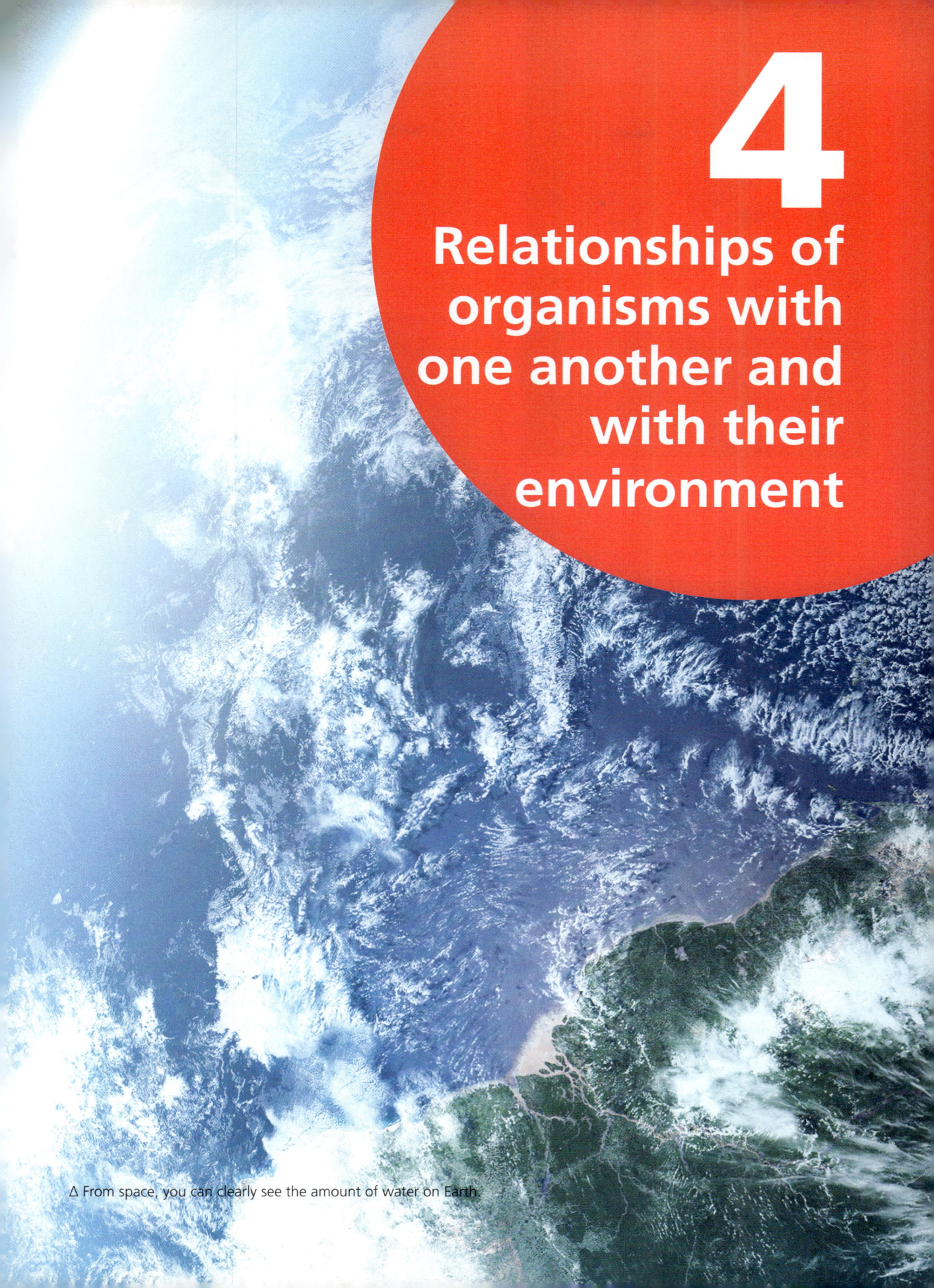

4

Relationships of organisms with one another and with their environment

△ From space, you can clearly see the amount of water on Earth.

Energy flow, food chains and food webs

△ Fig. 4.1 Alligators can survive for months without food.

INTRODUCTION

All animals need to eat in order to provide the fuel for respiration. Some animals, such as the common shrew, need to consume two or three times their body weight every day in order to survive. They live life quickly, on the hunt for food for most of the time, especially at night. In contrast, alligators only need to feed about once a week, and can live for months without food. They live much more slowly than shrews, waiting in ambush for prey to get close before attacking. Most animals eat on average somewhere between these extremes, although adult mayflies have no mouthparts and never eat. They live their brief few days using energy stored from earlier stages in their life cycle. Their only purpose is to reproduce, after which they die.

KNOWLEDGE CHECK

- ✓ Know that respiration is the release of energy from food molecules.
- ✓ Know that during photosynthesis, light energy is transferred by plants into chemical energy in sugars.
- ✓ Know that energy released by respiration is used for a range of purposes, including making new body tissue.
- ✓ Know that organisms that feed on one another can be displayed in a food chain that shows who eats what.
- ✓ Know that food chains within a habitat can be combined to produce a food web.

LEARNING OBJECTIVES

- ✓ Be able to state that the Sun is the main source of energy input to biological systems.
- ✓ Be able to describe the non-cyclical nature of energy flow.
- ✓ Be able to define the terms *food chain*, *food web*, *producer*, *consumer*, *herbivore*, *carnivore*, *decomposer*, *ecosystem*, *trophic level*.
- ✓ Be able to describe the gains and losses of energy between one trophic level and the next.

- ✓ Be able to draw, describe and interpret pyramids of biomass and numbers.
- ✓ **EXTENDED** Be able to explain why food chains usually have fewer than five trophic levels.
- ✓ **EXTENDED** Be able to explain why it is more efficient in energy terms to eat crop plants than to feed the plants to animals that we then eat.

THE ENERGY INPUT FOR BIOLOGICAL SYSTEMS

You already know that plants make their food (sugars) from carbon dioxide and water using light energy from the Sun. In systems terminology, sunlight is the energy input for plants.

Most food chains on the surface of the Earth begin with photosynthesising plants. This means that the Sun is the main input of energy into biological systems, or what we call **ecosystems**. An ecosystem is all the organisms in an area and their environment, interacting together. For example, there is an ecosystem within a lake, a desert or a decomposing log.

> **SCIENCE IN CONTEXT**
>
> ### ENERGY INPUT FROM THE SUN
>
> Because the earth curves, the amount of light energy from the Sun that falls on every square metre is greatest at the equator and decreases as you move towards the poles. The tilt of Earth's axis in relation to the Sun causes variation in the amount of sunlight energy received by high latitude regions at different times of the year. This why we have different seasons.
>
> These differences in the energy received has major effects on the ecosystems in each region. Parts of the world near the equator that receive sufficient rainfall, such as tropical rainforests, have a greater productivity of plants in a year than other regions. This greater productivity supplies more food for animals, leading to a greater productivity of animals. Some of these areas are the most biodiverse on the planet.
>
> The seasonal effects in high-latitude regions result in rapid plant growth in summer months and virtually no growth during the winter, although some of this effect is the result of lack of heat energy from the Sun as much as lack of light energy.

TROPHIC LEVELS

Food chains

You should be familiar with food chains from your earlier work. A **food chain** shows 'who eats what' in a habitat. For example, in Fig. 4.2, owls eat voles, voles eat grasshoppers, grasshoppers eat grass. (Remember, the arrows in a food chain show the direction in which the food passes.)

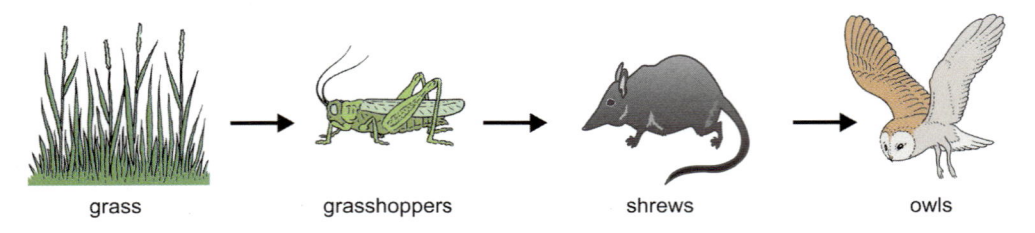

△ Fig. 4.2 An example of a food chain.

Each level in a food chain shows a separate **trophic level**, or level at which that species is feeding. So in the diagram there are:

- grass – The **producer** level, because grass is a plant and produces its own food using light energy during photosynthesis.
- grasshoppers – These are the primary **consumers**, so called because they eat the grass and 'primary' because they are the first eaters of other organisms in the food chain. This level may also be called **herbivores** because they eat plant material.
- voles – These are consumers too, but they are specifically secondary consumers because they eat the primary consumers. They are also called **carnivores**, because they eat meat.
- owls – These are also consumers, but they are specifically tertiary consumers because they eat the secondary consumers. They are also carnivores.

If anything ate owls, they would be quaternary consumers, but food chains are rarely this long. Animals at the highest trophic level in a food chain may also be called the top consumers.

All animals are consumers, because they eat other organisms to get their food. In contrast, plants are producers. Predators are any animals at secondary consumer level or above, because they kill other organisms in order to eat them. The organisms they kill are called their prey. Scavengers eat dead plants and animals, and omnivores eat both plants and animals.

Food chains do not show what happens to all the dead plant and animal material that isn't scavenged. This material decays as a result of the action of **decomposers**, such as fungi and bacteria. Fungi digest their food by secreting enzymes outside their hyphae; they then absorb the dissolved food materials (see Fungi). Many bacteria also do this. However, only some of the digested food materials are absorbed. The rest are released into the environment. Decomposers play an essential role in ecosystems (see Nutrient cycles).

◁ Fig. 4.3 The hyphae of this fungus are growing through the dead tree and secreting enzymes that cause the wood to break down into simpler chemicals.

SCIENCE IN CONTEXT

OTHER PRODUCERS

Not all producers are plants, and not all producers use light energy. There are species, mainly bacteria, that produce their own food without the presence of light energy from the Sun. Instead they get the energy they need for the formation of sugars from chemical reactions.

These bacteria are the source of food for food chains and webs that exist where there is no sunlight, such as deep in oceans and in deep underground caves. Be careful to avoid the statement that 'All life on Earth depends on the Sun', which is an oversimplification and not totally accurate.

Food webs

Looking more closely at the trophic levels in a food chain, it is rare to find an organism that is eaten by just one other species, or a predator that feeds on just one prey. A predator may also feed at more than one trophic level – an omnivore, for example, is a primary consumer when feeding on plants, but a secondary or tertiary consumer when eating other animals. So food chains within a habitat are linked together to form a **food web**. A food web is a better description of the feeding relationships in a habitat because it shows how living organisms are interdependent.

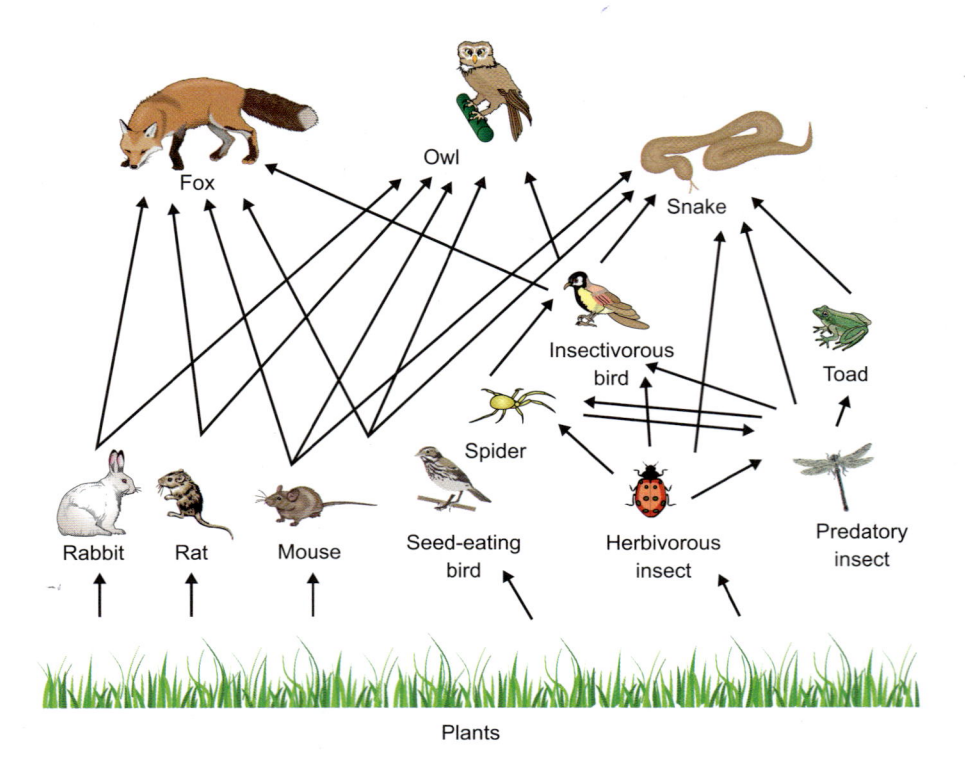

△ Fig. 4.4 A simplified food web.

Food webs still usually try to group the organisms according to their trophic level. For example, in Fig. 4.4 of a simplified food web, the rabbit, squirrel, mouse, seed-eating bird and herbivorous insect are all primary consumers and are placed just above the producer level.

There are usually many more species in a habitat than shown in the diagram, and linking them all in one food web can get confusing. So food web diagrams may focus on the relationships between key organisms rather than all of them. For example, they may only include the most numerous species, or focus on the most vulnerable species. This can be helpful if you want to use the food web to predict what would happen to the ecosystem if the food web changed in some way, such as by human activity.

For example, you could use the food web shown in Fig. 4.4 to predict what would happen if the plants were sprayed with an insecticide. This would kill the herbivorous insects and so reduce the amount of food available to all the animals that feed on them.

QUESTIONS

1. Use your own words to define the following terms: a) ecosystem, b) producer, c) consumer, d) decomposer.

2. Name the principal source of energy to an ecosystem. Explain your answer.

3. Distinguish between a food chain and food web.

4. Describe the usefulness and problems with drawing food webs.

Energy transfers in food webs

Food chains and food webs not only show who eats what; because 'food' is a store of chemical energy, they also show the flow of energy through the trophic levels, from producers to top consumers. You can investigate this transfer of energy using simple diagrams.

Pyramids of number

In many food chains, if you look at the number of organisms in each trophic level you will find far more producers than primary consumers, and far more primary consumers than secondary consumers. For example, a few tigers eat a larger number of antelopes, which eat a far larger number of grass plants. We can use these data to create a **pyramid of numbers** as shown in the diagram in Fig. 4.5. Each bar in the pyramid represents a different trophic level, arranged in order starting with producers at the bottom and ending with the top consumer at the top. The width of each bar is drawn to scale, representing the numbers of individuals in the trophic level.

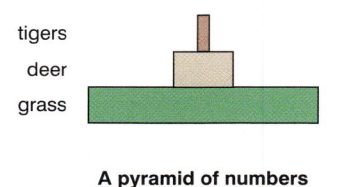

△ Fig. 4.5 A typical pyramid of numbers.

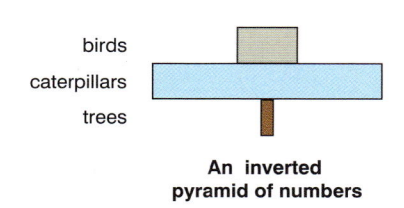

△ Fig. 4.6 Inverted pyramid of numbers.

These diagrams are called 'pyramids' because their shape is often wider at the base and narrower towards the top. However, this isn't always the case. Imagine one large tree on which hundreds of caterpillars feed, on which many birds feed. This pyramid of numbers produces a shape with a very narrow base. This is sometimes called an *inverted* pyramid of numbers.

Gathering data for a pyramid of numbers is relatively simple, because you just have to count the number of organisms in each trophic level within the area of observation.

Pyramids of biomass

It is not surprising that, in the example of tree/caterpillars/birds, you get an inverted pyramid of numbers because one tree is huge in comparison with the tiny caterpillars that feed on it. If you measure **biomass** (the mass of living material) in the organisms instead of number, you can avoid this problem.

The biomass of each trophic level is usually calculated as the average dry mass of one individual multiplied by the total number of individuals. This is usually within a given area, so the values will be in mass per unit area. If you draw these values to scale, you can produce a **pyramid of biomass**.

Fig. 4.7 shows what happens if you produce a pyramid of biomass for the tree/caterpillars/birds example from before. Using the biomass of the organisms produces the anticipated pyramid shape.

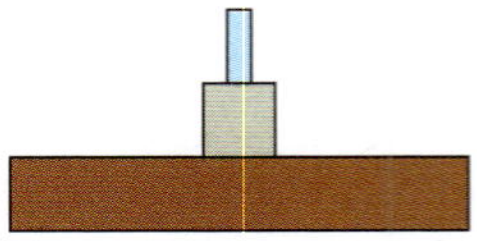

A pyramid of biomass

△ Fig. 4.7 A pyramid of biomass.

Note that dry mass is used when constructing pyramids of biomass. This is the mass without any the water in the body, because water is relatively heavy and water content in a body can vary a lot depending on its state of hydration. Measuring dry mass may involve killing at least one organism of each type, so that the tissues can be dried fully in an oven and the masses averaged to give an average biomass per individual. Tables of average biomass can be used in order to avoid killing any organisms, but this adds another level of estimation to the process, and so may decrease the reliability of the data in the diagram even further.

Pyramids of biomass produce problems of their own. The mass of the tree is actually a measure of the biomass accumulated over many years. This is often called the 'standing crop'. The insects living off it may have been produced and consumed in a matter of days and so do not accumulate biomass in the same way. This produces the correct shape for the pyramid but creates a different problem.

The growth of plankton in an area of sea was measured over a few days and a pyramid of biomass created (see Fig. 4.8). Again the pyramid is inverted. This is due to 'under-sampling' of the algae in the food web, which is caused by the relatively short life span of the algae compared with the longer-lived herbivorous organisms. Unlike the tree, algae do not produce a significant standing crop so do not provide a significant base layer for the habitat.

To help you understand this, think of a shop that sells fruit. In the morning, the shopkeeper fills the shelves with fruit, and during the day customers take fruit from the shelves to buy. The 'standing crop' of fruit on the shelves looks smaller in the afternoon than in the morning. Only if the shopkeeper has someone to continually restock the shelves will the standing crop be the same throughout the day.

Developing investigative skills

Some students were collecting data on the abundance of plants and snails on some rough wasteland so that they could construct ecological pyramids.

Using and organising techniques, apparatus and materials
❶ Draw a food chain for these organisms, and identify each trophic level.
❷ Write a plan for this investigation, to explain how they could gather reliable data for a pyramid of numbers.

Observing, measuring and recording
❸ Explain fully how they could convert the data for a pyramid of numbers into a pyramid of biomass.

Handling experimental observations and data
The students collected the following data for their pyramid of numbers.

sample site	1	2	3	4	5
number of plants	46	75	39	28	22
number of snails	4	8	5	1	2

❹ Calculate a mean number for each trophic level.
❺ Use your mean values to draw a pyramid of numbers.

The students calculated the following mean values of dry mass for one organism in each trophic level: plant 38 g, snail 6 g.

❻ Use these values to construct a pyramid of biomass for these organisms.
❼ Describe and explain the shape of both pyramids.

Pyramids of energy

Remember that each trophic level represents the amount of chemical energy in the organisms that are available to the organisms that feed on them when they digest their food. So you can create a **pyramid of energy**, in units of energy per area per time such as kJ/ha/yr to show the amount of energy in each trophic level. This avoids the problem you saw with pyramids of biomass, because it includes a value for time as well as area. If you look at pyramids of number, biomass and energy for the same food chain, we can see the effect of this.

A pyramid of numbers

A pyramid of biomass

A pyramid of energy

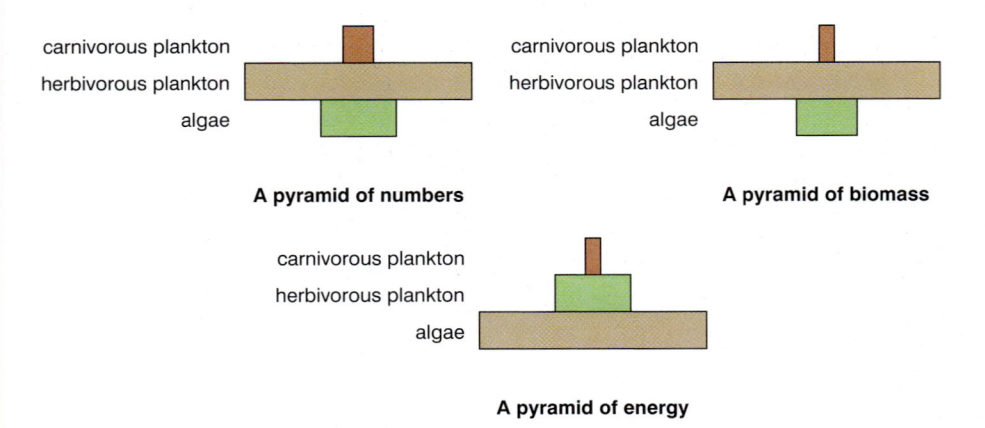

△ Fig. 4.8 Pyramids of number, biomass and energy for the same marine community. (Herbivorous and carnivorous plankton are tiny animals that float in the top layers of the open oceans.)

Although pyramids of energy produce a pyramid shape each time, they are more difficult to calculate because you need to measure the energy content of each population within the area. This involves burning the plant or animal tissue in a calorimeter to measure the energy it contains and then multiplying by the number of individuals. Alternatively, using estimates of the average fat and muscle content of an individual, you can estimate the average energy content in an individual. All these phases reduce the reliability of the calculated values.

SCIENCE IN CONTEXT: USING PYRAMIDS

In 1976 a scientist named J.O. Farlow attempted to construct a pyramid of energy for a community of dinosaurs that lived in part of North America about 85 million years ago. This community included several herbivorous dinosaur species as well as carnivorous tyrannosaurs. At the time that Farlow carried out his calculations, there was a major debate about whether dinosaurs were 'warm-blooded' (like modern mammals and birds) or 'cold-blooded' (like modern reptiles).

Farlow calculated the energy requirements of the herbivorous and carnivorous dinosaurs using estimates from modern mammals, and also from modern reptiles. Using energy values from mammals, he calculated that the carnivores in the community needed far more energy to support their needs than could be provided by the herbivores. However, using lizard energy values, he was able to produce a pyramid of energy of the right shape.

He used this as evidence to suggest that large dinosaurs were cold-blooded like modern lizards, although he admitted that the number of assumptions he had made in his calculations didn't prove that they were not warm-blooded like mammals. Since Farlow's time, other evidence has suggested more strongly that many dinosaur species were 'warm-blooded'.

The reason that a pyramid of energy is always a pyramidal shape is that each trophic level can only use a proportion of the energy in the food that they eat (or, in the case of plants, the food they make using the light energy they absorb). At each stage in the food chain, some of the energy within the trophic level is lost to the environment, as heat energy from respiration, or as chemical energy in waste substances that pass to the environment and the decomposers.

QUESTIONS

1. Define the terms a) pyramid of number, b) pyramid of biomass and c) pyramid of energy.
2. Explain, with an example, how you would produce a pyramid of numbers for a community with three trophic levels.
3. Explain why a pyramid of biomass may have an inverted shape.

ENERGY TRANSFER LOSSES

If you compare the amount of energy that enters a trophic level with the amount of energy available to the next trophic level, you can see that there are several sources of energy loss at each stage. The energy losses from plants and animals differ in some ways.

Energy losses in plants

The amount of light energy that falls on the Earth's surface varies at different times of day and year, and varies in different parts of the world (places near the equator receiving more light energy than places nearer the poles). On average, tropical areas receive between 3 and 5 kWh/m² per day (which is about the same energy as a one-bar electric heater left on for 3 to 5 hours).

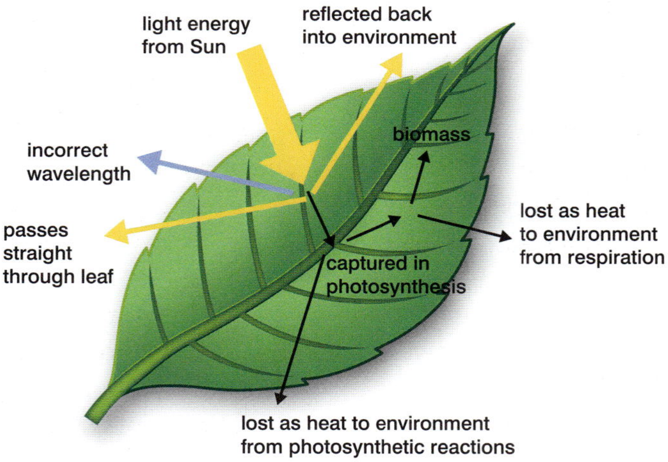

Plants use only a tiny proportion of this for many reasons, as shown in Fig. 4.9. It has been estimated that most plants only convert about 1 to 2% of the light energy that falls on them into chemical energy in biomass. This is the energy available to a herbivore that eats the plant.

△ Fig. 4.9 Energy gains and losses from a leaf.

Energy losses in animals

When an animal eats, the food is digested in the alimentary canal and the soluble food molecules are absorbed into the body. The undigested and unabsorbed food in the alimentary canal is egested as faeces.

Absorbed food molecules may be used for different purposes in the body:

- to produce new animal tissue or gametes for reproduction
- used as a source of energy for respiration
- converted to waste products in chemical reactions.

The chemical energy in the food molecules remains as chemical energy in body tissue, or in the waste products, such as urea, which are excreted to the environment.

When food molecules are broken down during respiration, the chemical energy in the molecules is converted to heat energy. This heat energy is lost to the environment by conduction and radiation. So only a proportion of the chemical energy in the animal's food is converted into chemical energy in its body tissues. This is what increases the animal's biomass.

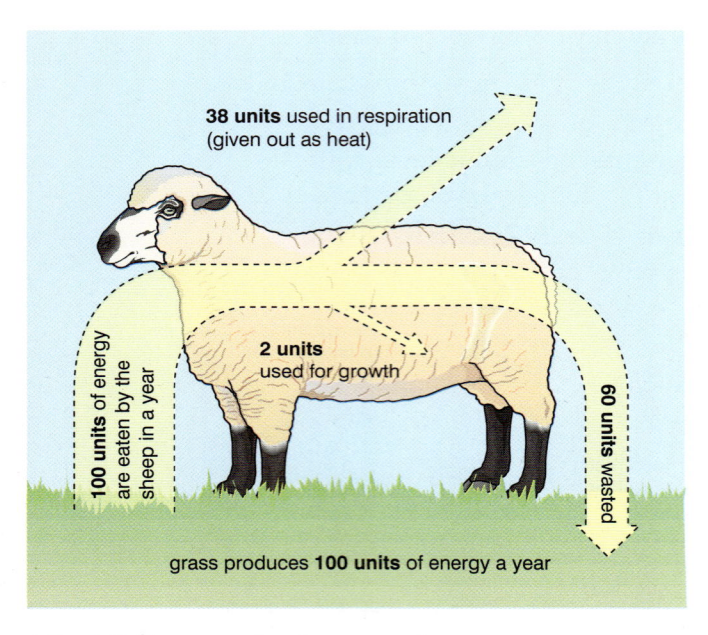

△ Fig. 4.10 The energy flow in a young sheep.

Energy transfer efficiency is the amount of energy available in body tissue of a trophic level compared with the amount in the previous level. Calculating the energy transfer efficiency between trophic levels involves the estimation of many values. This means that the transfer efficiencies you may find in textbooks and on the internet are only best estimates and must not be taken as exact.

In addition, many sources quote a value of 10% as the efficiency of energy transfer between any trophic level and the one above. Calculations of efficiency vary from about 0.2% to around 20% for different organisms in different ecosystems. This gives an *average* of 10%, but over such a large range this is not very reliable. It is better to prepare to explain how energy is gained and lost between trophic levels, in order to explain the shape of pyramids of energy and lengths of food chains, than to quote specific figures for energy transfer efficiency.

The fact that there are energy losses to the environment at each trophic level explains the shape of the pyramid of energy. It also explains the fact that food chains are rarely more than 4 or 5 organisms in length. Any organism expends energy when looking for food, and the more scattered the food, the more energy is lost in moving about to find it. Top consumers usually have to hunt over large distances to find enough food. If an organism fed exclusively on them, it would expend more energy hunting for its food than it would gain from eating it.

EXTENDED

Human food chains

Humans are **omnivores**, meaning we get our energy from eating both plant and animal tissues. This gives us choice of what we eat, but these choices have an impact on what we grow and how we use ecosystems.

Consider two food chains that include humans:

- crop plant (such as wheat grain) → herbivore (such as cow) → human
- crop plant (such as wheat grain) → human

If we consider these food chains in terms of energy flow, we can see that if humans eat the grain, there is much more energy available to them than if they eat the cows that eat the grain. This is because energy will be lost from the cows in the system. So it is energetically more efficient within a crop food chain for humans to be herbivores than to be carnivores.

In reality, we often use animals to gather plants that either we cannot eat (such as grass), or are too widely distributed for us to collect (such tiny algae in oceans, that form the food of fish that we eat).

END OF EXTENDED

QUESTIONS

1. Draw a flowchart to show the energy gains and losses from a plant leaf.
2. Draw a flowchart to show the energy gains and losses from a herbivore.
3. **EXTENDED** Sketch pyramids of biomass for the two food chains given above, and use these to explain why it is energetically more efficient for humans to eat the crops than feed them to animals that we then eat.

End of topic checklist

Key terms

biodiversity, biomass, carnivore, consumer, decomposer, food chain, food web, herbivore, omnivore, producer pyramid of biomass, pyramid of energy, pyramid of number, trophic level

During your study of this topic you should have learned:

○ That the Sun is the main source of energy input to biological systems.

○ How to describe the non-cyclical nature of energy flow.

○ Definitions of the terms:

- *food chain* – as a chart showing the flow of energy (food) from one organism to the next, beginning with a producer
- *food web* – as a network of interconnected food chains showing the energy flow through part of an ecosystem
- *producer* – as an organism that makes its own organic nutrients, usually using energy from sunlight, through photosynthesis
- *consumer* – as an organism that gets its energy by feeding on other organisms
- *herbivore* – as an animal that gets its energy by eating plants
- *carnivore* – as an animal that gets its energy by eating other animals
- *decomposer* – as an organism that gets its energy from dead or waste organic matter
- *ecosystem* – as a unit containing all of the organisms and their environment, interacting together, in a given area
- *trophic level* – as the position of an organism in a food chain, food web or pyramid of biomass, numbers or energy.

○ How to describe the gains and losses of energy between one trophic level and the next.

○ How to draw, describe and interpret pyramids of biomass and numbers.

○ **EXTENDED** How to explain why food chains usually have fewer than five trophic levels.

○ **EXTENDED** How to explain why it is more efficient in energy terms to eat crop plants than to feed the plants to animals that we then eat.

End of topic questions

Note: The marks awarded for these questions indicate the level of detail required in the answers. In the examination, the number of marks awarded to questions like these may be different.

1. The photograph shows lions feeding on the carcass of a zebra. When a lion catches a zebra, it will share the meat with other lions. Before the lion started chasing the zebra, the zebra has been feeding on grass.

   a) Is the lion a carnivore or herbivore? Explain your answer. **(2 marks)**

   b) At which trophic level does the zebra feed? **(1 mark)**

   c) Draw a food chain for the organisms shown in the photograph. **(2 marks)**

   d) Lions also feed on the herbivores gazelle and wildebeest. Use all these organisms to draw a food web for the African grassland. **(3 marks)**

2. In a community of organisms in a garden there are 5 lettuces. There are 40 caterpillars feeding on the lettuces until 2 thrushes (insectivorous birds) eat all the caterpillars.

   a) Draw a pyramid of number for this community. **(3 marks)**

   b) Describe the limitations of this pyramid. **(2 marks)**

   c) Describe the difficulty of preparing the data for a pyramid of biomass for these organisms. **(2 marks)**

3. Use the food web on in Food webs to predict what would happen to the following species if all the herbivorous insects were killed by insecticide. Explain your answers.

   a) predatory insects **(2 marks)**

   b) insectivorous birds **(2 marks)**

   c) mice **(2 marks)**

   d) snakes. **(2 marks)**

End of topic questions continued

4. The diagram shows a pyramid of numbers for a food chain early in the year. Later in the year the caterpillars change into butterflies and fly away to feed on flowers.

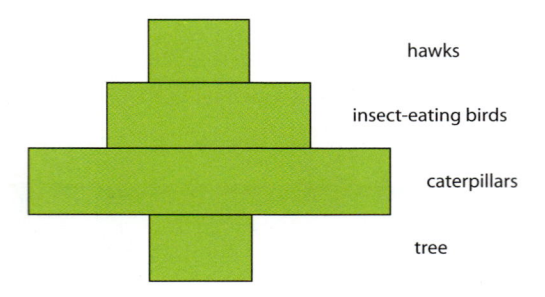

   hawks
   insect-eating birds
   caterpillars
   tree

   a) Which organisms are the primary consumers in this food chain? Explain your answer. **(2 marks)**

   b) Explain why this pyramid is not the usual pyramid shape. **(1 mark)**

   c) Suggest what this pyramid would look like for the same food chain in late summer. Explain your answer. **(2 marks)**

   d) The data in this pyramid were converted to a pyramid of energy. Suggest the shape of the pyramid of energy. Explain your answer. **(2 marks)**

5. **EXTENDED** Explain as fully as you can why a food chain is unlikely to include more than five trophic levels. **(6 marks)**

6. **EXTENDED** Food chains in high northern latitudes may be much shorter than food chains in tropical rainforests. Thinking only in terms of energy, try to explain this difference. **(4 marks)**

7. **EXTENDED** A farmer feeds wheat grain grown in his fields to his chickens to fatten them up for meat. Explain the error of this in terms of trophic energy efficiency. **(2 marks)**

8. Compare the difficulty of producing a pyramid of number, pyramid of biomass and pyramid of energy. **(6 marks)**

Nutrient cycles

INTRODUCTION

The light energy from the Sun, on which most of life on Earth depends, seems to be an inexhaustible supply. (In fact, in 4 to 5 billion years the Sun will fade and stop supplying that energy, but this is not a problem for life at the moment.) Light energy from the Sun is converted to chemical energy in the tissues of living organisms and then transferred to the environment as heat energy, which organisms can no longer use; this also is not a cause for worry. However, all the chemical substances that are found in living tissue, such as water, carbon and nitrogen, are limited to what is currently on Earth. So it is just as well that these substances are continually cycled between the living and non-living parts of ecosystems, otherwise life would end fairly rapidly.

△ Fig. 4.11 Sunlight provides the energy for all plant growth.

KNOWLEDGE CHECK
✓ Know that water is essential for many life processes.
✓ Know that carbohydrates, proteins and lipids all contain carbon.
✓ Know that proteins also contain nitrogen.
✓ Know that plants lose water to the environment through their leaves, in transpiration.
✓ Know that decomposers digest dead organic material releasing some of the products of digestion into the environment.

LEARNING OBJECTIVES
✓ Be able to describe the carbon and water cycles.
✓ **EXTENDED** Be able to describe the nitrogen cycle.
✓ **EXTENDED** Be able to discuss the effects of combustion of fossil fuels and the cutting down of forests on the oxygen and carbon dioxide concentrations in the atmosphere.

THE WATER CYCLE

Water (H_2O) is essential to life processes. Without water, most organisms usually die within a few days. Fortunately, the world's water does not get used up; it is constantly recycled.

Most of the world's water is in the oceans. Some **evaporates** into the atmosphere as water vapour. Water vapour is transported in the atmosphere by air movement until it **condenses** into liquid water droplets in clouds. It eventually falls as **precipitation**, such as rain or snow. Some returns straight to the ocean, but some falls on land. Much of this water drains into rivers and returns to the oceans.

Plants take up water through their roots and lose it through **transpiration** through their leaves. Within the plant, water is used for transport of dissolved substances in the phloem, for photosynthesis to make glucose and to keep cells turgid to support the plant. Animals get the water they need from their food, through respiration, and directly from the environment when they drink. They lose water to the environment through evaporation and through excretion of urine.

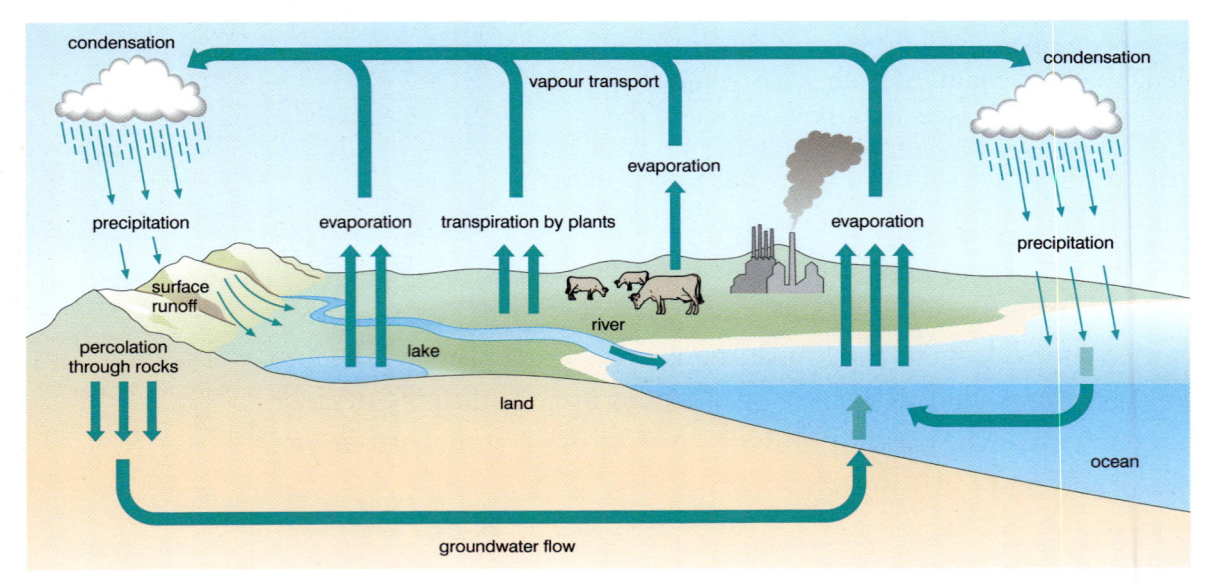

△ Fig. 4.12 A summary of the water cycle.

QUESTIONS

1. Define the following terms, and give an example of each from the water cycle:

   a) evaporation

   b) transpiration

   c) condensation

   d) precipitation.

2. Starting with water in a pond, describe two different routes that water may take through the water cycle before returning to the pond, one through living organisms and the other not through organisms.

3. Explain the importance of the water cycle for life.

THE CARBON CYCLE

Carbon is also cycled in different forms through the living and non-living parts of ecosystems at different stages of the **carbon cycle**. Carbon dioxide from the atmosphere is converted to complex carbon compounds in plants during photosynthesis. This is often called the 'fixing' of carbon by plants. Respiration in plants returns some of this fixed carbon back to the atmosphere as carbon dioxide. Carbon in the form of complex carbon compounds passes along the food chain. At each stage, some of this carbon is released as carbon dioxide to the atmosphere as the result of respiration.

When organisms die, their bodies decay as they are digested by decomposers. Some of the complex carbon compounds are taken into the bodies of the decomposers, where some may be converted to carbon dioxide during respiration. Carbon dioxide may also be released directly into the atmosphere during decay.

△ Fig. 4.13 Water, like sediment, excludes air, which prevents decay organisms from respiring. So dead plant material builds up over time, forming peat. Peat can be burned as a fuel, although this is being discouraged so that peat bog habitats can be protected.

QUESTIONS

1. Describe the role of the following in the carbon cycle:

    a) respiration

    b) photosynthesis

    c) decomposition.

2. In what form is carbon when it is in the following stages of the carbon cycle?

    a) Earth's atmosphere

    b) plant tissue

    c) fossil fuels

EXTENDED

Combustion and deforestation

If dead organic material is buried too quickly by sediment or water for decomposers to cause decay, and remains buried, then it may be converted to other complex carbon compounds. Peat is formed when mosses and other plants are buried in swampy ground for hundreds of years. Over many millions of years, where there were once huge forests growing in swampy regions, heat and pressure have turned the organic material to coal. Heat and pressure over many millions of years also produces oil from the decaying bodies of tiny marine organisms that were buried in sediment at the bottom of oceans. Peat, coal and oil are **fossil fuels**. The carbon from their complex carbon compounds is released into the air as carbon dioxide during **combustion**, when we burn the fossil fuels.

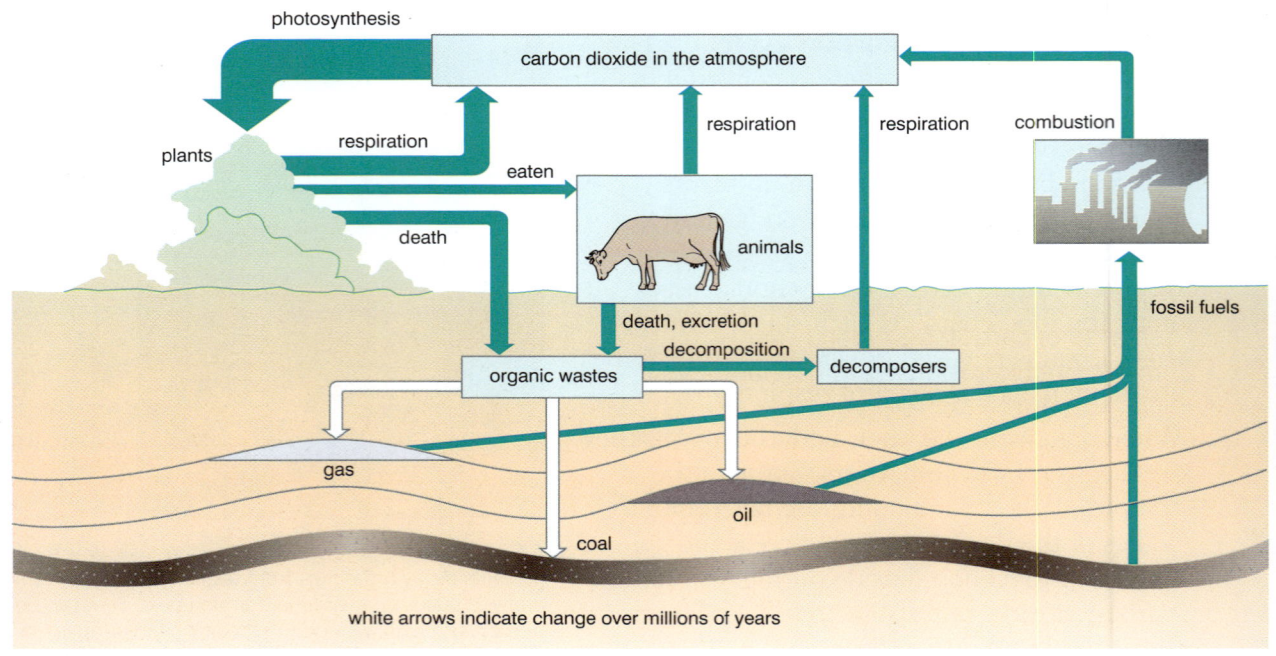

△ Fig. 4.14 A summary of the carbon cycle.

REMEMBER

Make sure you are certain what form carbon is in (carbon dioxide or complex carbon compounds such as carbohydrates) at each stage of the carbon cycle.

Deforestation is the permanent destruction of large areas of forests and woodlands. It usually happens in areas that provide quality wood for furniture and to create farming or grazing land.

Forests act as a major carbon store because carbon dioxide is taken up from the atmosphere during photosynthesis and used to produce the chemical compounds that make up the tree. When forests are cleared, and the trees are either burned or left to rot, this carbon is released quickly into the air as carbon dioxide. This rapidly increases the proportion of carbon dioxide compared with oxygen in the air surrounding the forest. In addition, the amount of oxygen removed from the local atmosphere by plants for photosynthesis may also drop, changing the balance between carbon dioxide and oxygen in the atmosphere locally.

▷ Fig. 4.15 Satellite images of the island of Sumatra show the extent of deforestation between 1992 (top photo, land mostly green) and 2001.

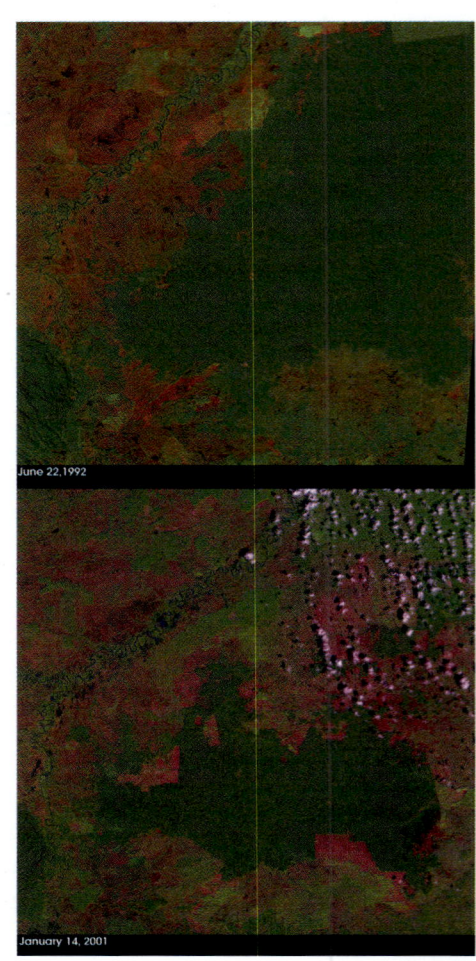

Large-scale deforestation, such as in Sumatra, releases so much carbon dioxide that it cannot be balanced by the amount taken from the atmosphere for photosynthesis. The additional carbon dioxide remains in the atmosphere.

> **SCIENCE IN CONTEXT**
>
> ### GETTING OUT OF BALANCE
>
> Over the past 10 000 years or so, as a result of photosynthesis and respiration and other physical processes, the exchange of carbon between organisms and the atmosphere caused little change in the amount of carbon dioxide in the atmosphere. On average, over one year about 120 billion tonnes of carbon dioxide are removed from the atmosphere by photosynthesis, and a similar amount returned by respiration.
>
> △ Fig. 4.16 Atmospheric carbon dioxide concentration from 1000 CE to recent times.
>
> During the past 250 years, however, human activity has added increasing amounts of carbon dioxide to the atmosphere. Today about 5.5 billion tonnes of carbon are added to the atmosphere every year through human activity, particularly through combustion of fossil fuels.
>
> Compared with 120 billion tonnes through natural processes, this may not seem a lot. However, there is no process to balance this extra carbon dioxide, so its concentration in the atmosphere is increasing. It is this additional carbon dioxide in the atmosphere that most people believe is causing global warming and climate change.

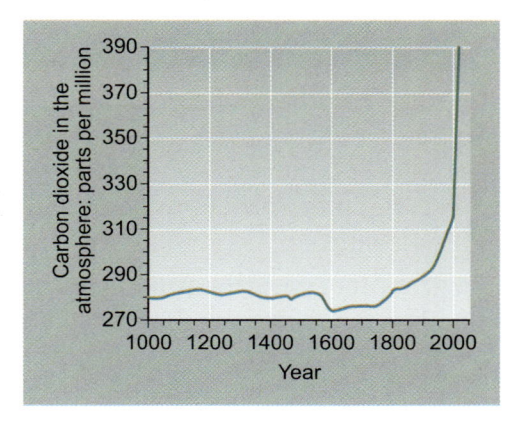

QUESTIONS

1. **EXTENDED** Explain the importance of combustion in the carbon cycle.
2. **EXTENDED** Describe the effect of large-scale deforestation on the oxygen and carbon dioxide concentrations of the atmosphere.

THE NITROGEN CYCLE

Living things need nitrogen to make proteins, which are used, for example, to make new cells during growth. Air is 79% nitrogen gas (N_2), but nitrogen gas is very unreactive and cannot be used by plants or animals.

The **nitrogen cycle** describes the way in which nitrogen passes between the living and non-living parts of an ecosystem. Animals take in nitrogen in the form of proteins when they eat plant tissue or animal tissue. They break down the proteins to amino acids in digestion and convert them into new proteins in their own body tissue. Since plants don't eat, they can't take their nitrogen in this form. Instead they absorb nitrogen in the form of nitrate ions (NO_3^-) using active transport from the soil water around their roots. They then convert these ions into the proteins they need.

Nitrate ions are present in the soil as a result of several processes, one non-living and the others as the result of living organisms. The non-living process is lightning, which generates large amounts of energy that cause atmospheric nitrogen to react with oxygen. This forms nitrate ions, which dissolve in rain and fall to the ground where they remain in soil water.

The biological processes that produce nitrate ions are much more important. The first begins with organic material containing complex nitrogen compounds, such as proteins, in the form of faeces or urine from animals, and as dead plant and animal tissue.

- This material decays as a result of decomposers, releasing nitrogen in the form of ammonium ions into the soil.
- Some bacteria in the soil can take in ammonium ions and produce nitrite ions (NO_2^-). Other soil bacteria can take in nitrite ions and produce nitrate ions.
- Both kinds of bacteria are called **nitrifying bacteria** because their action adds soluble nitrogen to the soil water.

The other biological process that makes nitrogen available to plants as nitrates also involves bacteria. These bacteria are unusual because they can convert nitrogen gas from the atmosphere directly into a form that plants can use. They are called **nitrogen-fixing bacteria**. Some of these bacteria live free in the soil and some grow in root nodules, small bumps on the roots, of some kinds of plants especially legumes (plants of the pea and bean family).

Nitrifying bacteria can only grow in aerobic conditions, when there is plenty of oxygen in the soil. In waterlogged conditions, air in the soil is pushed out, so these bacteria cannot grow. These conditions favour another type of bacteria, which can respire anaerobically. As they grow, they convert nitrates back to nitrogen gas, which escapes to the atmosphere. They are called **denitrifying bacteria**, because they remove nitrates from the soil and make it less fertile for growing plants.

△ Fig. 4.17 Some legume plants produce special nodules on their roots in which nitrogen-fixing bacteria live.

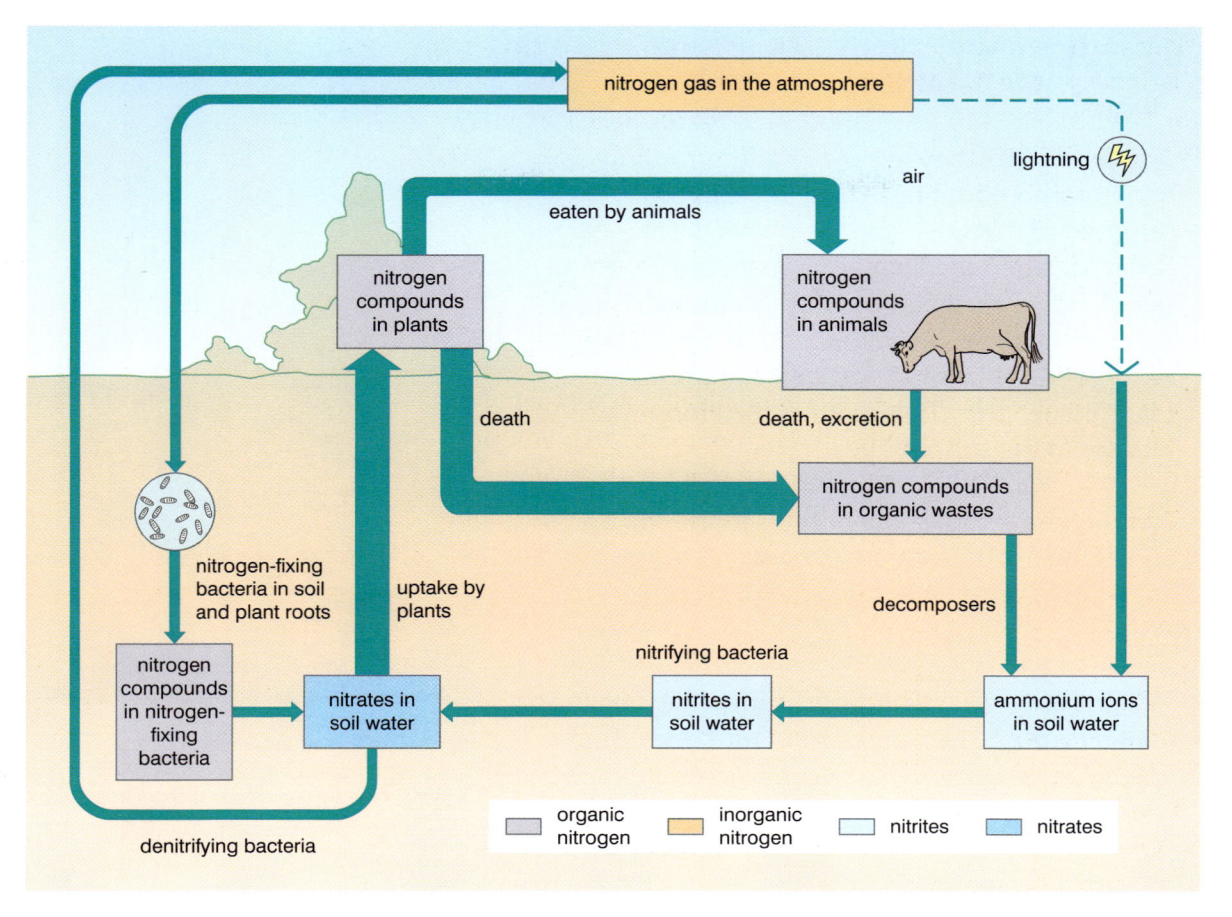

△ Fig. 4.18 A summary of the nitrogen cycle.

QUESTIONS

1. **EXTENDED** Describe what is meant by:
   a) *nitrifying bacteria*
   b) *nitrogen-fixing bacteria*
   c) *denitrifying bacteria*.
2. **EXTENDED** Explain the importance of nitrifying bacteria in the nitrogen cycle for the fertility of soils.
3. **EXTENDED** Explain the importance of decomposers in the cycling of nitrogen.

END OF EXTENDED

SCIENCE IN CONTEXT

MAINTAINING SOIL FERTILITY

As plants grow, they remove nitrates from the soil. In natural communities, those nitrates are returned to the soil when the plants die and decay, or when the animals that ate them die and decay. However, when we grow crops, we usually remove the plants from the ground, taking the nitrates (in the form of nitrogen compounds in plant tissue) away. This reduces the amount of nitrates in the soil each year.

Farmers can return some nitrates to the soil by spreading animal manure (faeces and urine mixed with straw) or by growing legume crops in some years.

Millions of tonnes of nitrogen-containing fertiliser are made each year by chemical processes. These are used to fertilise fields to provide food for the growing human population.

△ Fig. 4.19 Farmers sometimes plant legume crops in order to add nitrogen to the soil. The following year another crop, such as potatoes or cabbages, will grow better without additional fertiliser.

These processes often use energy from fossil fuels, which are non-renewable resources. There is concern that, if we do not find other ways of making these fertilisers, global food production may decrease as fossil fuels reserves are used up.

Developing investigative skills

Students were given some young wheat plants and some young legume plants growing in separate pots. The plants had been grown from seed. The seed for the legume plant had been inoculated with nitrogen-fixing bacteria so that it developed root nodules containing the bacteria. The students were also given two nutrient solutions for watering the plants: one that contained all nutrients and one that contained all nutrients except nitrogen.

Using and organising techniques, apparatus and materials

❶ Write a plan, using this equipment, to investigate the effect of nitrogen on the growth of legume and non-legume plants. Make clear how your investigation is designed to produce reliable results.

❷ Write a suitable prediction for your investigation.

Observing, measuring and recording

The table shows the results of the investigation carried out by the students.

Plants	Wheat		Legume	
Nutrient solution used for watering	All nutrients	Without nitrogen	All nutrients	Without nitrogen
Height at start (cm)	5.2	4.8	3.6	4.1
Height at end (cm)	20.6	13.6	18.1	19.3

❸ For each plant, calculate the percentage increase in height from the start to the end of the experiment.

❹ Describe the differences between the growth of each plant.

❺ Explain the differences in growth of the wheat plants.

❻ Explain the difference in results for the legume plants.

❼ Draw a conclusion for this experiment.

Handling experimental observations and data

❽ Identify any weaknesses in this method.

❾ Explain how the method should be improved so that a more reliable conclusion can be produced.

End of topic checklist

Key terms

carbon cycle, combustion, deforestation, denitrifying bacteria, fossil fuel, nitrifying bacteria, nitrogen cycle, nitrogen-fixing bacteria, water cycle

During your study of this topic you should have learned:

○ How to describe the roles of evaporation, transpiration, condensation and precipitation in the water cycle.

○ How to describe the roles of photosynthesis, respiration, decomposition and combustion in the carbon cycle.

○ **EXTENDED** How to describe the nitrogen cycle in terms of:

- the role of microorganisms in providing usable nitrogen-containing substances by decomposition and by nitrogen fixation in roots
- the absorption of these substances by plants and their conversion to protein
- followed by passage through food chains, death, decay
- nitrification and denitrification and the return of nitrogen to the soil or the atmosphere.

○ **EXTENDED** How to discuss the effects of combustion of fossil fuels and the cutting down of forests on the oxygen and carbon dioxide concentrations in the atmosphere.

End of topic questions

Note: The marks awarded for these questions indicate the level of detail required in the answers. In the examination, the number of marks awarded to questions like these may be different.

1. Imagine one water molecule in the ocean. Write bullet point notes as an outline for a children's story describing the journey of the water molecule through the water cycle. Include the processes of evaporation, condensation, transpiration and precipitation in your notes. **(7 marks)**

2. The graph shows the change in carbon dioxide concentration above a forest over two days, and the light intensity just above the top of the trees.

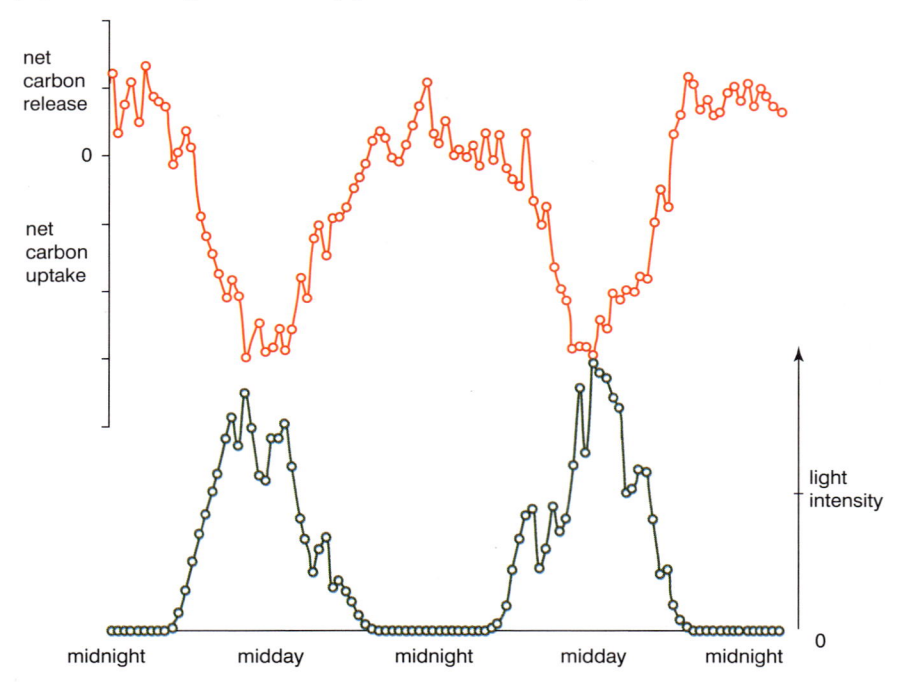

    a) Explain the changes in light intensity shown in the graph. **(2 marks)**

    b) Explain the changes in carbon dioxide concentration shown in the graph. (Remember there are more organisms than just the trees in the forest.) **(4 marks)**

End of topic questions continued

3. a) Explain why waterlogged soils, such as swamps and bogs, usually have very low concentrations of nitrates. **(3 marks)**

b) Sundew plants live in waterlogged soils of swamps and bogs. They have sticky hairs on some leaves that are special adaptations for catching insects. Once the insect is trapped, the leaf rolls up and enzymes are secreted to digest the animal. The digested liquid is absorbed by the plant. Suggest why these adaptations are important to the sundew. **(3 marks)**

4. In a tropical forest, the layer of dead leaves (called the leaf litter) on the forest floor is usually very thin at all times of the year. In temperate woodlands (where there are seasons of summer and winter), many trees drop their leaves in the autumn and grow new ones in the spring.

a) Tropical trees drop a few leaves at a time at any time of year. What happens to the leaves on the ground? Explain your answer as fully as possible. **(2 marks)**

b) The leaf litter in a temperate woodland is deep all through winter when it may be cold enough for snow, until it gets warm again in spring. Then the leaf litter disappears. Explain these observations as fully as you can. **(3 marks)**

5. Explain the importance of water cycle for life. **(7 marks)**

6. 'Without bacteria in the soil, there would be no plants and no animals.' Explain this statement. **(5 marks)**

Population size

INTRODUCTION

Norwegian lemmings have a reputation for committing mass suicide by jumping off cliffs in large numbers. Although this is not true, large numbers of lemmings do die while on migration in some years, but in other years there are so few lemmings that there is no migration. Their population numbers seem to go through a boom-and-bust cycle of about four years. When the population size is very large, many of the animals migrate together to seek places where there are fewer lemmings and more food. Scientists are still unsure why lemming population size changes so rapidly and so dramatically, though it may be due to availability of food.

△ Fig. 4.20 Lemmings on migration.

KNOWLEDGE CHECK

✓ Know that a healthy diet is needed for healthy growth and reproduction.
✓ Know that famine is caused by unequal distribution of food, droughts, flooding and war.
✓ Know that modern technology has resulted in increased food production.

LEARNING OBJECTIVES

✓ Be able to define the term *population* as a group of organisms of one species living in the same area at the same time.
✓ Be able to state the factors that affect the rate of population growth and describe their importance.
✓ Be able to identify the lag phase, exponential (log) phase, stationary phase and death phase in the sigmoid population growth curve for a population growing in an environment with limited resources.
✓ Be able to describe the increase in human population size and its social implications.
✓ Be able to interpret graphs and diagrams of human population growth.
✓ **EXTENDED** Be able to explain the factors that lead to the lag phase, exponential (log) phase and stationary phase in the sigmoid curve of population growth.

POPULATION GROWTH

A **population** is a group of organisms of the same species that live in the same environment at the same time. So we might talk about a population of mahogany trees in a forest, or a population of pill woodlice living in a rotting log.

When we talk about a human population we may mean a local population of people living in a town, or the global human population, which is all the people in the world. This is because humans are able to move around and mix together in a much more general way than most species.

The size of a population depends on various factors:

- the number of new individuals as a result of reproduction, that is, births
- the number of individuals lost from the population, that is, deaths
- for some populations, there may be **emigration**, as individuals move out to other areas, or **immigration**, as individuals join from other areas.

Births and immigration will increase the number of individuals, while deaths and emigration will decrease the number of individuals in the population. So any change in population size will depend on the balance between these:

population size = current population size + [births + immigration] − [deaths + emigration]

- If [births + immigration] is greater than [deaths + emigration], the population will increase in size.

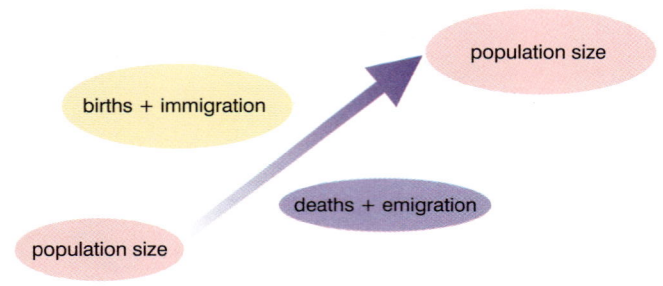

△ Fig. 4.21 Population increase is caused by more people coming 'in' (from births and immigration) than there are people going 'out' (by death or emigration).

- If [deaths + emigration] is greater than [births + immigration], the population will decrease in size.

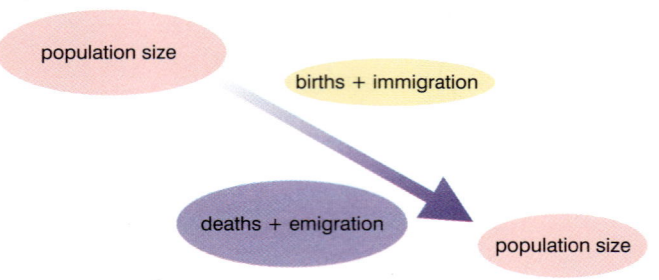

△ Fig. 4.22 Population decrease is caused by more people going 'out' (by death or emigration) than there are people coming 'in'.

- The greater the difference between [births + immigration] and [deaths + emigration], the more rapid the change in population size – that is, the rate of population growth. (A decrease in population size is often described as 'negative growth'.)

Any factor that affects the number of births or deaths, or the number immigrating or emigrating, will affect the rate of population growth.

Food supply

Food supply may affect several of the variables in the population size equation. When there is plenty of food, **birth rate** (the number of births in a particular time) may increase. Survival rate of the young may also increase, so **death rate** (the number of deaths in a particular time) may decrease. Emigration may also decrease because there is food for all, and immigration may increase as individuals arrive from areas where there is less food.

So, when food supply is good, the population will probably grow. And when food supply is poor, the population growth will more likely be negative and the population size will decrease.

Predation

The food supply for a predator is their prey. So, when there are plenty of prey, the population size of the predator will probably increase. This can have a negative effect on the population size of the prey species, particularly when the predator only feeds on one or two prey species. The death rate of the prey species will increase, which may cause a decrease in population size of the prey. That, in turn, means the food supply for the predator decreases, leading to starvation and/or reduced birth rate for the predator.

A well-known example of this cycle of predator and prey population change is shown by the snowshoe hare and lynx that live in northern Canada. The hare is a herbivore, and the lynx feeds mainly on hares. Studies using evidence of numbers of animals killed for their fur between 1845 and 1935 show regular cycles.

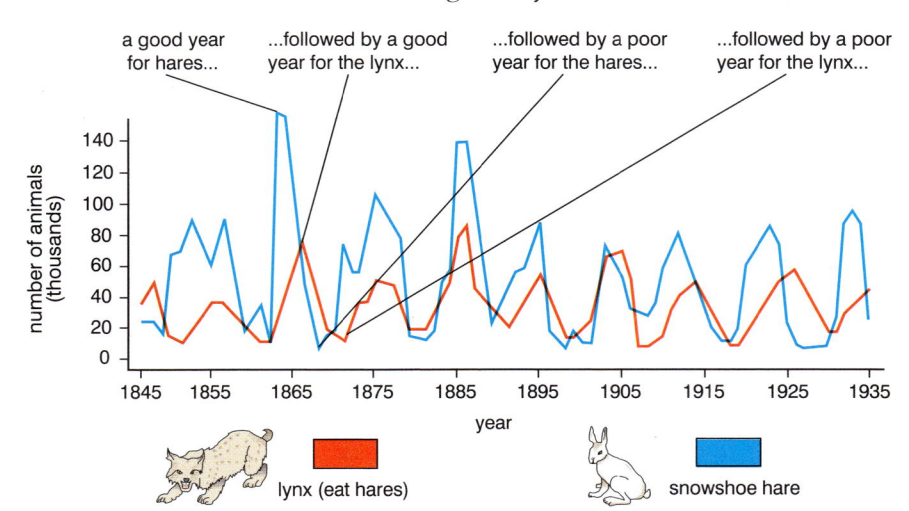

◁ Fig. 4.23 Population of snowshoe hares and lynxes.

These cycles were interpreted as the predator population size controlling the prey population size through predation, and in return prey population size controlling the predator population size as a result of food supply. Further research has suggested that the population size of the hares is controlled more by the availability of their food, than by the number of lynxes. However, the sudden immigration of a large number of predators could cause a rapid decrease in population size of a prey species.

> **REMEMBER**
>
> A predator–prey cycle like the one shown by hares and lynxes is extremely unusual in nature, because most predators feed on more than one or a few species of prey. If predator numbers are high, when one prey population decreases (and the prey became more difficult to find), the predator usually hunts a more common prey. So the change in prey population size is rarely as large as shown by the snowshoe hares.

Disease

Disease can increase the death rate. It can also reduce the birth rate by affecting mothers. Disease is more likely to be common when food supply is limited, because organisms that do not get enough to eat are not as healthy and so less able to fight off infection. The combination of limited food and disease can cause a decrease in population size and growth.

Infectious diseases need to be transmitted from one individual to another, so they will have a greater impact on death rate when the population is already large. They will have a more limited impact on a small population where the individuals are fairly widely spread in the area.

QUESTIONS

1. Define the term *population*.
2. Name two variables that increase population size and two variables that decrease population size.
3. Give an example of an environmental factor that can result in an increase in population growth, and explain why it has this effect.
4. Give examples of two environmental factors that can decrease population growth and explain why they have this effect.

SIGMOID POPULATION GROWTH CURVE

If you measure the growth of a population of microorganisms growing in a fermenter, you will get a curve similar to the one shown. Remember that the conditions within the fermenter are maintained to provide an ideal environment so that the microorganisms grow as rapidly as possible.

The shape of this curve gives it its name, a **sigmoid population growth curve**. This curve has four distinct phases, as shown on the graph.

- The level phase at the start (A) is known as the **lag phase**.
- The stage when population size is rapidly increasing (B), which is known as the **exponential phase** or **log phase**.
- This is followed by a stage where population size does not change (C), which is called the **stationary phase**.
- The final stage is where the population size decreases (D), which is known as the **death phase**.

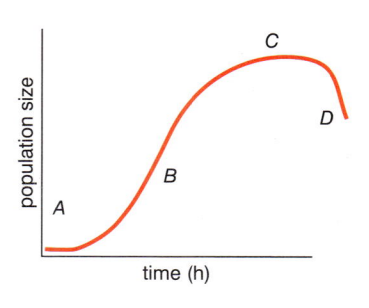

△ Fig. 4.24 A sigmoid population growth curve has four distinct phases, as described in the text.

EXTENDED

The change from one phase to the next in a sigmoid growth curve is a result of changing conditions within the environment in which the organisms are growing, caused by the organisms themselves.

- In the lag phase, the organisms are adapting to the environment, for example by making new enzymes to suit the nutrients in the food supply, before they can grow sufficiently to reproduce. Also there are so few individuals at the start that it takes a while for numbers to increase in a way that is measurable.
- In the exponential or log phase, the growth rate is **exponential**, meaning that the increase at any point is dependent on the number just beforehand, for example where population doubles over a particular interval. This is when the food supply is abundant, so birth rate is rapid and death rate small. The main factor limiting population growth at this point is the number of new individuals that can be produced, that is, the birth rate.
- In the stationary phase, population growth levels out. This occurs when some factor in the environment, such as a nutrient, becomes limited because it is not being replenished, increasing the death rate and decreasing the birth rate so that they become equal. Unless more nutrient is added at this point, the stationary phase will continue until the nutrient becomes severely limiting.
- In the death phase, the population decreases, either because food has become very scarce or something produced by the organisms (such as a metabolic waste) increases to a high enough concentration to become toxic. This causes the death rate to increase further and the birth rate to fall, so that death rate exceeds the birth rate.

Organisms in natural environments are unlikely to show a sigmoid population growth curve because they are affected by many other factors, such as changing physical conditions of temperature or light, predators and disease. They may also be able to migrate from the area when conditions get difficult.

END OF EXTENDED

QUESTIONS

1. Name the four phases of a sigmoid growth curve.
2. Give an example of where a population of organisms may have a sigmoid growth curve. Explain your answer.
3. **EXTENDED** Explain why a sigmoid growth curve shows the four phases that you named in Q1.

HUMAN POPULATION SIZE

Today there are around 7 billion people living on Earth. One hundred years ago there were about 1.7 billion people. In one hundred years from now estimates suggest there could be anything from 5.5 to 14 billion people. This variation depends on predictions for birth rate and death rate in different parts of the world.

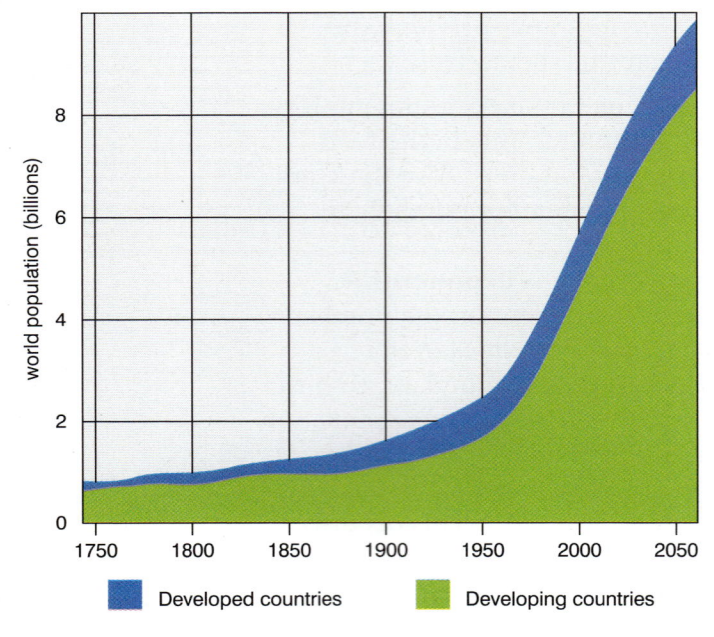

△ Fig. 4.25 The change in global human population size since 1750, and as predicted using average values for birth and death rate to 2050.

From the graph you can see that:

- the total growth in global human population seems to follow the lag and log phases of a sigmoid growth curve, although the lag phase was long
- the rate of growth in developed countries was greatest between about 1900 and 1950, and since then has become more constant
- the rate of growth in developing countries is still increasing rapidly
- the prediction from now to 2050 is an increase for all countries, but a slowly decreasing rate for the less industrialised countries.

If you think about the causes of the rapid increase in human population size, the key factors include:

- a more rapid increase in birth rate due to a greater abundance of food as a result of improved technology
- a decrease in death rate as a result of improved medicine, hygiene and health care.

Some predictions of future population size suggest that growth rate will slow down as birth rate falls in the less industrialised countries due to improved birth control.

REMEMBER

There are many ways of displaying human population growth, each with a different focus or message to tell. Look carefully at each chart or diagram and try to work out its 'message', and how that links to the social implications, both locally and globally, of the growth of the human population.

Another way of looking at population growth is to look at the age structure of a population. This can help show in more detail what is happening in a particular population, and help predict the impact of those changes.

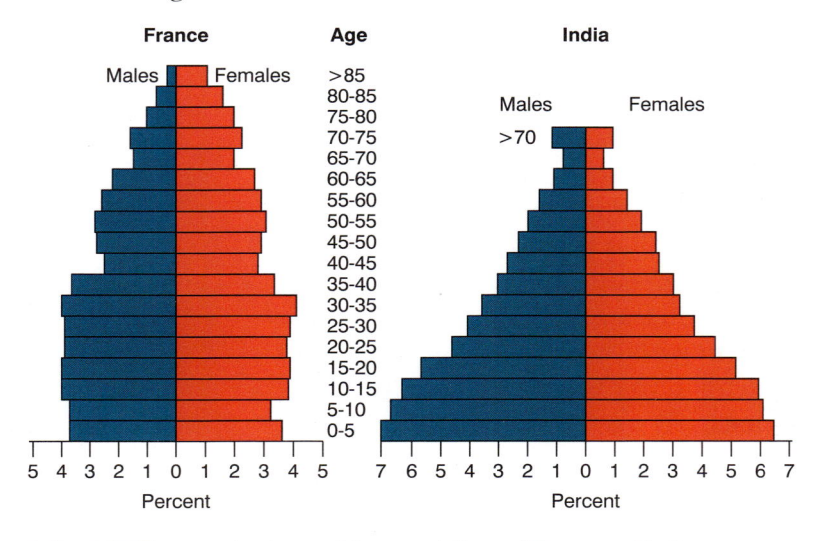

△ Fig. 4.26 The age structures of the populations of France and India.

Fig. 4.26 compares the age structure of the population living in France (an industrialised country) and in India (a less industrialised country). The shapes of these charts are quite typical of the two groups of countries. From the graphs you can see that:

- In France more people live beyond the age of 70 than in India. This is the result of better food, hygiene and medical care generally in France than in India.

- There are almost equal numbers of people in each five-year group up to the age of 40 in France, while there are many more younger people than older people in India. This is because improving food supply and medical care in India is increasing the number of children born, while in France the birth rate is much lower as couples choose to have fewer children and start their families at an older age.

> **SCIENCE IN CONTEXT**
>
> **PLANNING AND POPULATION SIZE**
>
> Population size and growth are important factors that are considered by governments, scientists and others when planning for the future. This includes planning for new roads, buildings, food supplies, school places and developments in health services, as well as predicting the impact of any changes on the environment and other species so as to minimise their effect and protect the world around us from our activities. (See also Conservation of species and habitats.)

Social implications of human population growth

Human population growth has implications on a global, regional and local scale, for the environment and for the way we will be able to live with each other (the **social implications**) in the future.

On a global scale, the main concerns are about being able to supply enough food for everyone, distributed so that nobody suffers from starvation or malnutrition.

On a country-wide or regional scale, where there is a large proportion of younger people, one concern is providing sufficient jobs and homes for them when they are old enough, as well as necessities of modern life such as transport or internet access. In countries where the birth rate has already fallen, and people live till they are much older than the normal working age, there is a concern with the country having enough money to support older people who no longer work, including the additional health care that they need.

On a local scale, many people are moving from farmlands to cities, where there may be a better chance of work, food and health care. If the population of a city increases too quickly, this can lead to large areas of slum development, where there is poor hygiene, poor quality buildings and little space. This increases the risk of rapidly spreading infectious diseases, as well as general unrest in people that can lead to violence.

◁ Fig. 4.27 A favela in the Morumbi neighbourhood in the city of Sao Paulo, Brazil. The low-rise slum buildings contrast against the expensive high-rise buildings.

QUESTIONS

1. Describe the rate of growth of the global human population over the past 150 years.

2. Explain why there are different predictions of global human population growth over the next 100 years.

3. Give two reasons why human population growth rate differs in different countries.

4. Give two examples of social implications of the human population growth in cities (include one positive and one negative implication in your answer).

End of topic checklist

Key terms

birth rate, death phase, death rate, emigration, exponential, immigration, lag phase, log phase, population, sigmoid population growth curve, social implications, stationary phase

During your study of this topic you should have learned:

○ The definition of the term *population* as a group of organisms of one species living in the same area at the same time.

○ About the factors that affect the rate of population growth and their importance:
- food supply
- predation
- disease.

○ How to identify the lag phase, exponential (log) phase, stationary phase and death phase in the sigmoid population growth curve for a population growing in an environment with limited resources.

○ **EXTENDED** How to explain the factors that lead to the lag phase, exponential (log) phase and stationary phase in the sigmoid curve of population growth.

○ How to describe the increase in human population size and its social implications.

○ How to interpret graphs and diagrams of human population growth.

End of topic questions

Note: The marks awarded for these questions indicate the level of detail required in the answers. In the examination, the number of marks awarded to questions like these may be different.

1. A population of mice live in field. Predict the change in population growth of the mice as a result of the following factors. Explain each of your answers.

    a) The crop in the field ripens, producing a lot of grain (a favourite food of the mice). (4 marks)

    b) The local fox families produce many offspring. (Foxes feed their young on mice.) (2 marks)

2. As a result of the threat of serious infectious diseases such as swine flu, scientists have modelled how these diseases are transmitted in human populations. Explain why these diseases could have a rapid effect on human population. (2 marks)

3. The graph shows global human population size since 1 CE, as predicted up to 2200 CE.

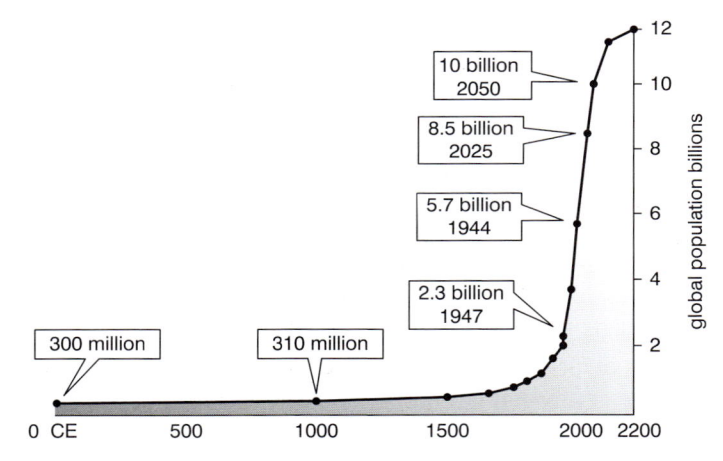

    [Graph showing global population in billions from 0 CE to 2200, with labels: 300 million, 310 million, 2.3 billion 1947, 5.7 billion 1944, 8.5 billion 2025, 10 billion 2050, rising to 12 billion]

    The graph shows that the growth in human population has been exponential up to about the year 2000 CE.

    a) What is meant by *exponential* growth? (1 mark)

    b) Suggest why the global human population has this shape up to about 2000 CE. (1 mark)

    c) Describe the predicted shape of the graph up to the year 2200 CE. (1 mark)

    d) Explain why the prediction gives this shape. (2 marks)

    e) Suggest one global and one local social implication of this increase in human population to 2200 CE. (2 marks)

Human influences on the ecosystem

INTRODUCTION

St Helena is an isolated island in the Atlantic Ocean. It was uninhabited until it was discovered in 1502. The human population of the island slowly increased to over 1000 in the 1700s, and around 4500 people live there now.

The first people to arrive on St Helena found many plants and animals that were unique to the island. As they cleared the dense tropical forest to make space for building, growing crops and keeping herd animals, many of these unique forest species became extinct.

△ Fig. 4.28 Before humans arrived on St Helena, this landscape would have been covered with dense tropical rainforest.

The introduction of animals that didn't naturally live there, such as cats, goats and rats, also had a devastating effect on wildlife. Cats catch and kill small animals and birds, and rats steal and eat eggs from bird nests. Many of the lower areas of St Helena near the sea are now completely bare of vegetation as a result of grazing by goats.

KNOWLEDGE CHECK

✓ Know that the Earth's atmosphere is affected by human activity (such as deforestation and combustion of fuels) and by natural processes (such as volcanoes).
✓ Know that development of the environment can be sustainable or non-sustainable.
✓ Know that organisms that cannot adapt to changing conditions fast enough may go extinct.

LEARNING OBJECTIVES

✓ Be able to outline the effects of humans on ecosystems, including tropical rainforests, oceans, rivers.
✓ Be able to list the undesirable effects of deforestation, including extinction, loss of soil, flooding, carbon dioxide build-up.
✓ Be able to describe the undesirable effects of overuse of fertilisers, including eutrophication of lakes and rivers.
✓ Be able to describe the undesirable effects of pollution, including water pollution, air pollution, global warming, pollution due to pesticides, pollution due to nuclear fall-out.
✓ Be able to describe the need for conservation of species and their habitats and natural resources (water and non-renewable materials including fossil fuels).

- ✓ **EXTENDED** Be able to discuss the effects of non-biodegradable plastics in the environment.
- ✓ **EXTENDED** Be able to discuss the causes and effects on the environment of acid rain, and the measures that might be taken to reduce its incidence.
- ✓ **EXTENDED** Be able to explain how increases in greenhouse gases (carbon dioxide and methane) are thought to cause global warming.
- ✓ **EXTENDED** Be able to explain how limited and non-renewable resources can be recycled, including paper recycling and sewage treatment.

REMEMBER

In your exams, you may be given other examples of polluted or conserved ecosystems than the ones discussed here. You will be expected to apply the principles you have learned here to the examples you are given.

THE EFFECTS OF AGRICULTURE

Agriculture is the cultivation of land for producing food, for humans and their animals. Ever since humans started to grow crops over 7000 years ago, huge areas of natural forests have been cleared to provide the space needed.

Deforestation

Deforestation is the permanent destruction of large areas of forests and woodlands. It usually happens in areas that provide quality wood for furniture, such as the tropical hardwood forests of Malaysia, or to create farming or grazing land (all over the world).

Forests act as a major carbon store because carbon dioxide is taken up from the atmosphere during photosynthesis and used to produce the chemical compounds that make up trees. When forests are cleared, and trees are either burned or left to rot, this carbon is released quickly as carbon dioxide. This rapidly increases the proportion of carbon dioxide compared with oxygen in the air surrounding the forest. The scale of deforestation in some forests, such as in Sumatra and the Amazon Basin, is so great that the amount of carbon dioxide released cannot be brought back into balance by photosynthesis (see Plant nutrition).

Deforestation also has an effect on the water cycle. Trees draw ground water up through their roots and release it into the atmosphere by transpiration. As trees are removed from the forest, the amount of water that can be held in an area decreases. This can affect the amount of water transpired by the plants, and so can affect the amount of rainfall (see Nutrient cycles).

Removing the protective cover of vegetation from the soil can also result in **soil erosion**. This is where the soil is washed away by heavy rain. The top layers of soil are the ones that contain the most nutrients, from the decay of dead vegetation, so soil erosion removes essential nutrients from the land. Soil nutrients are also lost by **leaching**, which is the soaking away of soluble nutrients in soil water because there are few plant roots in the soil to absorb the nutrients and lock them away in plant tissue. This loss of nutrients from the soil is permanent, and makes it very difficult for forest trees to regrow in the area, even if the land is not cultivated.

△ Fig. 4.29 This satellite image of a river estuary in Madagascar shows large amounts of soil in the water (orange). This is a result of deforestation in the surrounding area.

Loss of plant species due to deforestation will result in a loss of animal species in the same community, because of their feeding relationships in the food web. Many tropical rainforests are areas of high **biodiversity**, where many organisms live. They also contain many species found nowhere else because they have evolved together in an energy-rich and relatively unchanging environment. Destruction of tropical rainforests, such as in the Amazon Basin, is causing a high rate of extinction of species.

QUESTIONS

1. Explain these terms:

    a) *deforestation*

    b) *soil erosion*

    c) *leaching*.

2. Explain how deforestation may affect a) the water cycle, b) soil fertility, c) atmospheric carbon dioxide.

3. **a)** Define *biodiversity*.

    b) Explain why many people are concerned about the effect of deforestation on biodiversity.

Overuse of fertilisers

Fertilisers are chemicals that farmers use on fields to add nutrients, such as nitrates, that help the crops to grow better and so produce greater yields. However, if a farmer adds too much fertiliser to a field, the crop plants cannot absorb them. The excess nutrients will soak away in ground water into nearby streams and rivers. Also, if there is heavy rainfall soon after the fertiliser has been spread on a field, the nutrients will dissolve in the rain water and run off the surface of the field into streams and rivers.

Adding nutrients to water is called **eutrophication**. The nutrients in the water have the same effect on the plants in the water as they have on plants that grow on land, and encourage them to grow faster. This can have undesirable effects on other organisms in the water.

- If the plants and algae at the surface grow so much that they block light to plants that grow deeper in the water, the deep-water plants will die.
- This provides more food for decomposers, such as bacteria, which will increase in numbers rapidly.
- The bacteria respire more rapidly in order to make new materials for growth and reproduction. Respiration takes dissolved oxygen from the water, reducing the oxygen concentration of the water.
- Other aquatic (water-living) organisms find it increasingly difficult to get the oxygen they need from the water for respiration.
- If the amount of oxygen dissolved in the water falls too low, many organisms, particularly active animals such as fish, will die.
- The decay of dead organisms in the water provides more nutrients, so more bacteria grow, respire and take more oxygen from the water.
- Eventually, most of the large aquatic plants and animals in the water may die.

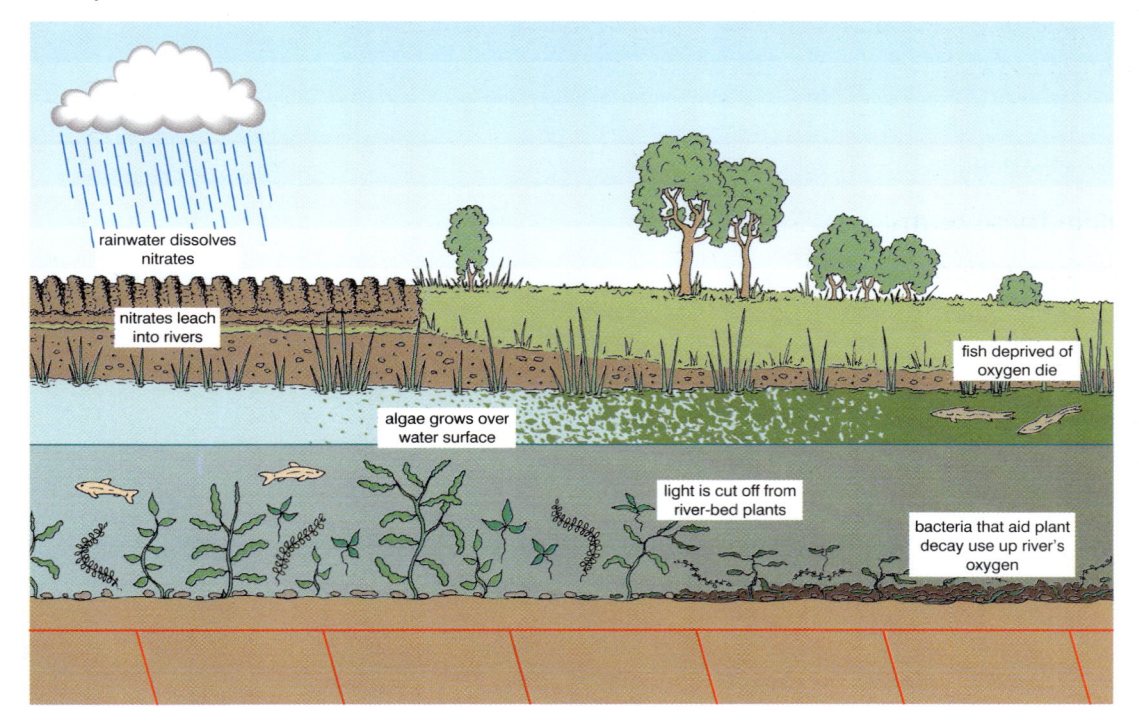

△ Fig. 4.30 Eutrophication can lead to the death of water organisms.

REMEMBER

Eutrophication is often wrongly defined as the pollution of water and death of aquatic organisms. This is incorrect – eutrophication is simply the adding of nutrients. It comes from the Greek word *eutrophia*, meaning healthy or adequate nutrition. Adding nutrients that the ecosystem can use normally may be an advantage, but adding them in excess may lead to the death of aquatic organisms as a result of the depletion of dissolved oxygen in the water. So excess nutrients can cause pollution.

△ Fig. 4.31 Large-scale algal growth can be seen in satellite photos. This algal bloom occurred in the Baltic Sea in 2010 as a result of fertilisers from the surrounding land.

QUESTIONS

1. Explain what we mean by *eutrophication*.
2. Give two reasons why the use of artificial fertiliser on a field could cause eutrophication of a nearby stream.
3. Describe how eutrophication can lead to the death of fish in a stream.

POLLUTION

Pollution is the adding of substances to the environment that cause harm. For example, the over-use of fertilisers on fields can lead to eutrophication of surrounding waterways, causing a reduction in oxygen concentration of the water and the death of organisms. We say that the fertilisers polluted the water.

SCIENCE IN CONTEXT: POLLUTION INDICATORS

When pollution happens as a result of eutrophication, the oxygen content in the water falls. This can change the community of organisms living in the water.

Some organisms are adapted to living in high-oxygen aquatic environments. For example, mayfly larvae are commonly found in clear freshwater streams that contain a high concentration of oxygen in the water. Other organisms are adapted to living in water that contains a low concentration of oxygen. For example, bloodworms (the larval stage of *Chironomus* midges) contain haemoglobin, which helps them absorb oxygen from water that contains only a little oxygen.

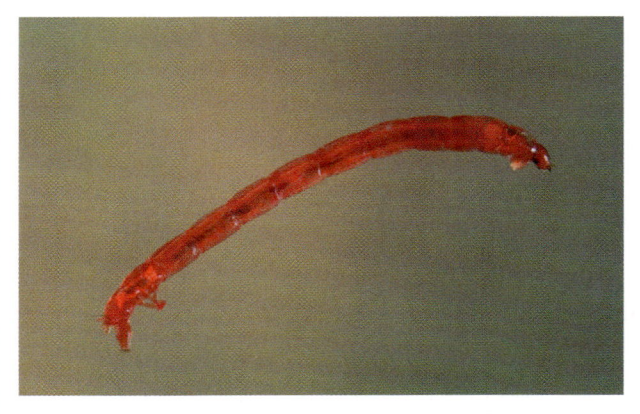

◁ Fig. 4.32 The red colour of this worm is caused by haemoglobin. Like the haemoglobin in our blood, this helps the worm to absorb oxygen from its surroundings.

Mayfly larvae and bloodworms are *bioindicators*, because their presence tells us what the range of concentration of oxygen concentration in the water is likely to be without having to measure it directly. They are useful because they can tell us very quickly how healthy or polluted the water is.

Water pollution

Sewage is human waste, faeces and urine, which we all produce and need to dispose of. Faeces and urine contain high concentrations of many nitrogen-containing substances, making them good sources of nutrients for plants and microorganisms. In fact, farmers often use animal waste as manure to spread on their fields to increase crop production.

In areas where many people live, sewage disposal is a big problem. Many cities have sewage management systems to carry the sewage in pipes to treatment centres where it can be broken down, or to take the sewage far away from the city.

△ Fig. 4.33 There are some fast-growing cities in the world where there are no proper systems for removing and treating sewage.

When untreated sewage is added to water, the nutrients in it dissolve into the water. This leads to eutrophication of the water and, like the addition of artificial fertilisers to water, can increase plant and algal growth, resulting in an increase in bacterial growth, a fall in oxygen concentration in the water, and death of organisms in the water. In addition, human waste contains many bacteria, some of which can cause infection leading to vomiting, diarrhoea, fever and even death.

Water pollution can be caused by other substances. Many industries produce liquid waste that is easiest to dispose of into water systems. In most places there are strict laws about what can and cannot be released into rivers, streams and lakes. Some substances, like sewage, lead to eutrophication. Others are toxic, such as the metals copper, mercury and lead and some organic chemicals. These must be cleaned completely from any waste water that drains into water systems. This is to prevent the toxins being absorbed by plants and animals and passing into food chains where they can cause damage.

SCIENCE IN CONTEXT: THE MINAMATA POLLUTION DISASTER

One of the worst water pollution incidents in modern history occurred in Japan between about 1932 and 1968. A chemical factory released a mercury-containing chemical into Minamata Bay in waste water from the industrial processes. In the bay, shellfish and fish absorbed the mercury. When people ate the shellfish and fish, the mercury was absorbed and stored in their bodies. Over 10 000 people were harmed by the mercury, including damage to muscles, nerves, vision and speech, and over 2000 people died as a direct result.

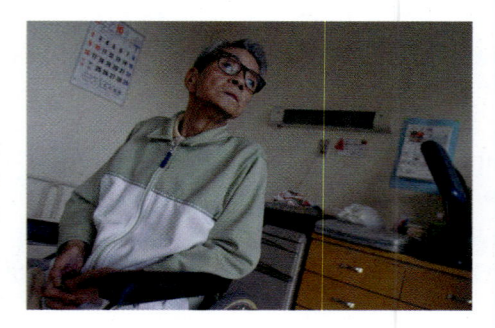

△ Fig. 4.34 Taisuke Mitarai, 69 years old, a patient suffering from Minamata disease, Minamata, Japan.

QUESTIONS

1. Draw a flow diagram to show how sewage can cause eutrophication leading to water pollution.
2. Describe one other danger from sewage in water systems, apart from eutrophication.

AIR POLLUTION

Sulfur dioxide

Since the Industrial Revolution began in northern Europe in the 1700s, humans have burned increasing quantities of fossil fuels (such as coal, oil and natural gas) to provide energy for industrial processes. Burning fossil fuels gives off many gases, including sulfur dioxide (SO_2).

Sulfur dioxide is an acidic gas which is highly soluble in water, forming sulfuric acid. Acid damages cells and delicate tissues directly. If breathed in, the sulfur dioxide can dissolve in the moisture lining the lungs and damage the delicate tissues of the alveoli. This can lead to breathing problems for life.

EXTENDED

Acid rain

Sulfur dioxide in the air can combine with water droplets in clouds to form sulfuric acid. When the water droplets fall as rain, it is more acidic than usual, so we call it **acid rain**. Burning fossil fuels gives off many gases, including **sulfur dioxide** and various **nitrogen oxides**.

Acid rain can cause damage directly to living organisms. It damages the leaves of plants, interfering with photosynthesis and normal growth. Animals with soft skins, such as fish and amphibians (frogs and toads), may have their skins damaged by the acid rain falling in the ponds, lakes and rivers where they live. Single-celled organisms, such as protoctists, are even more likely to be damaged.

Acid rain can also cause damage indirectly. In soil, it can cause some mineral ions to dissolve into the soil water, which is called **leaching**. Some of these ions, such as aluminium ions, are poisonous to some organisms. Other ions may be more easily washed out of the soil, away from plants that need them, so that plant growth is reduced.

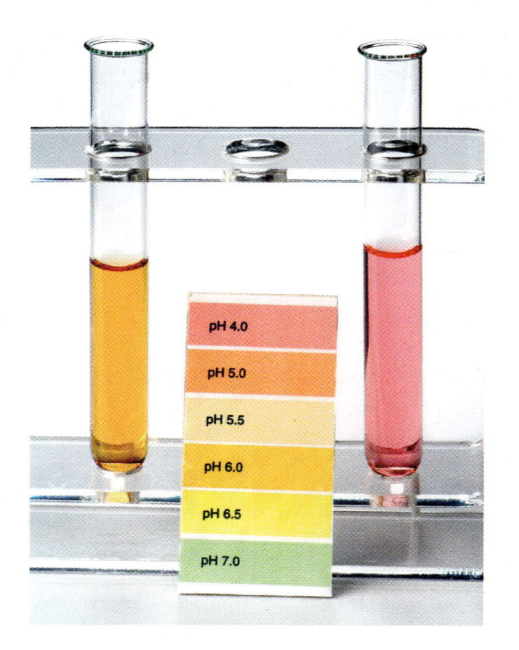

△ Fig. 4.35 The pH of normal rain water (tube on the right) is about 5.6, but acid rain can have a pH of less than 3.0.

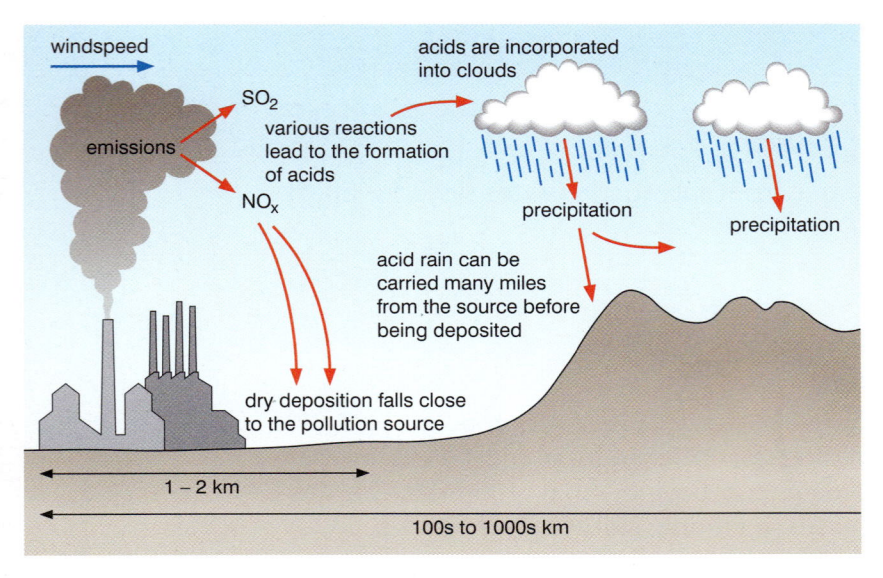

△ Fig. 4.36 Acid rain can be transported many kilometres from where the acidic gases were released into the air.

The interdependency of organisms in an ecosystem means that damaging some organisms in a food web will have an impact on others. Species that depend on those damaged by acid rain will either suffer or move away from the area. However, some species can tolerate acidity better than others, so they will benefit by having more space to live in.

◁ Fig. 4.37 This species of lichen can only grow where there is no pollution in the air. Other species can grow where there the air is polluted.

Developing investigative skills

You can investigate the effect of acid on the germination of seeds by adding acid to the water used to water growing seedlings.

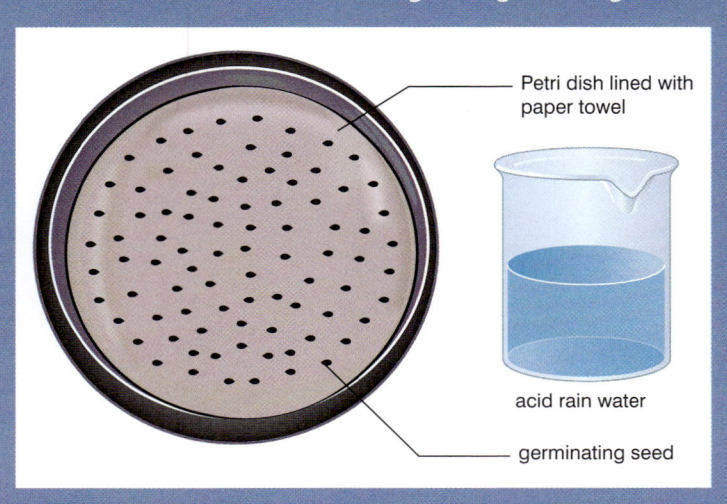

△ Fig. 4.38 Apparatus needed for the investigation into the effect of acid rain on germinating seeds.

Using and organising techniques, apparatus and materials

❶ Write a plan for an investigation on the effect of different acidic pHs on the germination of seeds.

Observing, measuring and recording

❷ Describe any hazards with your plan and how you should protect against them.

Handling experimental observations and data

The graph shows the results from an investigation into the effect of different pHs on the germination of wheat seeds.

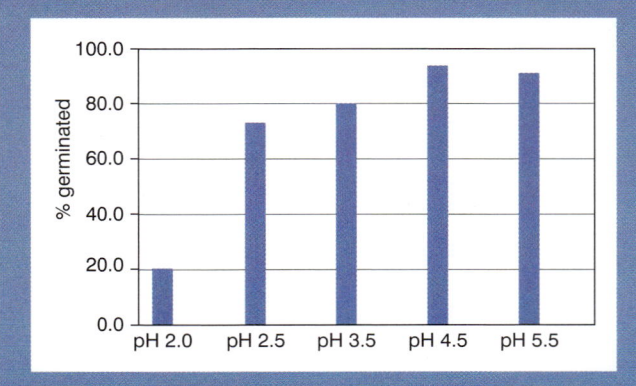

△ Fig. 4.39 Results of investigation.

❸ Use the graph to draw a conclusion about the effect of acid rain on wheat.

Reducing sulfur dioxide emissions

Most of the fossil fuels that we burn are either used in industry, particularly for the generation of electricity, or in vehicle engines. Over the past few decades, efforts have been made in many countries to reduce emissions of sulfur dioxide.

- Sulfur dioxide is removed ('scrubbed') from the gases given off from combustion as they pass up the chimneys of factories and power stations, so that it is not released into the atmosphere.
- Sulfur compounds are removed from petrol and diesel fuels before they are burned in vehicle engines.
- These efforts have resulted in a decrease in sulfur dioxide in the air, but more needs to be done to solve this problem completely.

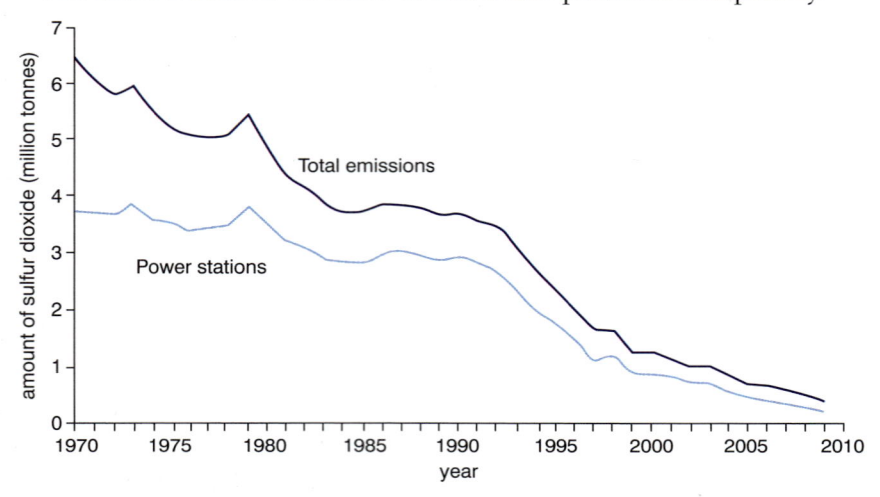

△ Fig. 4.40 Sulfur dioxide emissions from the UK.

END OF EXTENDED

QUESTIONS

1. Explain why sulfur dioxide in the air is a form of pollution.
2. **EXTENDED** a) Explain how human activity contributes to acid rain.

   b) Explain why acid rain is not just a problem for places where there is a lot of industry.
3. **EXTENDED** Describe the different ways in which acid rain can damage organisms and ecosystems.

POLLUTION BY GREENHOUSE GASES

There are many gases in the Earth's atmosphere, but one group plays an important role in keeping the Earth's surface warm. These gases are called the **greenhouse gases** and they include:

- carbon dioxide – produced naturally from respiration of living organisms (see Nutrient cycles)

- methane – produced during the decay of organic material, such as in swamps, and in the digestion of food in the alimentary canal.

Human activities are increasing the proportion of some greenhouse gases in the atmosphere.

- Carbon dioxide is increasing as a result of combustion of fossil fuels (see Air pollution).
- Methane is increasing as a result of the increasing numbers of people and farm animals, and from the release of the gas from artificial wetlands such as rice paddy fields.

Animals that 'chew the cud', such as sheep, goats, cattle and buffaloes, produce large quantities of methane. In countries, such as New Zealand, where many sheep and cattle are kept, over 30% of annual methane emissions may come from these animals.

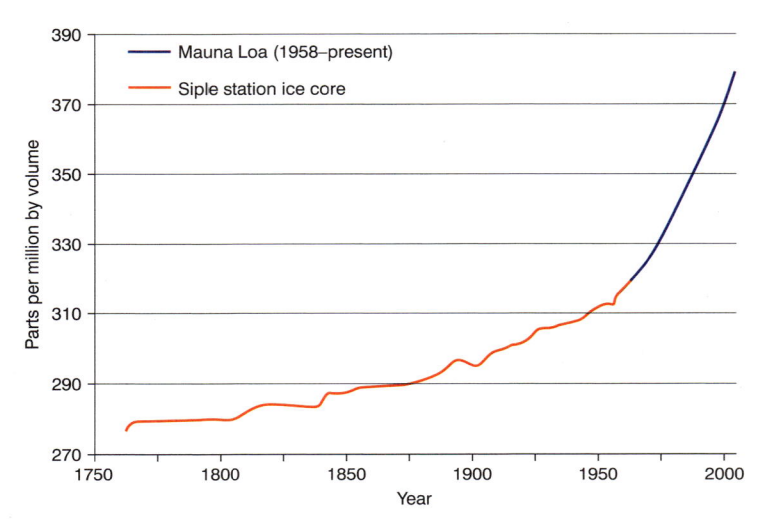

◁ Fig.4.41. Change in the concentration of carbon dioxide in the Earth's atmosphere between 1750 and 2004.

Average global surface temperature has also been rising over the past few centuries. Global temperatures can vary over a wide range due to many natural factors, such as the amount of radiation received from the Sun, which varies due to a predictable but complex cycle. However, many scientists are certain that the recent increases in temperature are caused by increased emissions of greenhouse gases from human activity, resulting in what is commonly called **global warming**.

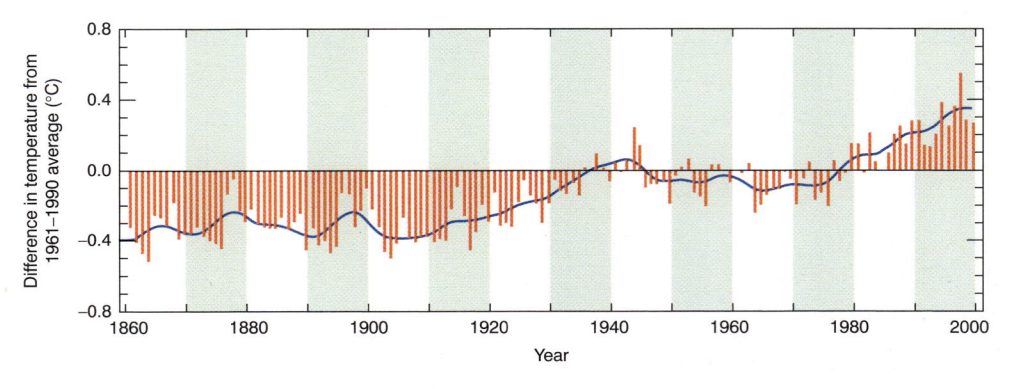

△ Fig. 4.42 Variation of the Earth's surface temperature, calculated as the difference from the average temperature between 1961 and 1990.

Many features of our climate are the result of differences in local surface temperatures in different places, such as the speed of winds or amount of precipitation. Predictions from computer modelling of the effects of global warming suggest that different parts of the world, at different times, may experience an increase in:

- the number and strength of storms
- drought
- flooding as a result of increased rainfall and rising sea levels
- hotter summers and warmer winters
- cooler, wetter summers and colder winters.

These changes will not only affect humans, but whole ecosystems, potentially increasing the rate at which other species become extinct. This will have greater impacts as a result of the interdependency of organisms through food webs in communities.

EXTENDED

The greenhouse effect

Short-wave radiation from the Sun warms the ground which gives off heat as longer-wave radiation. Much of this radiation is stopped from escaping from the Earth by the greenhouse gases and returns to warm the Earth's surface. This is known as the **greenhouse effect**.

The greenhouse effect is responsible for keeping the Earth warmer than it otherwise would be. The greenhouse effect is normal, and important for life on Earth. Without it, it is estimated that the surface of the Earth would be about 33 °C cooler than it is now. All water on the surface of the Earth would be frozen, and very little life could exist in these conditions.

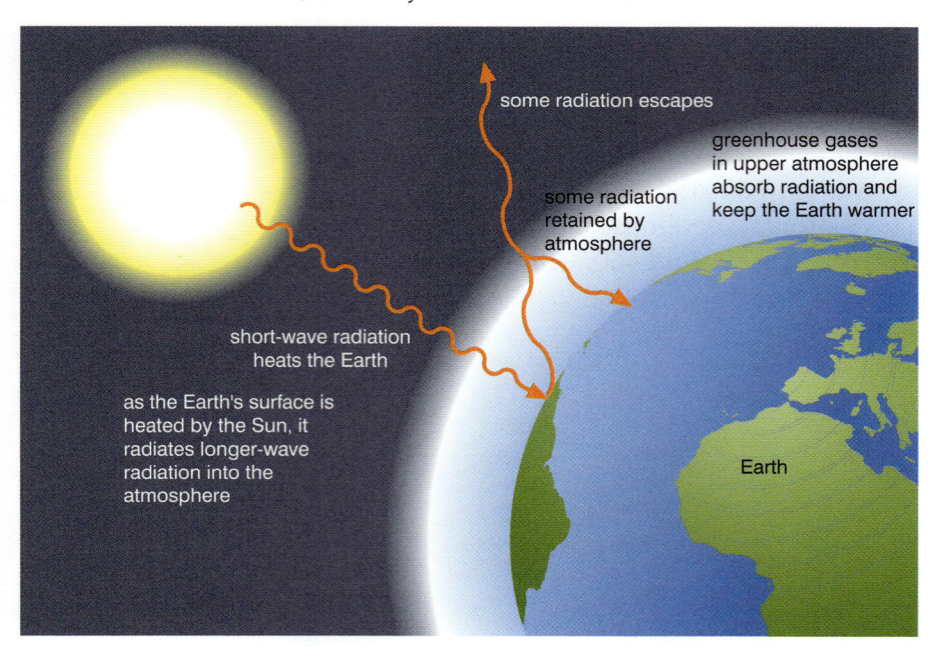

△ Fig. 4.43 The greenhouse effect on Earth.

Additions of greenhouse gases to the atmosphere as a result of human activity are thought to increase the greenhouse effect, producing what some call an **enhanced greenhouse effect**. This is pollution because, in causing global warming, it could damage many species and the environment.

SCIENCE IN CONTEXT — LIFE ON MARS OR VENUS?

△ Fig. 4.44 People once thought that Martians lived on Mars, but scientific exploration suggests there is probably no life on the planet.

The importance of the greenhouse effect can be seen by comparing the conditions on the surfaces of Earth, Venus and Mars. Mars has a relatively thin atmosphere, and although this is composed mainly of carbon dioxide, the effect of this as a greenhouse gas is limited. The average surface temperature on Mars is about −55°C, varying from about 27 °C during the day at the equator to −143 °C at the poles at night. In contrast, Venus has a much denser atmosphere than Earth, which consists mainly of carbon dioxide with clouds of sulfur dioxide. These gases create a very strong greenhouse effect, heating the surface of Venus to an average of around 460 °C. It is not surprising, therefore, that Earth is the only planet in our Solar System where we know life has evolved.

> **REMEMBER**
>
> Be very careful not to confuse the natural greenhouse effect, which is essential for life on Earth, with the enhanced greenhouse effect and global warming, which are the result of the release of additional greenhouse gases from human activity.

END OF EXTENDED

QUESTIONS

1. Give examples of natural causes and human causes of emissions of the following gases:

   a) carbon dioxide

   b) nitrous oxide

   c) methane.

2. Give three examples of consequences of global warming.

3. **EXTENDED** Distinguish between the greenhouse effect and the enhanced greenhouse effect.

POLLUTION DUE TO PESTICIDES

A pesticide is a chemical used to kill pests. Pesticides include:

- **insecticides** used to kill insect pests that damage a crop
- **herbicides** used to kill plants that compete with a crop (that is, a weed).

Using insecticides

Plants do not provide food just for humans. A wide range of other animals and insects will eat crop plants if they are not protected. These animals damage the plants by eating parts of their leaves (such as locusts), or sucking out sap (such as aphids). Damaging the plants reduces their ability to make food and produce new tissue, so they don't grow as well and so don't produce as great a crop yield. We call these animals **pests**, because they are a problem to us.

△ Fig. 4.45 Locusts are pests of many plant species. They damage plants by eating their leaves and other parts. A swarm of locusts may destroy crops over large areas.

Traditionally insect pests were controlled by hand, picking the pests off the plants, or by using domesticated animals, such as chickens, to eat them. Today, in the huge crop fields of commercial farming, pests are mainly controlled by using chemical poisons that kill insects. These insecticides are sprayed onto the plants and eaten by the insects when they feed on the plants. The chemicals kill or harm the insects, which reduces the damage they cause, and so increases crop growth and yield.

Although insecticides are useful, there are problems with using them. Some insecticides kill not only the pest species but also other insect species in the community. This can have two drawbacks.

- Killing lots of different types of insect in an area reduces the amount of food available for any animals that specialise on eating insects, such as insectivorous birds. This affects other organisms in the food web, because of interdependence.
- If the other species killed are predators of the pest then, once the insecticide has been washed away by rain, it is possible for the pest species to return and increase in number even more rapidly, causing even more damage to the crop.

SCIENCE IN CONTEXT

THE PROBLEM WITH DDT

Some insecticides can cause other problems higher up the food chain through bioaccumulation, when the toxin is stored in tissue. A good example of this is DDT, an insecticide used widely in the 1950s and 1960s. DDT is stored in fatty tissues in animals. A small amount in the body may have no noticeable effect on the animal, so predators that eat insects treated with DDT may not be obviously harmed. However, the more insects they eat, the more DDT that is stored in their tissues – the DDT accumulates. Predators that feed on these animals will absorb much higher doses of DDT than are in the environment, and store the DDT in their tissues. At high doses DDT is toxic to larger organisms too. In birds it can also cause eggs to be laid with thinner shells than normal. The shells break more easily, killing the developing chicks inside. In the 1960s it became clear in the US and Europe that numbers of birds of prey, which are top consumers in food webs, were decreasing rapidly as a result of poisoning by DDT and eggshell thinning. DDT was then banned for use in agriculture in the US and Europe.

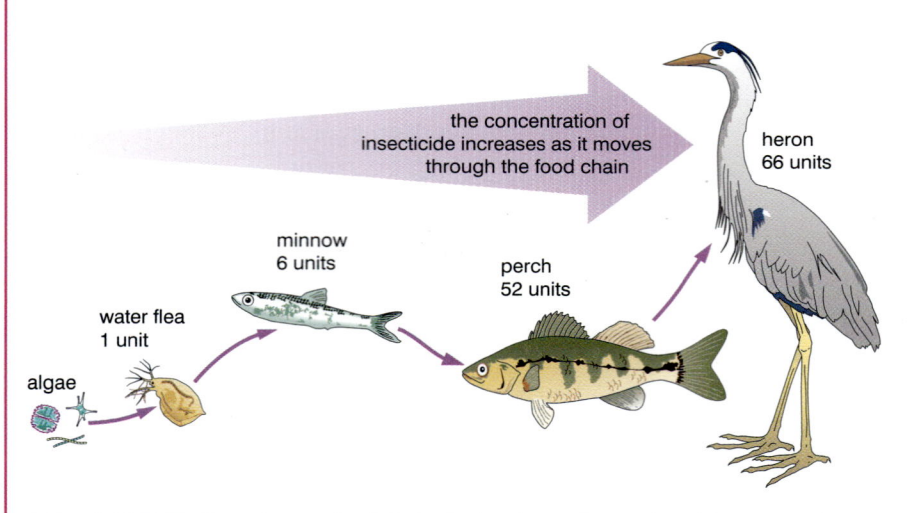

the concentration of insecticide increases as it moves through the food chain

heron 66 units
perch 52 units
minnow 6 units
water flea 1 unit
algae

△ Fig. 4.46 DDT affected many food chains before it was banned. In this one, the figures give the relative concentration of DDT in each trophic level.

A growing problem with insecticides is that pest species are evolving resistance to the chemicals. This is because any individuals that survive the use of an insecticide are more resistant than those that are killed, so the offspring of those surviving individuals carry the genes for resistance. Farmers have responded by using greater amounts of insecticide, which only increases its damaging effects of on the environment.

Using herbicides

Herbicides can be used in a field to kill plants that are not the crop. They are usually sprayed onto a field where they selectively kill weeds, leaving the crop plants unaffected. This clears any plants that might compete with the crop plants for nutrients, water and light. So the crop plants can grow more rapidly and produce a greater yield.

The use of herbicides may help increase yield, but they can cause damage to the environment.

- Many weeds are important food plants for a wide range of insect species, so removing all the weeds reduces the food and shelter available for them. Some of the insects are pests of the crop, so removing the weeds might be helpful. However, some of the insect species may be predators of insect pest species. Clearing the weeds makes it easier for insect pest species to increase in number more quickly. Some farmers now leave a strip of weed plants around their crops to encourage the increase in predator species.
- Some herbicides quickly break down in the soil to simple substances that are no danger to the environment. However, some include highly dangerous substances that can poison soil organisms and anything that eats them. These herbicides are now banned in most countries, but as they don't break down easily, some areas still contain toxic concentrations in the soil.

QUESTIONS

1. Define the term *pesticide* in your own words.
2. Describe, with examples, the advantages of using pesticides on crops.
3. Give one example of a problem caused by using the following on a crop:

   a) an insecticide

   b) a herbicide.

POLLUTION FROM NUCLEAR FALL-OUT

Nuclear fall-out contains radioactive particles that get into the environment from accidental leakage of radioactive materials, such as from a nuclear power station or processing plant, or as a result of an explosion involving nuclear material.

If particles are thrown high into the atmosphere by an explosion, they may be transported by winds over long distances before they fall to the ground. Larger particles will fall out more quickly, and so nearer to the site of the explosion, than smaller particles which may travel hundreds of miles.

RADIATION FROM CHERNOBYL

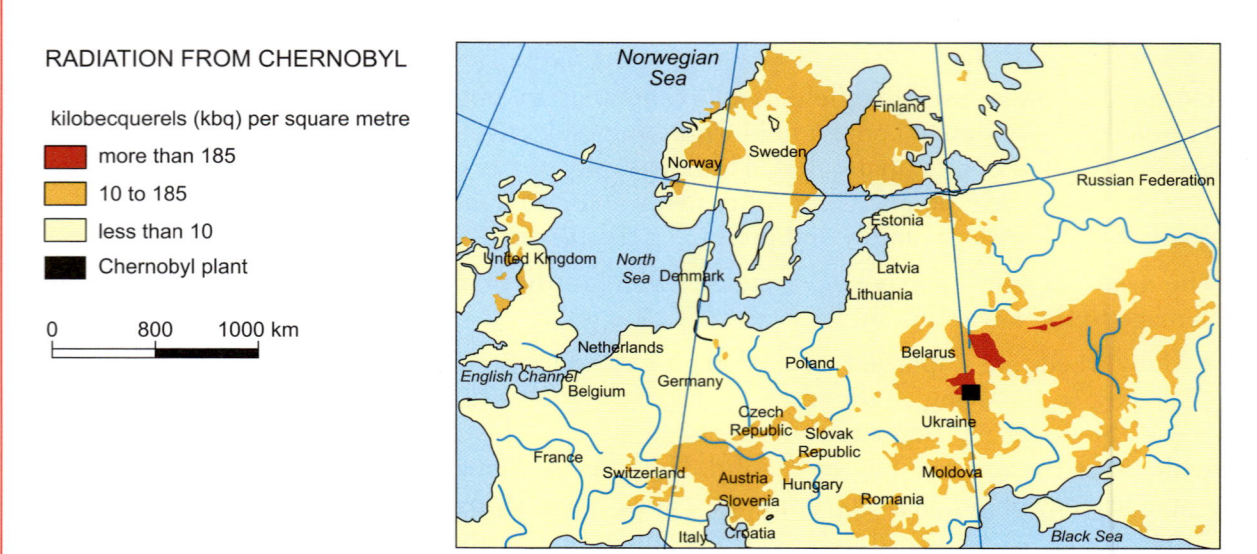

kilobecquerels (kbq) per square metre
- more than 185
- 10 to 185
- less than 10
- Chernobyl plant

△ Fig. 4.47 Nuclear fall-out from the explosion at the Chernobyl nuclear reactor in 1986.

Many kinds of radioactive particles may be released in an explosion like the one at the Chernobyl reactor in the former Soviet Union in 1986. Some particles quickly decay and become non-radioactive, so are unlikely to cause harm. Others keep their radioactivity for longer and may cause damage if they touch living material, such as by burning tissue or causing cancers. Radioactive particles of elements that are normally stored in tissue, such as iodine, cause greater problems because they can be passed along the food chain in animal tissue. This can result in increasing amounts of radioactivity in each trophic level of the food chain, leading to a greater risk of damage in animals higher up the food chain. The most notable effect on human health after the Chernobyl explosion was an increase in the proportion of children and young people with thyroid cancers since 1986.

Since the Chernobyl explosion, safety mechanisms and procedures on nuclear reactors have been greatly improved. The Japanese tsunami in 2011 damaged three working reactors at Fukushima. Automatic shutdown systems closed off the reactors, but the reactors then overheated because the cooling systems were damaged. Some radiation was released into the environment, but it is estimated that this was only about one-tenth of the amount that was released from Chernobyl. The type of radiation released was also less damaging than from Chernobyl. A few workers at the reactors were killed by the tsunami, but none from radiation. The estimated number of cancer cases that will be caused by the disaster is only between 0 and 100. As a result of Fukushima, many countries have reviewed their procedures for controlling reactors after a natural disaster.

QUESTIONS

1. Explain what is meant by *nuclear fall-out*.
2. Give two examples of causes of nuclear fall-out.
3. Explain the spread of nuclear fall-out from the Chernobyl explosion shown in the map above.
4. Describe one danger from nuclear fall-out.

EXTENDED

Pollution by plastics

Plastics are used to make a huge number of products these days. Many of these plastics are **non-biodegradable**, which means that they cannot be decayed by the action of decomposers such as fungi and bacteria. If left in the environment, they may remain there unchanged for tens, even hundreds of years.

One way of dealing with plastic refuse has been to place it in landfill tips. The problem is that, once covered over, that land can be only be used for a few purposes. Due to the risk of poisonous chemicals leaking, it cannot be used for growing crops or grass for herd animals. It takes several decades before it can be used for building on, after the ground has settled and no decay gases are being produced.

Plastic that gets into water systems may end up out at sea, where it collects in huge areas commonly called 'garbage patches' where the oceans circulate. The plastic is a form of pollution because:

- plastic bags may be swallowed by animals such as turtles that mistake them for one of their prey, jellyfish
- plastic nets and ropes may entangle organisms so that they die of starvation
- the plastic may be broken down to release toxins that affect organisms in the area
- the plastic may break down into small particles, called nurdles, that are swallowed by animals that mistake them for food.

Research has shown that at least 267 species worldwide have been affected by plastic marine debris.

◁ Fig. 4.48 Rubbish from the sea collecting on shores.

END OF EXTENDED

QUESTIONS

1. **EXTENDED** Describe two problems with the decay of non-biodegradable plastics.

2. **EXTENDED** Explain why dumping non-biodegradable plastics in landfill tips is being restricted.

3. **EXTENDED** Explain why non-biodegradable plastics are causing pollution in oceans.

CONSERVATION OF SPECIES AND HABITATS

Conservation is the protection of species and their habitats, so that they can continue to survive and reproduce successfully. If the number in a population of organisms falls too low, they may have difficulty finding mates and the population will die out. If this happens to all the populations of these organisms, the species will become **extinct**.

The IUCN (International Union for the Conservation of Nature) Red List shows which species are endangered or at risk of extinction, and is used to help conservationists prioritise which species to protect.

△ Fig. 4.49 The populations of all species of tiger in the wild are now so small that people think they will all become extinct in the next decade or so.

Species may be conserved by breeding them in safe places such as zoos (for animals) and botanic gardens (for plants); but if their habitats are not conserved as well, there is little point in trying to return these species to the wild.

A successful example of conservation is the survival of the Hawaiian goose. When the explorer James Cook visited the island of Hawaii in 1778, there was an estimated population of around 25 000. As a result of hunting the geese for food, and the introduction by people of predators such as pigs and cats, by 1952 there were only about 30 birds left. A few breeding pairs were brought back to the UK in the 1950s, where they bred successfully. Since then more birds have been bred in zoos and wildfowl sanctuaries around the world. Some have also been successfully re-introduced to national parks on an island near Hawaii where they are protected.

Conservation may be carried out for a range of reasons:

- the species or habitat has *ecological value*: this means it is important for the local ecosystem or on a larger scale, such as protection of tropical rainforests for their effect on the carbon cycle and water cycle, and for their biodiversity
- the species or habitat has *economic value*: this means it is worth money to us, such as maintaining wild fish stocks that we eat, or protecting plant species that may supply us with useful medicines, or tourists may wish to visit the area
- the species or habitat may have *aesthetic value*: these are species or places that we enjoy because they are beautiful, and in the case of habitats, may give us space to relax.

Some species or habitats may have more than one kind of value, which helps to make them more important to us.

QUESTIONS

1. Explain what is meant by *conservation*.
2. Give one example of the conservation of a species.
3. Explain why habitats need conservation.

CONSERVATION OF NATURAL RESOURCES

We use many resources from the Earth. Some, such as food, are renewable; others, such as the fossil fuels of coal, oil and gas, are non-renewable. These are the resources that cause most concern since, once we have used them all, we will not be able to produce any more.

Some resources that may seem unlimited, such as fresh water and wood, are limited by the rate at which they can be produced. So this means they also need conserving and managing carefully.

Water

Around two-thirds of the Earth is covered in water, but most of this is salt (sea) water. If we drink this, or use it on our crops or give it to our animals, it will cause problems because the concentration of solutes in the salt water is greater than in living cells. Drinking salt water can cause living cells to lose water by osmosis, and so causes dehydration.

Living organisms need fresh water. Fresh water is not equally distributed across the Earth's surface. Some parts of the world may receive metres of rain each year, other parts may only receive a few millimetres of precipitation in a decade. People need water not only to drink, to water crops and for their animals, but also for cooking and washing. It is not surprising that many people live close to a source of fresh water, such as a river or lake.

△ Fig. 4.50 A drought in Somaliland, Africa.

As the human population grows, the need for fresh water increases. This is partly because there are more people using the same fresh water sources, but also because people spread out into surrounding areas where fresh water may not be as plentiful.

Polluting water with sewage, chemicals and fertilisers reduces the amount of fresh water available to us, as well as to other organisms.

> **SCIENCE IN CONTEXT**
>
> ## RUNNING OUT OF FRESH WATER
>
> Early in the 21st century it was calculated that humans use about 60% of the available fresh water on the Earth every year. That was when the global human population was around 6 billion. At the time the projected human population for 2050 was 10 billion, which raises concerns about the availability of fresh water not only for all the humans and our needs, but also for all the wild organisms that share the water with us.

EXTENDED

Recycling water

In many places, sewage (human waste) is disposed of by washing it away into pipes using water. The pipes carry the liquid to sewage treatment plants where the organic waste can be removed. The water is cleaned so that it can be returned to the water system without causing eutrophication.

The treatment includes several stages.
- First stage: The waste water (including sewage) is screened to remove inorganic material such as sand, gravel and grit.
- Second stage: The waste water is passed through treatment beds where microorganisms feed on the organic material and break it down to smaller molecules, such as carbon dioxide, water and mineral ions.
- Third stage: The water is treated with chemicals to remove dissolved minerals that are toxic. Chlorine is added to kill microorganisms.
- The water may receive extra processing to make it pure enough for use as drinking water, but at this point is safe to return to water systems.

Δ Fig. 4.51 Sewage treatment. The solid organic material that is collected from the water may be used as a fertiliser on fields, or in a biodigester where it is broken down by microorganisms to release methane. The methane can be used as a fuel gas.

END OF EXTENDED

Fossil fuels

The fossil fuels coal, oil and gas, supply most of our energy needs. They are burned in power stations to generate electricity and used to power vehicles for transport. In addition, they are an important source of chemicals for the plastics and other manufacturing industries.

Fossil fuels are non-renewable resources: we use them at a far greater rate than it takes to make them. It is difficult to predict how much longer the fossil fuel resources we know about will last, because it depends on how rapidly we use them.

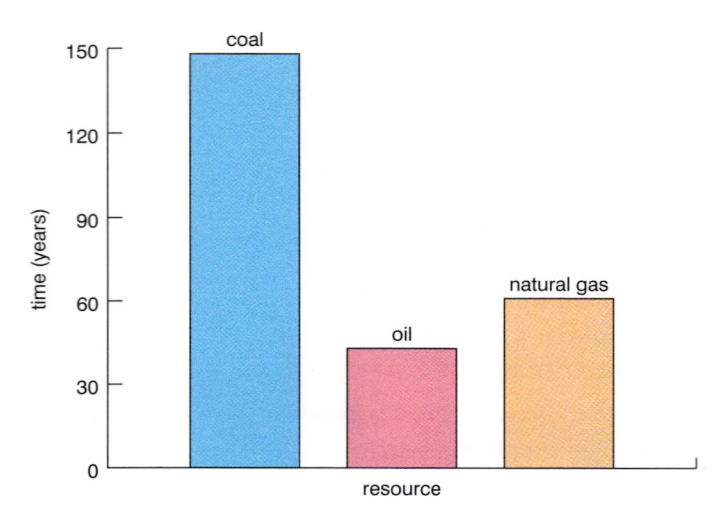

△ Fig.4.52 At the current global rates of use, fossil fuels might only last this long. However, there may be other sources of these fuels that we have not yet discovered.

We use these resources not only as fuels, but as the raw materials for other processes. For example, almost all plastics start with oil as a raw material. Even if we reduce our rate of use, fossil fuels will eventually run out, so we need other sources of energy and raw materials to replace them.

EXTENDED

Paper recycling

Paper is made by pulping the wood from whole trees. This requires energy for preparing the wood, by stripping off the bark, and then crushing it. The fibres are then soaked with chemicals to help separate the wood fibres, and to bleach the wood to make the paper white. The wood pulp that is left is used to make paper.

Used paper can also be recycled to produce more paper. Printed paper needs to have the ink removed first. Then the paper is mashed up in water to make pulp again, and the pulp is used to make more paper. Recycled paper is not as strong or as bright white as new paper, but it is useful for many purposes.

Although wood is a renewable resource, there is a limit to how quickly it can grow. And wood is used for many purposes other than paper, such as for building. So recycling paper means that more new wood is available for those other purposes, or that the land used for growing trees could be used for growing something else, such as crops, or left as forest to protect habitats for other organisms.

Recycling paper also:
- saves energy, because less energy is needed to produce the pulp from old paper than from trees
- reduces the amount of chemicals used, particularly the bleaching chemicals that can cause water pollution
- reduces the amount of paper dumped in landfill tips.

END OF EXTENDED

QUESTIONS

1. Explain why fresh water resources need to be conserved.
2. Explain why fossil fuels need to be conserved.
3. **EXTENDED** Describe how water is recycled.
4. **EXTENDED** Explain why recycling paper makes economic sense.

End of topic checklist

Key terms

acid rain, deforestation, enhanced greenhouse effect, eutrophication, global warming, greenhouse effect, greenhouse gases, leaching, pollution, sewage, soil erosion

During your study of this topic you should have learned:

○ About the effects of humans on ecosystems, with emphasis on examples of international importance:
- tropical rainforests
- oceans
- important rivers.

○ About the undesirable effects of deforestation, including:
- extinction
- loss of soil
- flooding
- carbon dioxide build-up.

○ About the undesirable effects of overuse of fertilisers, including eutrophication of lakes and rivers.

○ About the undesirable effects of pollution, including:
- water pollution by sewage and chemical waste
- air pollution by sulfur dioxide
- air pollution by greenhouse gases (carbon dioxide and methane) contributing to global warming
- pollution due to pesticides including insecticides and herbicides
- pollution due to nuclear fall-out.
- EXTENDED About the effects of non-biodegradable plastics in the environment.
- EXTENDED About the causes and effects on the environment of acid rain, and the measures that might be taken to reduce its incidence.
- EXTENDED How to explain how increases in greenhouse gases (carbon dioxide and methane) are thought to cause global warming.

○ About the need for conservation of:
- species and their habitats
- natural resources (water and non-renewable materials including fossil fuels).
- EXTENDED How to explain how limited and non-renewable resources can be recycled, including:
- paper recycling
- sewage treatment.

End of topic questions

Note: The marks awarded for these questions indicate the level of detail required in the answers. In the examination, the number of marks awarded to questions like these may be different.

1. **a)** Give two causes of large-scale deforestation. **(1 mark)**

   b) Describe the effects of large-scale deforestation on the organisms in the region. **(3 marks)**

   c) Explain why after deforestation, even if the land is left undisturbed so the vegetation regenerates naturally, plant growth will be much slower than before. **(3 marks)**

2. Rivers and lakes that are used for water supplies may be monitored to make sure the water in them is safe for use.

   One way of monitoring is to measure the amount of oxygen that is used by the water (the biochemical oxygen demand) over a period of 5 days.

   a) Why does the concentration of oxygen decrease in the water? Explain your answer. **(2 marks)**

   b) In this test would polluted water use more oxygen than unpolluted water? Explain your answer. **(2 marks)**

   Another way of monitoring the water is to sample the small organisms that live in it. Some species, such as worms, are better adapted for living in water that has a low oxygen concentration. Other species, such as mayfly larvae, need a high concentration of oxygen in the water.

   c) Which of the two species above would be more common in polluted water? Explain your answer. **(2 marks)**

   d) What does *adapted* mean? **(1 mark)**

   e) Why might sampling the organisms be a better measure of the long-term health of the water than measuring the oxygen demand of the water? **(2 marks)**

3. In parts of Europe, farmers now use satellite technology to help them see which parts of a field need additional fertiliser, and how much fertiliser they need.

   a) Why do farmers add fertiliser to their fields? **(1 mark)**

   b) What might happen if a farmer adds too much fertiliser to a field? **(1 mark)**

   c) Explain as fully as you can how this could result in water pollution. **(4 marks)**

   d) Explain the advantage of this use of technology. **(2 marks)**

End of topic questions continued

4. a) Explain the meaning of the term *pollution*. **(1 mark)**

 b) Name two human activities that are major sources of sulfur dioxide in the atmosphere. **(2 marks)**

 c) Explain how sulfur dioxide can lead to acid rain. **(2 marks)**

 d) Sketch a diagram to show how sources of sulfur dioxide in one region could result in acid rain in another region. **(1 mark)**

 e) Explain why sulfur dioxide causes pollution. **(2 marks)**

5. The graph shows the results of surveys of populations of birds on farmland in the UK between 1966 and 2009, compared with the 1966 level (set at 100 for easier comparison). The farmland birds were split into two groups: generalist species that are found on a range of other habitats as well as farmland, and specialists that live and breed almost exclusively on farmland habitats.

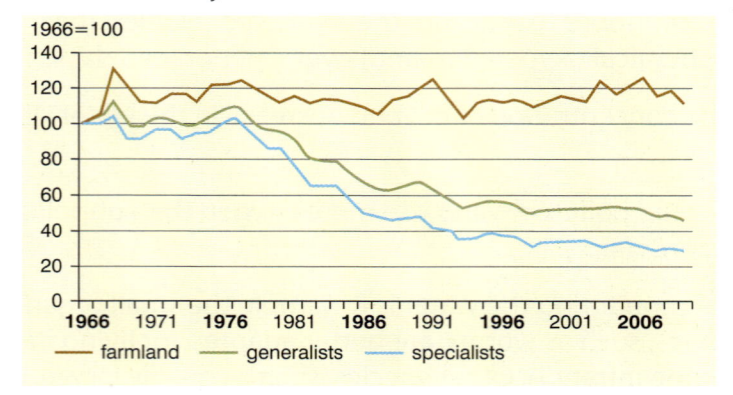

Of the five species of specialist seed eaters, all decreased in number up to 1986, after which one (goldfinches) increased while the rest continued to decrease. Of the three species of insect-eaters, all species decreased in number to about 1996 after which numbers seem to have stabilised.

 a) Describe what the graph shows. **(2 marks)**

 b) Suggest reasons for the difference between changes in population size for generalist and specialist species. **(2 marks)**

 c) Goldfinches are farmland birds that have increased in number since about 1986 possibly because people are feeding more birds in their gardens over winter. Suggest one possible cause for the continuing decrease in other species. **(1 mark)**

 d) In 1988 farmers in the European Union were encouraged to set aside some of their land for wildlife each year. That arrangement stopped in 2008. Explain why there is even greater concern now for specialist farmland bird species. **(2 marks)**

6. Explain as fully as you can why scientists think that human activities are leading to global warming. **(3 marks)**

TEACHER'S COMMENTS

a) i) Correct.

ii) The answers given for carbon dioxide and nitrous oxide are detailed and correct.

For methane, the student has not appreciated that the graph has been drawn to the scale on the right hand axis, which ranges from 0–2000 ppb. This has meant that although the trends have been described, the values of methane concentration are incorrect.

It is important to check scales carefully when reading data from graphs. The answer should therefore be:

'The concentration of methane has shown a very slow, slight upward trend from 0 to around 1750, ranging from 650 ppb to around 750 ppb.'

'Then after a slight dip, a steep increase to around 1925 ppb in 2005.'

Exam-style questions

Note: The questions, sample answers and marks in this section have been written by the authors as a guide only. The marks awarded for these questions indicate the level of detail required in the answers. In the examination, the number of marks awarded to questions like these may be different.

Sample student answer

EXTENDED Question 1

This question is about the enhanced greenhouse effect and global warming.

a) The graph shows the concentration of greenhouse gases in the air from the Year 0 to 2000.

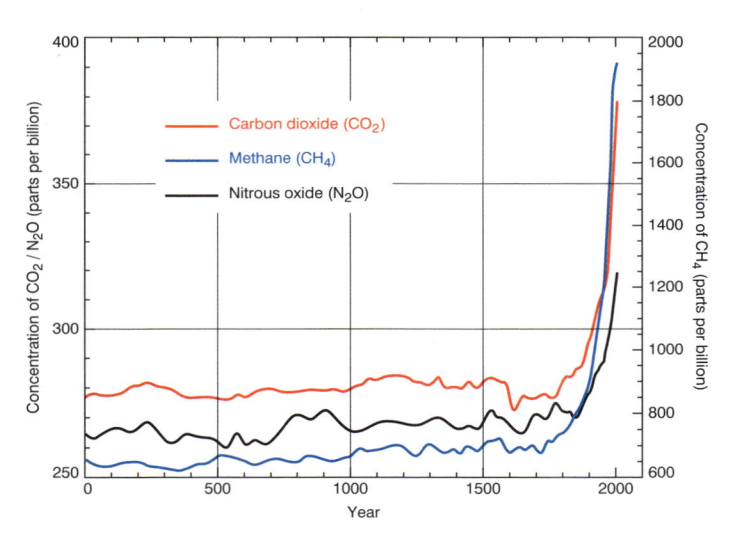

i) Which greenhouse gas was present in the highest concentration in the air in 2000? (1)

Methane ✓ ①

ii) Describe the trends in the changes of the concentration of each greenhouse gas from 0 to 2005. (6)

The concentration of carbon dioxide has been fairly stable at around 280 parts per billion from 0 to 1600. ✓ ①

Then, after a dip, it shows a steep increase to 380 ppb in 2005. ✓ ①

Exam-style questions continued

The concentration of nitrous oxide has shown a little fluctuation from 0 to around 1800, ranging from 265-275 ppb. ✓ ①

But there has been a steep increase to around 320 ppb in 2005. ✓ ①

The concentration of methane has shown a very slow, slight upward trend from 0 to around 1750, ranging from 255 ppb to around 260 ppb. ✗

But then a steep increase to around 390 ppb in 2005. ✗

iii) How has human activity contributed to the change in the concentration of carbon dioxide in the air? (2)

Carbon dioxide production has increased from the burning of fossil fuels in transport, heating and cooling, and in manufacture. ✓ ①

iii) The student has written a good answer for the contribution of the burning of fossil fuels to the increase in carbon dioxide. These all refer to the burning of fossil fuels, however, and the student could have picked up the second mark by referring to deforestation.

b) i) The student has picked up two marks, but for the third mark, has not mentioned the fact that sulfur hexafluoride has the longest lifetime – a greenhouse gas that's around for a shorter time will make less of a contribution to the greenhouse effect.

b) The table gives information on several greenhouse gases.

Gas	Chemical formula	Lifetime (years)	Global Warming Potential*
Carbon dioxide	CO_2	Variable	1
Methane	CH_4	12	21
Nitrous oxide	N_2O	114	310
CFC-11	CCl_3F	45	3 800
CFC-12	CCl_2F_2	100	8 100
Sulfur hexafluoride	SF_6	3 200	23 900

*The **Global Warming Potential (GWP)** is a measure of how much heat a greenhouse gas traps in the atmosphere relative to that trapped by the same mass of carbon dioxide. A GWP is calculated over a time interval. The values in the table are for a 100-year time scale.

From: IPCC/TEAP (2005) *Special Report on Safeguarding the Ozone Layer and the Global Climate System: Issues Related to Hydrofluorocarbons and Perfluorocarbons* [Metz, B., et al. (eds.)]. Cambridge University Press.

i) Which greenhouse gas contributes most to global warming? Explain your answer. (3)

Sulfur hexafluoride ✓ ①

It has the highest GWP. ✓ ①

ii) Explain how greenhouse gases result in the greenhouse effect and global warming. (6)

Shortwave radiation from the Sun passes through the Earth's atmosphere and warms the ground. ✓ ①

The warmed Earth gives off longer wave radiation that is prevented from leaving the earth by greenhouse gases in the atmosphere. ✓ ①

The trapping of the radiation leads to the Earth warming up, which is called the greenhouse effect. ✓ ①

The greenhouse effect is important, because without it, the temperature on the Earth would be 33°C lower – the Earth would be uninhabitable. ✓ ①

But increases in greenhouse gases as a result in human activity is leading to a significant warming of the Earth called global warming. ✓ ①

(Total 18 marks)

13/18

> **ii)** This is a good answer, but the student has not mentioned the 'enhanced greenhouse effect'. The final marking point could have been extended:
>
> 'But increases in greenhouse gases as a result in human activity is leading to the enhanced greenhouse effect.'
>
> 'This is leading to a significant warming of the Earth called global warming.'

Exam-style questions continued

Question 2

The table gives information on how the land area covered by forest has changed from 1990 to 2005.

Country	Area covered by forest/millions of hectares		Area of forest lost from 1990 to 2005/%
	1990	2005	
Bolivia	109.9	58.7	46.6
Brazil	851.5	477.7	
Colombia	113.9	60.7	
French Guiana	9.0	8.1	
Peru	125.5	68.7	
Suriname	16.3	14.8	
Venezuela	91.2	47.7	

Data from http://rainforests.mongabay.com/deforestation_alpha.html

a) **i)** Calculate the area of forest lost for each country, as a percentage of the area in 1990. The first one has been done for you. (6)

  ii) During the time period 1990 to 2005, in which country is there:

   the greatest deforestation? (1)

   the least deforestation? (1)

  iii) Suggest two reasons for deforestation. (2)

b) List the effects of deforestation. (5)

(Total 15 marks)

Question 3

The food web below shows the relationship of some of the organisms on a rocky shore.

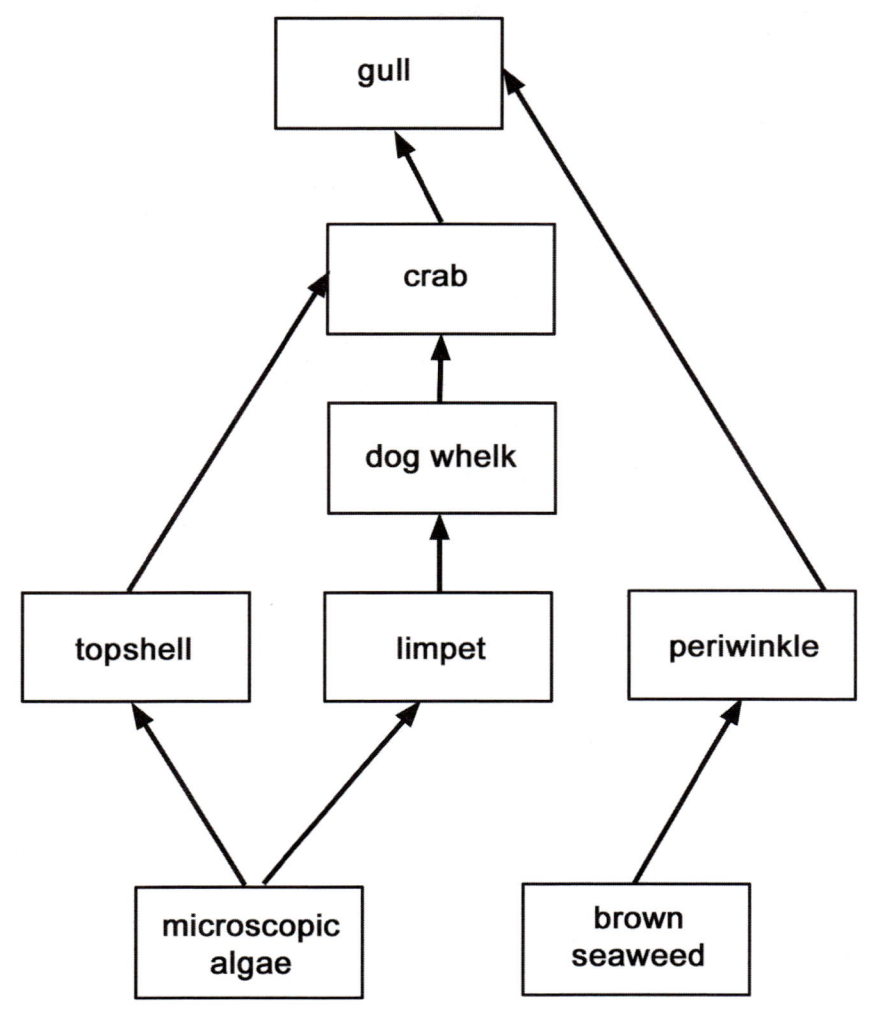

a) In the food web, state which organisms are:

  i) producers. (2)

  ii) primary consumers. (3)

  iii) secondary consumers. (3)

b) State an alternative name for a primary consumer. (1)

c) In an ecosystem, the numbers of crabs is severely reduced.

  i) State **one** reason for the reduction of an organism in an ecosystem. (3)

  ii) Describe the impact on dog whelks, limpets and gulls in the food web. (5)

Exam-style questions continued

d) A student investigated the distribution of a species of brown seaweed and periwinkles down a rocky shore. Her results are shown here.

Distance below high water on seashore/ metres	Distribution of organisms	
	Observed distribution of brown seaweed	Density of periwinkles/ mean number of periwinkles per m²
0	Absent	0
10	Rare	16
20	Occasional	52
30	Abundant	156
40	Abundant	128
50	Occasional	44
60	Rare	12

Suggest two reasons for the distribution of the periwinkles. (2)

(Total 19 marks)

Question 4

Nutrients are cycled in nature.

a) The passage below describes the stages in the carbon cycle.

Use suitable words to complete the sentences in the passage. (9)

_____ from the atmosphere is converted to complex carbon compounds in _____ by the process of _____ . This is often called carbon _____

Plants are then often eaten by _____ , which build up their own complex carbon compounds.

The process of _____ in both plants and animals, returns some of this carbon back to the atmosphere as _____ .

When organisms die, their bodies decay as they are worked on by _____ . Some of the complex carbon compounds are taken into the bodies of these organisms, where some may be converted to carbon dioxide during their _____ .

b) By what other process does carbon dioxide enter the air? (1)

(Total 10 marks)

Question 5

Modern technology is used to increase the yields of crop plants.

a) One method used to increase crop yields is to apply fertilisers. List the effects, in sequence, when nitrates are washed from the fields and pollute water. (4)

b) Pesticides are often applied to crops.

   i) Explain why pesticides are applied to crops. (5)

   ii) Describe two negative impacts of pollution by one type of pesticide. (2)

(Total 11 marks)

Question 6

The graph shows the actual and projected changes in human populations in developed and developing countries from 1750.

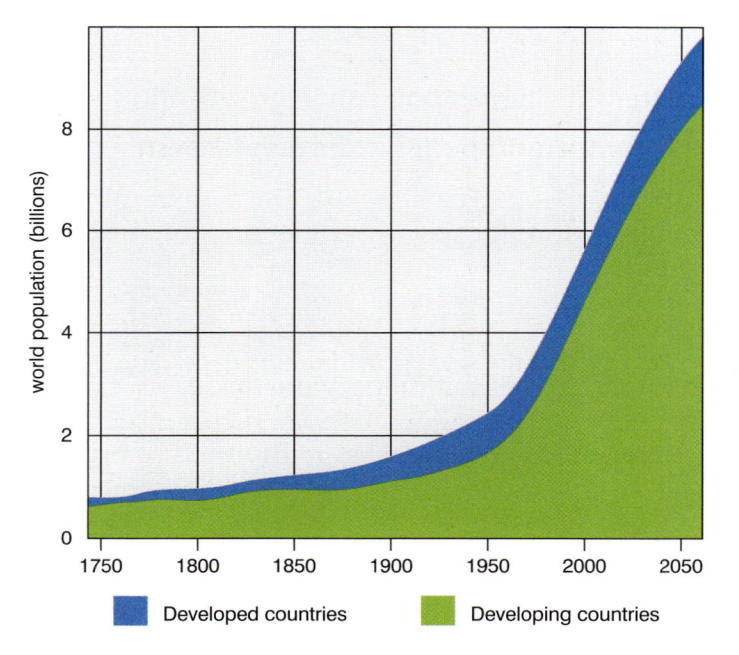

a) Describe the change in population growth from 1750 to 2050. (8)

b) Explain reasons for:

   i) Time periods where the population growth is slow. (2)

   ii) Rapid increases in human population size. (2)

Exam-style questions continued

c) The graph shows human population size and estimated extinctions of organisms.

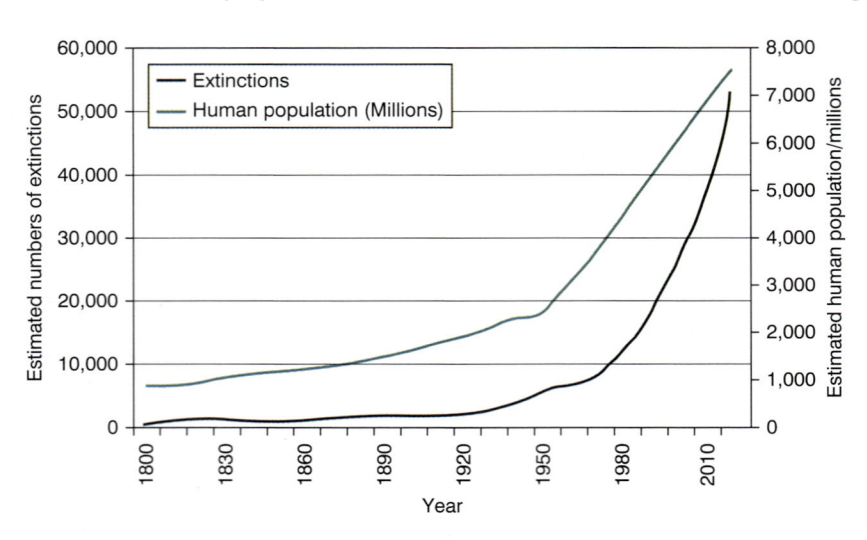

i) Describe the patterns in human population growth and species extinctions. (2)

ii) Explain why the numbers of species extinctions are estimated and not actual numbers. (2)

iii) State **five** possible reasons for species extinction. (5)

(Total 21 marks)

EXTENDED Question 7

In an ecosystem, the following measurements were made.

Organism	Number in ecosystem	Biomass of organisms/g
Oak trees	1	500 000
Aphids	100 000	100
Ladybirds	200	10

a) Draw a food chain to illustrate the feeding relationships of the three organisms. (3)

b) Using a sketch, compare the feeding relationships in the food chain using a pyramid of numbers and a pyramid of biomass. (2)

c) Explain one advantage and one disadvantage of using a pyramid of biomass to illustrate feeding relationships. (2)

d) In observations of the ecosystem, ladybirds were seen to be fed on by spiders, and spiders fed on by blackbirds.

i) Draw a food chain to illustrate these feeding relationships. (1)

ii) Explain fully why food chains longer than this are rare. (5)

(Total 13 marks)

EXTENDED Question 8

The diagram shows the nitrogen cycle. Name the organisms missing from the diagram. (4)

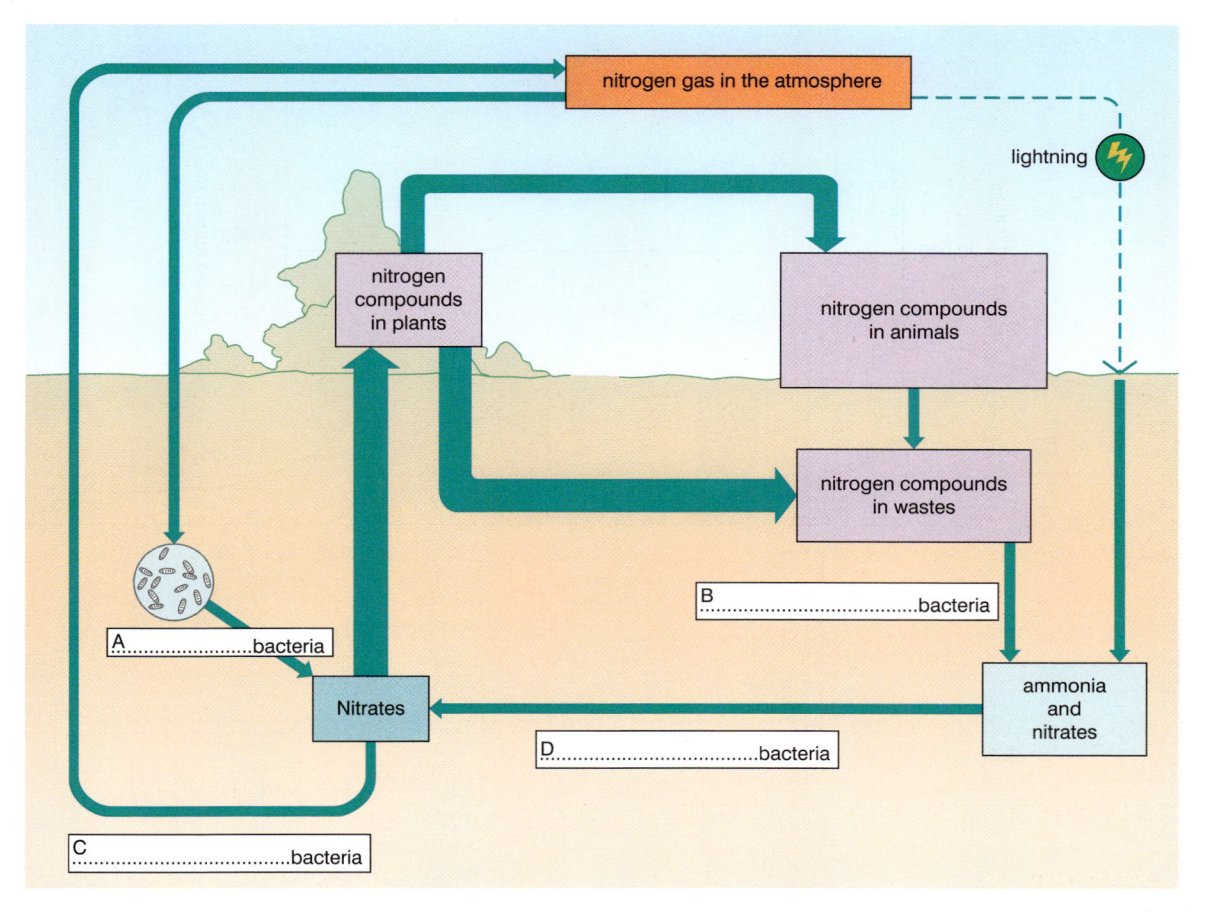

A .... bacteria
B .... bacteria
C .... bacteria
D .... bacteria

(Total 4 marks)

Exam-style questions continued

EXTENDED Question 9

The effect of sewage into a river was monitored over a number of years. The results are shown in the graphs.

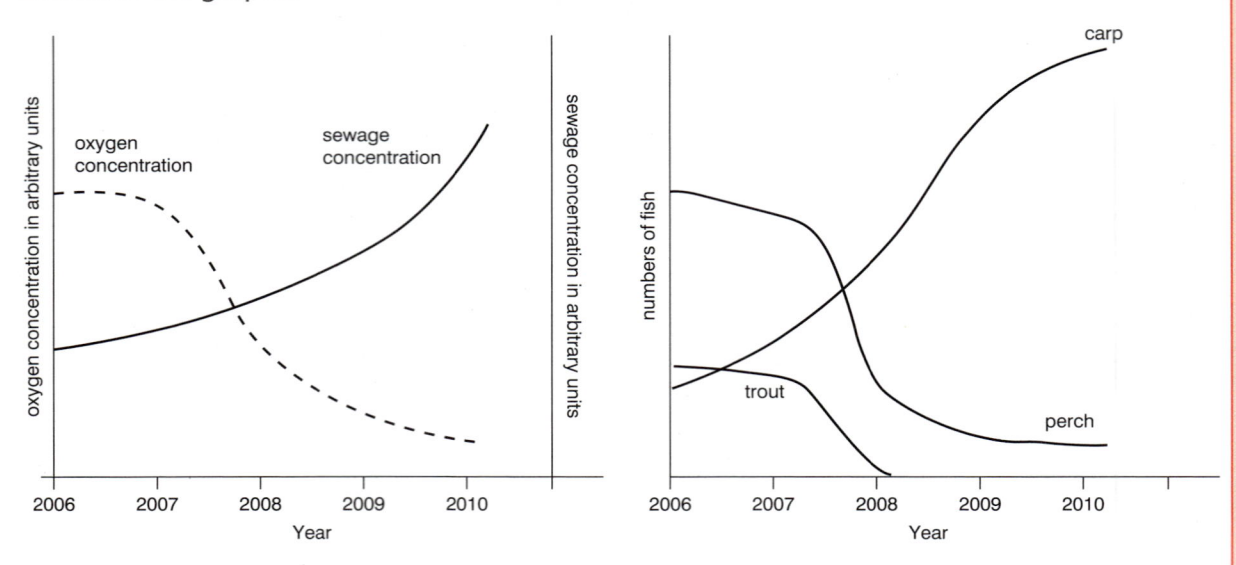

a) i) Describe the trends in sewage and oxygen concentration between 2006 and 2010. (2)

ii) Explain why sewage had this effect on the oxygen concentration in the water. (3)

b) Describe and explain the effects of the changing oxygen concentration on fish populations. (6)

(Total 11 marks)

EXTENDED Question 10

Population pyramids are used to show numbers of a population within certain age ranges. Population pyramids for India and the United Kingdom for 2010 are shown here.

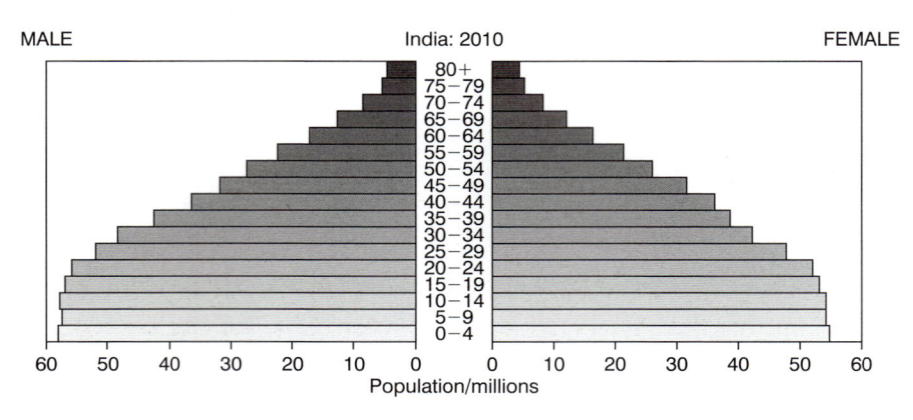

United Kingdom: 2010

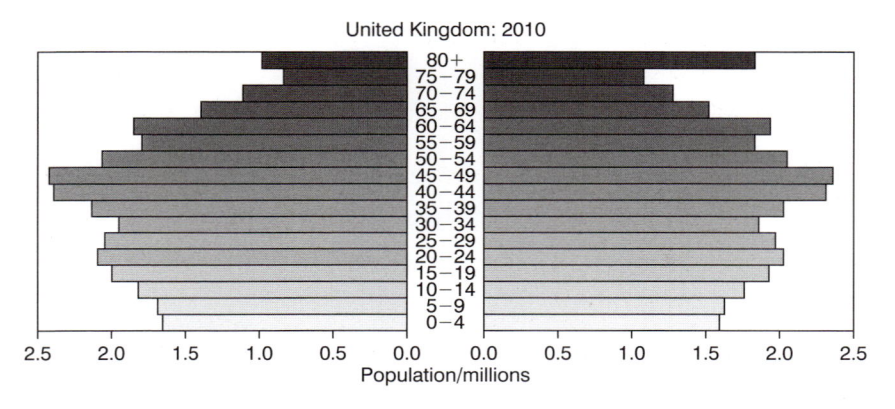

From: www.nationmaster.com

a) Compare the populations of India and the UK in 2010. (4)

b) Explain how the shape of the pyramids indicate that India can be described as a 'developing country' and the UK a 'developed country'. (3)

c) Suggestion one reason for:

   i) an indent in a population pyramid (1)

   ii) a bulge in a population pyramid. (1)

d) The graph shows the impact, and the projected impact of AIDS on the human population in sub-Saharan Africa.

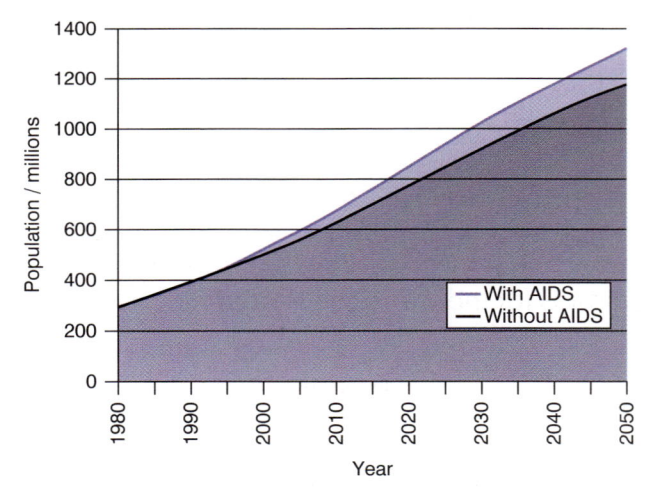

   i) Describe the impact of AIDS on population growth. (3)

   ii) Outline how HIV affects the human immune system. (2)

   iii) Projections of human populations of sub-Saharan Africa cover the next 40 years. Name three factors that could produce a marked change these projections. (3)

(Total 17 marks)

Doing well in examinations

INTRODUCTION

Examinations will test how good your understanding of scientific ideas is, how well you can apply your understanding to new situations and how well you can analyse and interpret information you have been given. The assessments are opportunities to show how well you can do these.

To be successful in exams you need to:

- ✓ have a good knowledge and understanding of science
- ✓ be able to apply this knowledge and understanding to familiar and new situations
- ✓ be able to interpret and evaluate evidence that you have just been given.

You need to be able to do these things under exam conditions.

OVERVIEW

Ensure you are familiar with the structure of the examinations you are taking. Consult the relevant syllabus for the year you are entering your examinations for details of the different papers and the weighting of each, including the papers to test practical skills. Your teacher will advise you of which papers you will be taking.

You will be required to perform calculations, draw graphs and describe, explain and interpret biological ideas and information. In some of the questions the content may be unfamiliar to you; these questions are designed to assess data-handling skills and the ability to apply biological principles and ideas in unfamiliar situations.

ASSESSMENT OBJECTIVES AND WEIGHTINGS

For the Cambridge IGCSE examination, the assessment objectives and weightings are as follows:

- ✓ A: Knowledge with understanding (50%)
- ✓ B: Handling information and problem solving (30%)
- ✓ C: Experimental skills and investigations (20%).

The types of questions in your assessment fit the three assessment objectives shown in the table.

Assessment objective	Your answer should show that you can…
A Knowledge with understanding	Recall, select and communicate your knowledge and understanding of science.
B Handling information and problem solving	Apply skills, including evaluation and analysis, knowledge and understanding of scientific contexts.
C Experimental skills and investigations	Use the skills of planning, observation, analysis and evaluation in practical situations.

EXAMINATION TECHNIQUES

To help you maximise your chances in exams, there are a few simple steps.

Check your understanding of the question.

- ✓ **Read the introduction to each question carefully before moving on to the questions themselves**.
- ✓ Look in detail at any **diagrams, graphs** or **tables**.
- ✓ Underline or circle the **key words** in the question.
- ✓ **Make sure you answer the question that is being asked** rather than the one you wish had been asked!
- ✓ Make sure you understand the meaning of the '**command words**' in the questions.

REMEMBER

Remember that any information you are given is there to help you to answer the question.

EXAMPLE

- ✓ '**Give**', '**state**', '**name**' are used when recall of knowledge is required, for example you could be asked to give a definition or provide the best answers from a list of options.
- ✓ '**Describe**' is used when you have to give the main feature(s) of, for example, a biological process or structure.
- ✓ '**Explain**' is used when you have to give reasons, e.g. for some experimental results or a biological fact or observation. You will often be asked to 'explain your answer', i.e. give reasons for it.

- ✓ **'Suggest'** is used when you have to come up with an idea to explain the information you're given – there may be more than one possible answer, no definitive answer from the information given, or it may be that you will not have learned the answer but have to use the knowledge you do have to come up with a sensible one.
- ✓ **'Calculate'** means that you have to work out an answer in figures.
- ✓ **'Plot'** and **'Draw a graph'** are used when you have to use the data provided to produce graphs and charts.

Check the number of marks for each question
- ✓ Look at the **number of marks** allocated to each question.
- ✓ Look at the **space provided** to guide you as to the length of your answer.
- ✓ Make sure you include at least as many points in your answer as there are marks.
- ✓ Write neatly and keep within the space provided.

REMEMBER
Beware of continually writing too much because it probably means you are not really answering the questions. Do not repeat the question in your answer.

Use your time effectively
- ✓ Don't spend so long on some questions that you don't have time to finish the paper.
- ✓ Check how much time you have left regularly.
- ✓ If you are really stuck on a question, leave it, finish the rest of the paper and come back to it at the end.
- ✓ Even if you eventually have to guess at an answer, you stand a better chance of gaining some marks than if you leave it blank.

ANSWERING QUESTIONS
Multiple choice questions
- ✓ Select your answer by placing a cross (not a tick) in the box.

Short-and long-answer questions
- ✓ In short-answer questions, **don't write more than you are asked for**.
- ✓ You may not gain any marks, even if the first part of your answer is correct, if you've written down something incorrect later on or which contradicts what you've said earlier. This may give the impression that you haven't really understood the question or are guessing.

- ✓ In some questions, particularly short-answer questions, answers of only one or two words may be sufficient, but in longer questions you should aim to use **clear scientific language**.
- ✓ Present the information in a logical sequence.
- ✓ Don't be afraid to also use **labelled diagrams** or **flow charts** if it helps you to show your answer more clearly.

Questions with calculations

- ✓ **In calculations always show your working**.
- ✓ Even if your final answer is incorrect you may still gain some marks if part of your attempt is correct.
- ✓ If you just write down the final answer and it is incorrect, you will get no marks at all.
- ✓ Write down your answers to as many **significant figures** as are used in the numbers in the question (and no more). If the question doesn't state how many significant figures then a good rule to follow is to quote three significant figures.
- ✓ Don't round off too early in calculations with many steps – it's always better to give too many significant figures than too few.
- ✓ You may also lose marks if you don't use the correct **units.** In some questions the units will be mentioned, e.g. calculate the mass in grams; or the units may also be given on the answer line. If numbers you are working with are very large, you may need to make a conversion, e.g. convert joules into kilojoules or millimetres into metres.

Finishing your exam

- ✓ When you've finished your exam, **check through** your paper to make sure you've answered all the questions.
- ✓ Check that you haven't missed any questions at the end of the paper or turned over two pages at once and missed questions.
- ✓ Cover over your answers and read through the questions again and check that your answers are as good as you can make them.

REMEMBER

You will be asked questions on investigative work. It is important that you understand the methods used by scientists when carrying out investigative work.

More information on carrying out practical work and developing your investigative skills are given in the next section.

Developing experimental skills

INTRODUCTION

As part of your Biology course, you will develop practical skills and have to carry out investigative work in science.

This section provides guidance on carrying out an investigation.

The experimental and investigative skills are divided as follows:

1. Using and organising techniques, apparatus and materials
2. Observing, measuring and recording
3. Handling experimental observations and data
4. Planning and evaluating investigations.

1. USING AND ORGANISING TECHNIQUES, APPARATUS AND MATERIALS

Learning objective: to demonstrate and describe appropriate experimental and investigative methods, including safe and skilful practical techniques.

Questions to ask:

How shall I use the equipment and chemicals safely to minimise the risks – what are my safety precautions?

✓ When writing a Risk Assessment, investigators need to be careful to check that they've matched the hazard with the concentration of a chemical used. Many acids, for instance, are corrosive in higher concentrations, but are likely to be irritants or of low hazard in the concentration used when working in biology experiments.

✓ Don't forget to consider the hazards associated with all the chemicals and biological materials, even if these are very low.

✓ You may be asked to justify the precautions taken when carrying out an investigation.

How much detail should I give in my description?

✓ You need to give enough detail so that someone else who has not done the experiment would be able to carry it out to reproduce your results.

How should I use the equipment to give me the precision I need?

✓ You should know how to read the scales on the measuring equipment you are using.

✓ You need to show that you are aware of the precision needed.

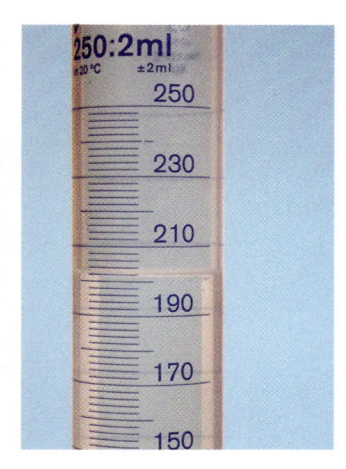

△ Fig. 6.1 The volume of liquid in a burette must be read to the bottom of the meniscus. The volume in this measuring cylinder is 202 cm^3 (ml), not 204 cm^3.

EXAMPLE 1

This is an extract from a student's notebook. It describes how she carried out an experiment to investigate the production of carbon dioxide by yeast at different temperatures.

What are my safety precautions?

a) Chemicals.

> *I have looked up the hazards associated with the chemical I am using:*
>
> *Glucose (solutions from 0.05 to 0.25 M): LOW HAZARD*
>
> *Although it is only a low hazard, it is still best to wear eye protection when using the solutions, especially as some of the liquids will be hot. It is also important to handle all chemicals carefully, and wipe up any spills of liquid.*

COMMENT

The student has used a data source to look up the chemical hazards.

b) Organisms.

> *I found my information on yeast from the Fisher Scientific website:*

Dried yeast may cause eye, skin, and respiratory tract irritation. It is expected to be a low hazard for usual handling. If the dust is inhaled, you should 'remove from exposure and move to fresh air immediately'.

I will handle the powdered yeast carefully when making up my suspension, trying to avoid making any dust.

COMMENT
The student has used a data source to look up the biological hazards.

c) Equipment

I must be careful when using the water bath not to get water near the electrical sockets.

I need to handle the glassware (conical flask, gas syringe and glass tubing) carefully. In particular, I need to protect my hands with a towel (or glove) when linking together the glass delivery tubes from the rubber bung in the conical flask to the gas syringe with rubber tubing. The tubing needs to be lubricated with water and I need to hold my hands close together to limit the movement of glass if a break occurs.

COMMENT
The student has suggested some sensible precautions.

How much detail should I give in my description?

The student's method is given below:

1. *A solution of 0.25 M glucose was made up by dissolving 45.00 g of glucose in water to make a litre of solution.*

2. *100 cm^3 of the glucose solution was transferred to each of six conical flasks.*

3. *The conical flask was transferred to a water bath at 20 °C. It was left for a few minutes to reach the temperature.*

4. *1.00 g of yeast was then added, and the mixture swirled to mix in the yeast. The stop clock was started.*

5. *The bung, on which I had placed tubing connecting it to the gas syringe, was placed in the conical flask.*

6 Every minute, the volume of carbon dioxide in the gas syringe was recorded.

7 The average volume of carbon dioxide produced per minute was calculated.

8 The experiment was repeated three times and the average rate of carbon dioxide production calculated.

9 The investigation was then carried out at 10 °C, 30 °C, 40 °C, 50 °C and 60 °C.

10 A graph was drawn of the average rate of carbon dioxide production over temperature.

COMMENT

The method is detailed and well written. The student has appreciated that it is important for the sugar solution to reach the temperature being investigated before the yeast is added.

2. OBSERVING, MEASURING AND RECORDING

Learning objective: to make observations and measurements with appropriate precision, record these methodically, and present them in a suitable form.

Questions to ask:

How many different measurements or observations do I need to take?

✓ Sufficient readings have been taken to ensure that the data are consistent.

✓ It is usual to repeat an experiment to get more than one measurement. If an investigator takes just one measurement, this may not be typical of what would normally happen when the experiment was carried out.

✓ When repeat readings are consistent they are said to be **repeatable**.

Do I need to repeat any measurements or observations that are anomalous?

✓ An **anomalous result** or **outlier** is a result that is not consistent with other results.

✓ You want to be sure a single result is accurate. So you will need to repeat the experiment until you get close agreement in the results you obtain.

- ✓ If an investigator has made repeat measurements, they would normally use these to calculate the arithmetical mean (or just mean or average) of these data to give a more accurate result. You calculate the mean by adding together all the measurements, and dividing by the number of measurements. Be careful though; anomalous results should not be included when taking averages.

- ✓ Anomalous results might be the consequence of an error made in measurement. But sometimes outliers are genuine results. If you think an outlier has been introduced by careless practical work, you should omit it when calculating the mean. But you should examine possible reasons carefully before just leaving it out.

- ✓ You are taking a number of readings in order to see a changing pattern. For example, measuring the volume of gas produced in a reaction every 10 seconds for 2 minutes (so 12 different readings). It is likely that you will plot your results onto a graph and then draw a **line of best fit**.

- ✓ You can often pick an anomalous reading out from a results table (or a graph if all the data points have been plotted, as well as the mean, to show the range of data). It may be a good idea to repeat this part of the practical again, but it's not necessary if the results show good consistency.

- ✓ If you are confident that you can draw a line of best fit through most of the points, it is not necessary to repeat any measurements that are obviously inaccurate. If, however, the pattern is not clear enough to draw a graph then readings will need to be repeated.

How should I record my measurements or observations – is a table the best way? What headings and units should I use?

- ✓ A table is often the best way to record results.
- ✓ Headings should be clear.
- ✓ If a table contains numerical data, do not forget to include units; data are meaningless without them.
- ✓ The units should be the same as those that are on the measuring equipment you are using.
- ✓ Sometimes you are recording observations that are not quantities. Putting observations in a table with headings is a good way of presenting this information.

EXAMPLE 2

How many different measurements or observations do I need to take?

A student cut a number of cylinders of tissue from a potato and weighed them, and recorded the mass of each cylinder. Six dishes were set up with each dish containing a different concentration of sucrose. Four potato cylinders were placed into each dish. After one hour, the cylinders were removed, blotted dry and reweighed. The student then calculated the percentage change in mass for each cylinder. The results are shown below.

Concentration of sucrose/M	Percentage change in mass of potato cylinders/g				Average percentage change in mass	Texture of potato cylinders (qualitative)
	Experiment 1	Experiment 2	Experiment 3	Experiment 4		
0.0	+31.4	+33.7	+31.2	+32.5	+42.9	Firm
0.2	+20.9	+33.4	+22.8	+21.3	+21.7	Firm
0.4	−2.7	−1.8	−1.9	−2.4	−2.2	Slightly soft
0.6	−13.9	−12.8	−13.7	−13.6	−13.5	Soft
0.8	−20.2	−19.7	−19.3	−20.4	−19.9	Floppy
1.0	−19.9	−20.3	−21.1	−20.3	−20.4	Very floppy

△ Table 6.1 Results for Example 2.

In this table of results:
- ✓ the description of each measurement is clear.
- ✓ the units are given.
- ✓ the data are recorded to the same number of decimal places, and decimal points are aligned.
- ✓ calculations of means are recorded to the appropriate number of significant figures.

The student has recorded four measurements for each concentration investigated.

With the exception of the cylinder in Experiment 2 in a concentration of 0.2 M sucrose (highlighted in the table), the repeats show good consistency.

Do I need to repeat any measurements or observations that are anomalous?

The result from the cylinder in Experiment 2 in a concentration of 0.2 M sucrose is not consistent with the other results for this concentration. It an anomalous result and is highlighted in the table. The student has not included this result in the calculation of the mean for this concentration.

EXAMPLE 3

How should I record my measurements or observations?

Here are some results of food tests carried out by an investigator:

Food substance tested	Colour change on heating with Benedict's solution
10% glucose solution	blue → green → yellow → orange → red
Biscuit	blue → greenish blue
Grape	blue → green → yellow → orange → red
Honey	blue → green → yellow → orange → red-brown
Potato	remained blue

△ Table 6.2 Results of some food tests.

Note the clear table headings. The right hand column doesn't simply say 'colour' but refers to colour *changes*.

COMMENT

Don't forget that some investigations might benefit from including both numerical data and observations, e.g. in the osmosis experiment in Example 2, the student also found it useful to include information on the firmness of the potato cylinders at the end of the experiment.

3. HANDLING EXPERIMENTAL OBSERVATIONS AND DATA

Learning objectives: to analyse and interpret data to draw conclusions from experimental activities which are consistent with the evidence, using biological knowledge and understanding, and to communicate these findings using appropriate specialist vocabulary, relevant calculations and graphs.

Questions to ask:

What is the best way to show the pattern in my results? Should I use a bar chart, line graph or scatter graph?

✓ Graphs are usually the best way of demonstrating trends in data.
✓ A bar chart or bar graph is used when one of the variables is a **categoric variable**, for example when one of the variables is the type of leaf, or species of organism.
✓ A line graph is used when both variables are continuous, e.g. time and temperature, time and volume.
✓ Scattergraphs can be shown to show the intensity of a relationship, or degree of *correlation*, between two variables.
✓ Sometimes a line of best fit is added to a scatter graph, but usually the points are left without a line.

When drawing bar charts or line graphs:

✓ Choose scales that take up most of the graph paper.
✓ Make sure the axes are linear and allow points to be plotted accurately. Each square on an axis should represent the same quantity. For example, one big square = 5 or 10 units; not 3 units.
✓ Label the axes with the variables (ideally with the independent variable on the x-axis).
✓ Make sure the axes have units.
✓ If more than one set of data is plotted use a key to distinguish the different data sets.

If I use a line graph, should I join the points with a line or a smooth curve?

✓ When you draw a line, do not just join the dots!
✓ Remember there may be some points that don't fall on the curve – these may be incorrect or anomalous results.
✓ A graph will often make it obvious which results are anomalous and so it would not be necessary to repeat the experiment
✓ If following the biological rhythms of an organism over a period of time, you should join the data points, point-to-point.

Do I have to calculate anything from my results?

✓ It will be usual to calculate means from the data.
✓ Sometimes it is helpful make other calculations, before plotting a graph (see Example 2). Other types of calculation include:
✓ the energy content of food *per gram* when burning a sample of food
✓ the volume of water taken up by a plant from the distance moved by a bubble in a potometer (volume = $\pi r^2 \times$ distance)
✓ the density of plants, e.g. as plants per m^2, or frequency of plants or animals, from sampling using a square frame called a quadrat
✓ Investigators also look for numerical trends in data, for example, the doubling of a reaction rate every 10 °C; the doubling of numbers of microorganisms every 20 minutes.
✓ Sometimes you will have to make some calculations before you can draw any conclusions.

Can I draw a conclusion from my analysis of the results, and what biological knowledge and understanding can be used to explain the conclusion?

✓ You need to use your biological knowledge and understanding to explain your conclusion.
✓ It is important to be able to add some explanation which refers to relevant scientific ideas in order to justify your conclusion.

EXAMPLE 4

What is the best way to show the pattern in my results?

A student did an experiment to compare the loss of water from leaves of three different species of tree – hazel, lime and oak.

He measured the mass of 10 leaves of similar size and hung the leaves on a line. After three hours, he removed the leaves and measured the masses of the leaves again and calculated the average loss of water in grams per hour.

Species	Average loss of water/g per hour
Apple	0.30
Hazel	0.05
Oak	0.01

△ Table 6.3 Results of Example 4.

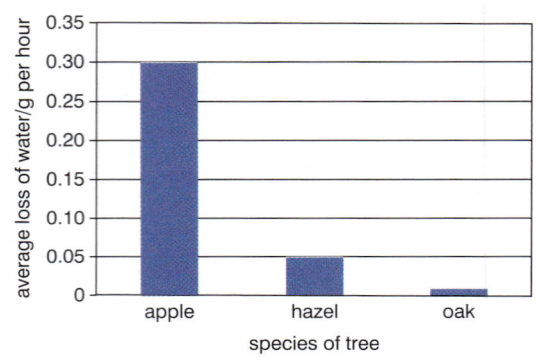

△ Fig. 6.2 Bar chart showing water loss from different leaves.

A bar chart or bar graph is used to display the data in this instance, as the type of leaf is a categoric variable.

EXAMPLE 5

What is the best way to show the pattern in my results?

A student investigated the effect of different light intensities on photosynthesis in pondweed.

He measured the oxygen collected over a number of days.

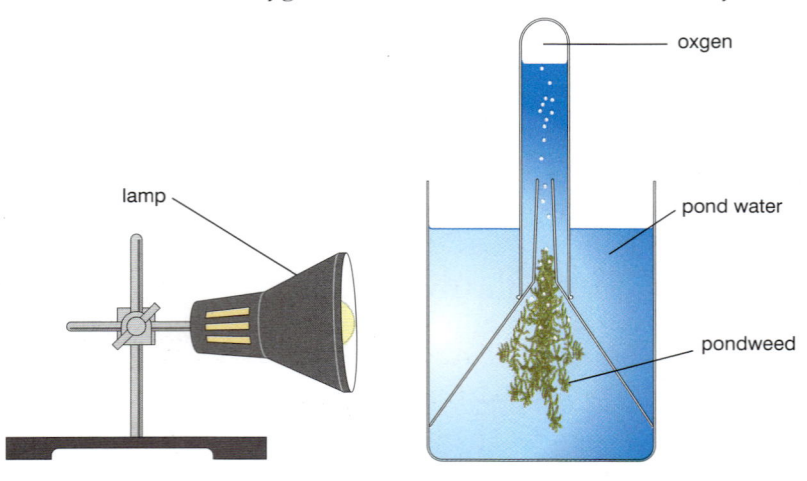

△ Fig. 6.3 Apparatus for Example 5.

A line graph is needed as both the volume of gas and time are continuous variables.

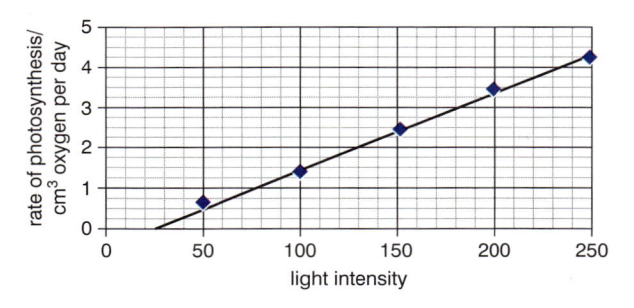

△ Fig. 6.4 Experimental results of Example 5.

If I use a line graph, should I join the points with a line or a smooth curve?

In this case the data fit a straight line most closely. You need to look at the shape that the points make to help you decide how to join them.

EXAMPLE 6

What is the best way to show the pattern in my results?

A student investigated the effect of applying different amounts of fertiliser to plants in the school grounds. He used a scatter graph to display the results he collected.

A scatter graph is most appropriate for displaying these data because each point represents one plant and is unrelated to another point. The results show a trend that should be described in the conclusion.

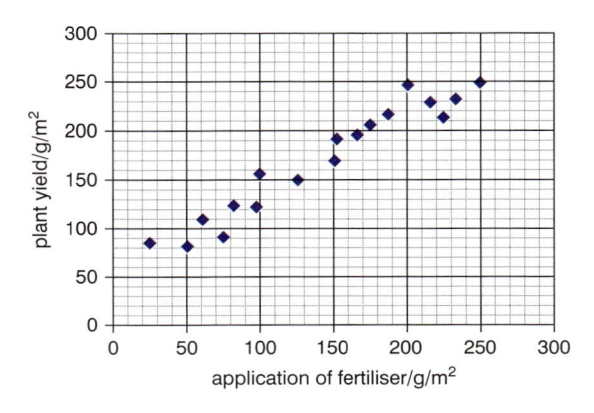

△ Fig. 6.5 Experimental results for Example 6.

EXAMPLE 7

If I use a line graph, should I join the points with a line or a smooth curve?

In an investigation on the activity of an enzyme at different temperatures, an investigator obtained a set of results that shows different phases.

Photographic film is made from a sheet of plastic coated with light-sensitive silver particles bonded by the protein gelatine. When the gelatine is broken down by a protease enzyme, the silver particles fall off, and the film becomes clear.

Photographic film was cut into five strips of equal size, and each strip was placed into a test tube. An identical volume of protease was added to the tubes, and each tube was kept at a different temperature.

The amount of time taken for each strip of film to become clear was measured and recorded.

Temperature/°C	Average time taken for breakdown of gelatine/s
4	3450
13	667
25	175
30	130
40	133
50	7140

△ Table 6.4 Table of results for Example 7.

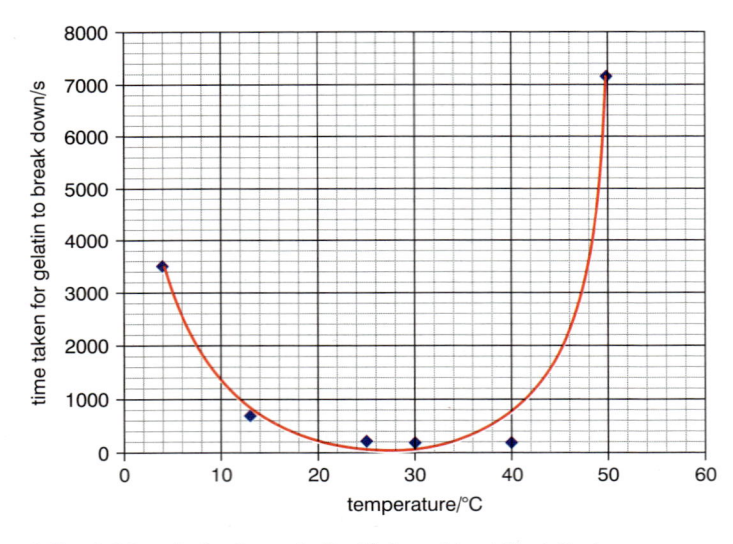

△ Fig. 6.6 Results for Example 7 with line of best fit plotted.

Do I have to calculate anything from my results?

The trend of how enzyme activity is affected by temperature is not well illustrated by the graph. It is shown better if the *rate of reaction* is calculated and plotted against temperature. The rate is the inverse of the time taken to break down the gelatine, i.e. 1 ÷ the time taken.

Temperature/°C	Average time taken for breakdown of gelatine/s	Rate of breakdown of gelatine (= 1/time)/s^{-1}
4	3450	0.00029
13	667	0.00150
25	175	0.00571
30	130	0.00769
40	133	0.00752
50	7140	0.00014

△ Table 6.5 Table of results with rate column added.

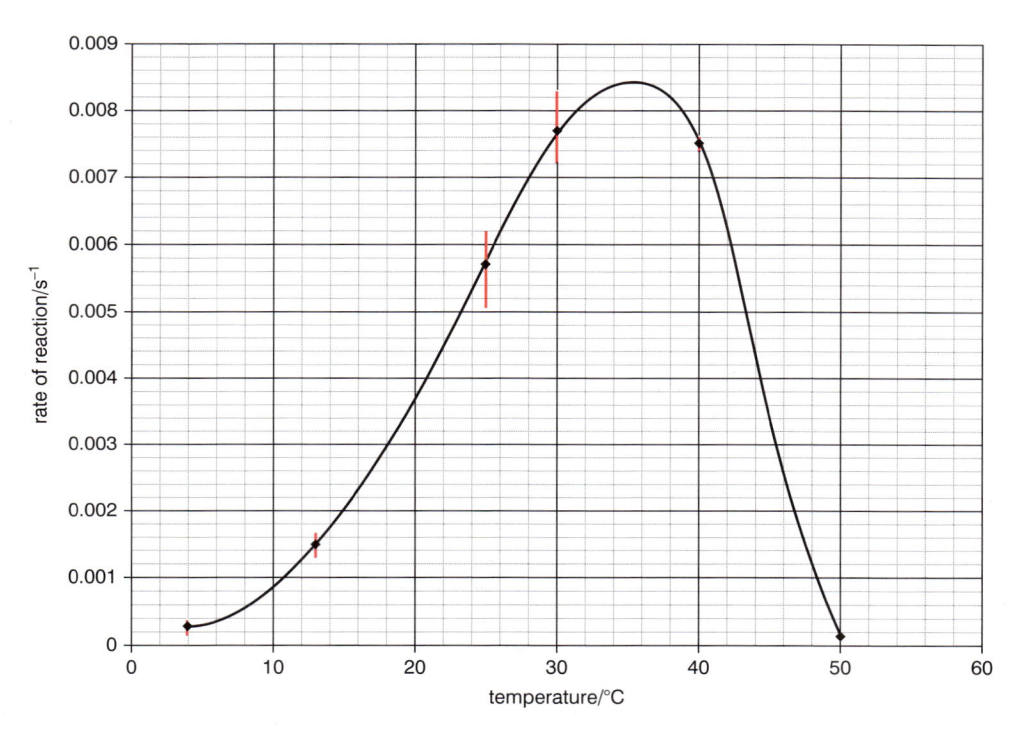

△ Fig. 6.7 Rate of reaction for Example 7.

Note that on this graph, the student has also drawn range bars to show how the data points are arranged around the mean.

Rates can also be calculated by looking at the change in steepness/gradient of a line graph, or part of a graph.

Can I draw a conclusion from my analysis of the results?

The student wrote:

> *Enzymes work best at a particular temperature. My graph suggests that the optimum temperature for protease is around 35 °C. At lower temperatures, enzymes work slowly because the molecules have less energy and move around more slowly, so there are fewer successful collisions between enzyme and substrate (gelatine) molecules.*

Enzymes work by a lock and key mechanism, with the substrate fitting into the enzyme. At temperatures that are too high, the structure of an enzyme will be changed so that it will not work. This change is permanent and the enzyme is said to be denatured.

COMMENT
This is a good, concise conclusion and links the data to the mechanism of enzyme action.

4. PLANNING AND EVALUATING INVESTIGATIONS
4a Planning
Learning objective: to devise and plan investigations, drawing on biological knowledge and understanding in selecting appropriate techniques.

Questions to ask

What do I already know about the area of biology I am investigating, and how can I use this knowledge and understanding to help me with my plan?

- ✓ Think about what you have already learned and any investigations you have already done that are relevant to this investigation.
- ✓ List the factors might affect the process you are investigating.

What is the best method or technique to use?

- ✓ Think about whether you can use or adapt a method that you have already used.
- ✓ A method, and the measuring instruments, must be able to produce **valid** measurements. A measurement is valid if it measures what it is supposed to be measuring.

You will make a decision as to which technique to use based on:

- ✓ The accuracy and precision of the results required.

Investigators might require results that are as accurate and precise as possible but if you are doing a quick comparison, or a preliminary test to check a range over which results should be collected, a high level of accuracy and precision may not be required.

- ✓ The simplicity or difficulty of the techniques available, or the equipment required; is this expensive, for instance?
- ✓ The scale, for example using standard laboratory equipment or on a micro-scale, which may give results in a shorter time period.
- ✓ The time available to do the investigation.
- ✓ Health and safety considerations.

What am I going to measure?

✓ You need to decide what you are going to measure.

✓ You need to choose a range of measurements that will be enough to allow you to plot a graph of your results and so find out the pattern in your results.

✓ You might be asked to explain why you have chosen your range rather than a lower or higher range.

✓ The factor you are investigating is called the **independent variable**. A **dependent variable** depends on the value of the independent variable that you select.

How am I going to control the other variables?

✓ These are **control variables**. Some of these may be difficult to control. This may be especially difficult if you are carrying out an ecology investigation in the field, where varying factors such as light and temperature are impossible to control.

✓ You must decide how you are going to control any other variables in the investigation and so ensure that you are using a fair test and that any conclusions you draw are valid.

What equipment is suitable and will give me the accuracy and precision I need?

✓ The *accuracy* of a measurement is how close it is to its true value.

✓ Precision is related to the smallest scale division on the measuring instrument that you are using, e.g. when measuring the distance moved by the bubble in Method 3, a rule marked in millimetres will give greater precision that one divided into centimetres only.

✓ A set of precise measurements also refers to measurements that have very little spread about the mean value.

✓ You need to be sensible about selecting your devices and make a judgement about the degree of precision. Think about what is the least precise variable you are measuring and choose suitable measuring devices. There is no point having instruments that are much more precise than the precision you can measure the least precise variable to.

What are the potential hazards of the equipment, chemicals, organism and technique I will be using and how can I reduce the risks associated with these hazards?

✓ Investigators find out about the hazard associated with chemicals using CLEAPSS Student Safety Sheets or a similar resource. Information on biological hazards can be found in the CLEAPSS Laboratory Handbook.

✓ Be prepared to suggest safety precautions when presented with details of a biology investigation.

EXAMPLE 8

You have been asked about the design and planning of an investigation on carbon dioxide production by yeast under different conditions.

What do I already know?

You may have already learned about the role of yeast in the production of alcohol/beer. In the equation for anaerobic respiration of yeast, you can see that in the absence of oxygen, yeast uses sugars for (anaerobic) respiration, producing carbon dioxide and alcohol (ethanol).

As sugar is a reactant in the process, conditions that might affect the production of carbon dioxide on the process include the *concentration of sugar*, and the *type of sugar* may also be a factor.

The chemical reactions involved in this process are controlled by the yeast's enzymes, so you would expect this process to be affected by *temperature*.

Conditions that might be expected to affect the process are therefore:

✓ concentration of sugar
✓ type of sugar
✓ temperature.

You might be questioned on how to measure the effect these factors, or one of these factors, has on carbon dioxide production.

All of these factors are independent variables. The amount of carbon dioxide produced is the dependent variable because it is affected by the independent variables – concentration of sugar, type of sugar and temperature.

What is the best method or technique to use?

An investigator needs to set up an experiment so that they can measure the carbon dioxide produced by the respiring yeast.

Several methods are available to produce valid measurements.

Method 1

A simple way is to add some yeast to a sugar solution in a conical flask, as shown in the diagram below and count the bubbles produced in given time periods, e.g. every minute, or over a period of time, e.g. one hour.

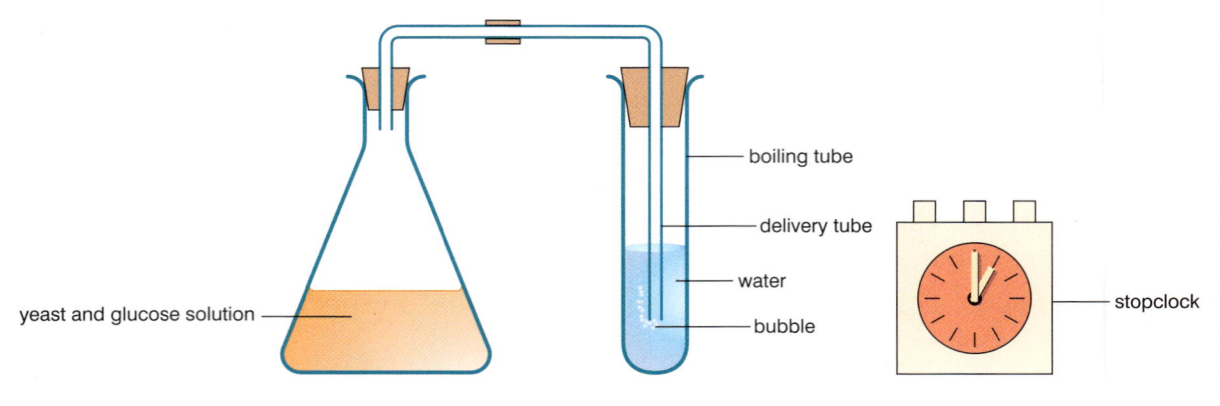

△ Fig. 6.8 Apparatus for method 1.

Method 2

An alternative method is to set up the yeast and sugar solution as before, but this time connect the glass tube to a gas syringe. This time, you can measure the volume of carbon dioxide produced in given time periods, e.g. every minute, or over a period of time, e.g. one hour.

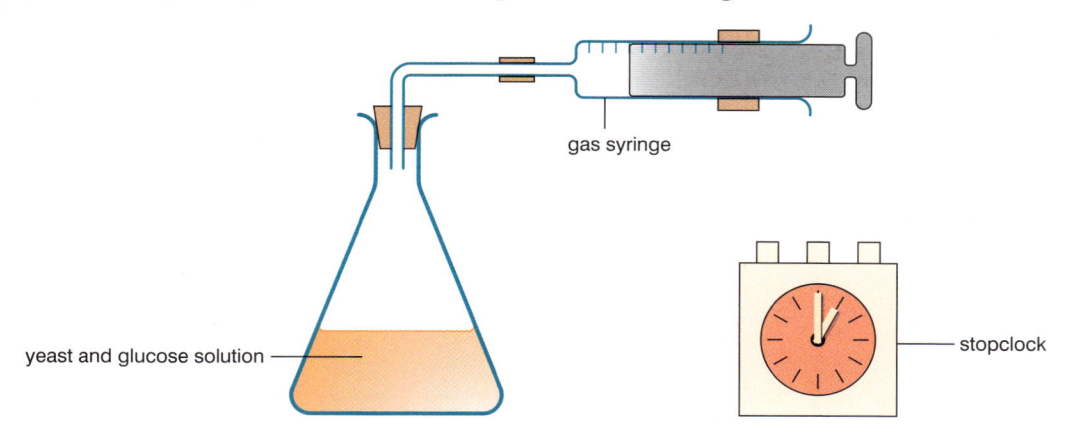

△ Fig. 6.9 Apparatus for method 2.

Method 3

Another method is based on a smaller scale. The yeast and sugar solution is placed in a syringe, which is connected to a pipette that contains a bubble of water. The movement of the bubble is measured in given time periods, e.g. every minute, or over a period of time, e.g. one hour. It is less suitable for the temperature investigation, however, as it would be inadvisable to immerse the syringes in water baths.

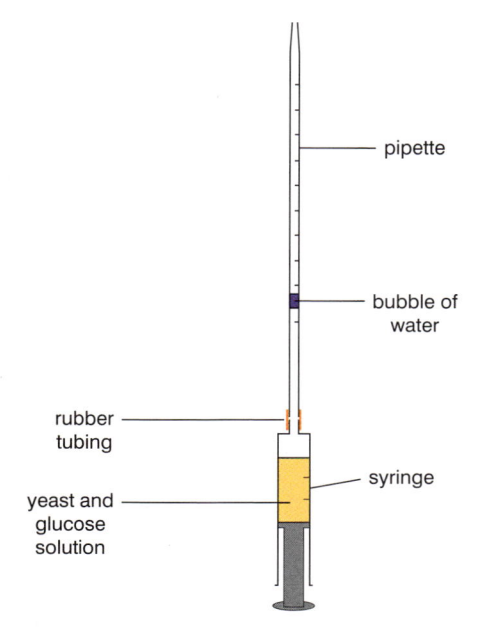

△ Fig. 6.10 Apparatus for method 3.

Choosing the method:

- ✓ Accuracy and precision: methods 2 and 3 would be preferred when accurate and precise results are needed. Method 1 cannot be used to actually measure volume of carbon dioxide, but would be best for a preliminary test or a quick comparison.
- ✓ Micro-scale: method 3 might be most suitable here.
- ✓ Time available: method 3 might be most suitable, as smaller volumes are involved and the investigation could be carried out more quickly.
- ✓ Health and safety considerations are less relevant here.

What am I going to measure?

You need to make a measure of the amount of carbon dioxide produced using different concentrations of sugar, different types of sugar, or at different temperatures.

The **independent variables** are the factors under investigation: the concentration of sugar, the type of sugar and the temperature.

The **dependent variables** are the measurements made: the number of bubbles per minute, the volume of carbon dioxide per minute, or the volume moved by the bubble.

Different concentrations of sugar

A sensible range of sugar concentrations to use might be based on concentrations that yeast might encounter in nature, or might be used to produce different beers in a brewery. The investigator here chose to use concentrations ranging from 0 M to 0.25 M (at intervals of 0 M, 0.05 M, 0.10 M, 0.15 M, 0.20 M and 0.25 M), as a preliminary test showed that these concentrations would give suitable results in the time period allocated. Concentrations chosen were no higher as the investigator realised that these might have an osmotic effect; water could be drawn from the yeast cells by osmosis, and not function.

The investigator plotted a graph using the results from the six different concentrations to look for a pattern in those results.

Different types of sugar

Here, the investigator decided to use sugars commonly found in nature, or in malted barley in brewing, that yeast might use for respiration. These included fructose, glucose, lactose, maltose and sucrose.

Different temperatures

It would be sensible to choose a range of temperatures that yeast would encounter in nature or in a brewery. Respiration in yeast is a series of enzyme-controlled reactions. In most cases, enzymes work best around 40 °C, and cease to function above around 60 °C. It was decided, therefore, to measure carbon dioxide production at five different temperatures (10 °C, 20 °C, 30 °C, 40 °C, 50 °C and 60 °C). Again, this would be sufficient to allow the investigator to plot a graph of results and so find any pattern in the results.

How am I going to control the other variables?

The investigator must ensure that any differences in carbon dioxide production must be the result of, in the different investigations, sugar concentration, type of sugar and temperature, and not the result of some other factor. In other words, it must be a fair test and produce valid measurements.

So, the investigator must decide what other factors could affect the experiment and try to keep these constant. These are the control variables.

Factor under investigation/ independent variable	Factors to be kept constant				
	Yeast concentration	Sugar concentration	Type of sugar	Temperature	Length of investigation
Sugar concentration	Yes	No – vary	Yes	Yes	Yes
Type of sugar	Yes	Yes	No – vary	Yes	Yes
Temperature	Yes	Yes	Yes	No – vary	Yes

△ Table 6.6 Variables in Example 8.

EXAMPLE 9

In this experiment, the production of heat from germinating seeds is being investigated. Some seeds have been placed in an insulated flask (Thermos flask) and their temperature measured over a period of time.

But a rise in temperature alone would not be sufficient to demonstrate that it is the respiring seeds that are causing this rise. Some other factor could be involved. So an identical experiment is set up, but this time with seeds that have been killed. So any change in the temperature must be the result of the living seeds' respiration. Both sets of seeds are also sterilised with disinfectant so that any temperature rise can't be down to the growth of microorganisms on the seeds.

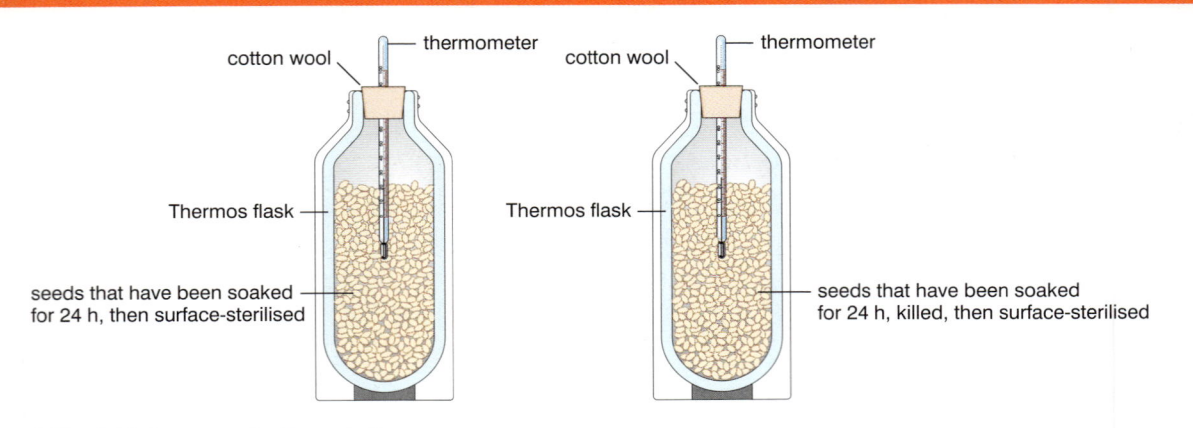

△ Fig. 6.11 Apparatus for Example 9.

What equipment is suitable and will give me the accuracy and precision I need?

You now know what you will need to measure and so can decide on your measuring devices.

Referring back to Example 8:

Measurement	Quantity required	Equipment
Mass of yeast	1.00 g	Balance measuring up to two decimal places
Volume of sugar solution	100 cm^3	100 cm^3 volumetric flask
Temperature	10–60 °C	Thermometer, with 1 °C precision
Time	One minute intervals	Stop watch (1 s precision)

△ Table 6.7 Suitable equipment for experiment.

You will need to be sensible about selecting your equipment and make a judgement about the degree of accuracy. The accuracies of equipment need to be comparable. It would be not appropriate to measure the yeast that was put in every conical flask accurately without measuring the volume of sugar solution poured onto each flask to a similar level of accuracy.

What are the potential hazards of the equipment and how can I reduce the risks?

In Example 8, the chemical hazards are as follows:

Fructose solution	LOW HAZARD
Glucose solution	LOW HAZARD
Lactose solution	LOW HAZARD
Maltose solution	LOW HAZARD
Sucrose solution	LOW HAZARD
Yeast, dried	LOW HAZARD

These indicate that there are no specific hazards the investigator needs to be aware of. However, when handling *any* chemicals, it would be sensible to wear eye protection.

In terms of the equipment and technique, the major hazards are:

✓ handling hot liquids (at 50 °C and 60 °C).

✓ when connecting tubing together.

✓ possible contact between water and electrical sockets.

4b Evaluating

Learning objective: to evaluate data and methods.

Questions to ask:

Do any of my results stand out as being inaccurate?

✓ You need to look for any anomalous results or outliers that do not fit the pattern.

✓ You can often pick this out from a results table (or a graph if all the data points have been plotted, as well as the mean, to show the range of data). The investigator has not included this result in the calculation of the mean for this concentration. It may be a good idea to repeat this part of the practical again, but it is not necessary if the results show good consistency.

What reasons can I give for any inaccurate results?

✓ When answering questions like this it is important to be specific. Answers such as 'experimental error' will not score any marks.

✓ It is often possible to look at the practical technique and suggest explanations for anomalous results.

✓ When you carry out the experiment you will have a better idea of which possible sources of error are more likely.

✓ Try to give a specific source of error and avoid statements such as 'the measurements must have been wrong'.

Your conclusion will be based on your findings, but must take into consideration any uncertainty in these introduced by any possible sources of error. You should discuss where these have come from in your evaluation.

Error is a difference between a measurement you make, and its true value.

The two types of errors are:

✓ random error

✓ systematic error.

With **random error**, measurements vary in an unpredictable way. This can occur when the instrument you are using to measure lacks sufficient precision to indicate differences in readings. It can also occur when it is difficult to make a measurement. If two investigators measure the height of a plant, for instance, they might choose different points on the compost, and the tip of the growing point to make their measurements.

With **systematic error**, readings vary in a controlled way. They're either consistently too high or too low. One reason could be down to the way you are making a reading, for example taking a burette reading at the wrong point on the meniscus, or not being directly in front of an instrument when reading from it.

What an investigator *should not* discuss in an evaluation are problems introduced by using faulty equipment, or by using the equipment inappropriately. These errors can, or could have been, eliminated, by:

- ✓ checking equipment
- ✓ practising techniques before the investigation, and taking care and patience when carrying out the practical.

Overall, was the method or technique I used accurate enough?

- ✓ If your results were good enough to provide a confident answer to the problem you were investigating the method probably was good enough.
- ✓ If you realise your results are not precise when you compare your conclusion with the actual answer it may be you have a **systematic error** (an error that has been made in obtaining all the results). A systematic error would indicate an overall problem with the experimental method.
- ✓ If your results do not show a convincing pattern then it is fair to assume that your method or technique was not precise enough and there may have been a **random error** (i.e. measurements vary in an unpredictable way).

If I were to do the investigation again, what would I change or improve upon?

- ✓ Having identified possible errors it is important to say how these could be overcome. Again you should try and be absolutely precise.
- ✓ When suggesting improvements, do not just say 'do it more accurately next time' or 'measure the volumes more accurately next time'.
- ✓ For example, if you were measuring small volumes, you could improve the method by using a burette to measure the volumes rather than a measuring cylinder.
- ✓ Other errors arise from accuracy of measurement. And investigations can also often be improved by extending the range (e.g. temperature, time, pH, etc.) over which it is carried out.

Do any of my results stand out as being inaccurate?

In Example 2, the results from the cylinder in Experiment 2 in a concentration of 0.2 M sucrose are not consistent with the other results for this concentration. The **anomalous result** or **outlier** is highlighted in the table.

EXAMPLE 10

What reasons can I give for any inaccurate results?

In this example, the pH of yoghurt was monitored during its production using a data logger. The data logger gave very precise, consistent readings. First, it had to be set, or *calibrated*, using solutions of known pH (called buffers).

In Example 2, it's possible that the potato cylinder had not been blotted dry properly before its mass was measured.

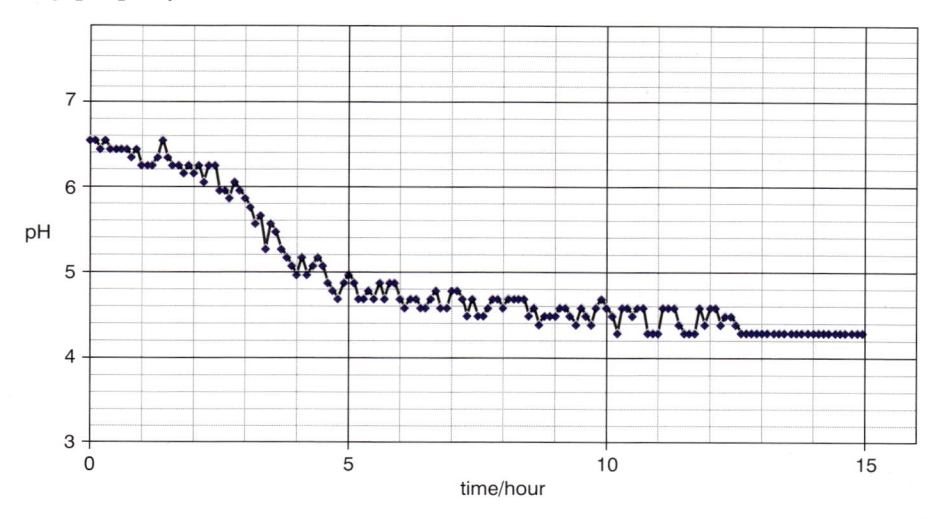

△ Fig. 6.12 Experimental results for Example 10.

EXAMPLE 11

In an investigation on the energy content of food, an investigator used the apparatus below to find the energy content of a piece of pasta.

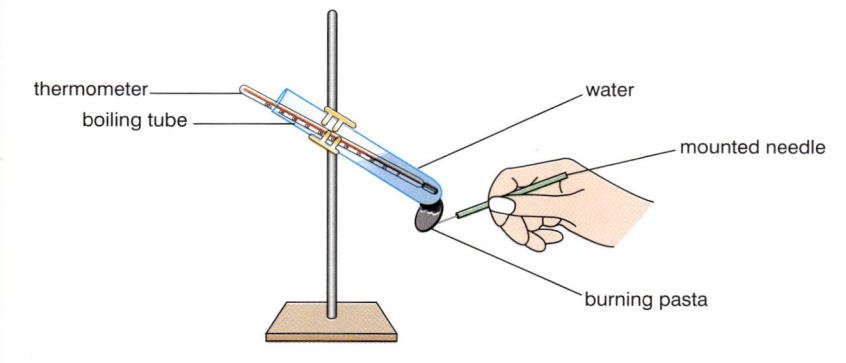

△ Fig. 6.13 Apparatus for Example 11.

Was the method or technique I used accurate enough?

The value obtained by the investigator was 1272 joules of energy per gram of pasta. This value was much lower than the value of 13 440 joules per gram of pasta printed on the packet.

The main problem with this method is that it relies on the transfer of all the chemical energy in the food to heat energy, which is used to warm the water. But using this equipment, the transfer is nowhere near 100 percent. Reasons for this are:

- ✓ conversion of the chemical energy to other forms of energy
- ✓ incomplete combustion of the food
- ✓ poor transfer of heat from the burning pasta to the water.

Measurements of the temperature rise may also be inaccurate because the heat energy is not evenly distributed through the water.

If I were to do the investigation again, what would I change or improve upon?

It is possible to improve the accuracy of energy values of food in the school laboratory by modifying the method above or using different equipment. The transfer of heat to the container holding the water (in this case, a boiling tube) can be improved, and the container could also be insulated. Many investigators use a device called a bomb calorimeter to overcome these problems. The food is burned inside the bomb calorimeter.

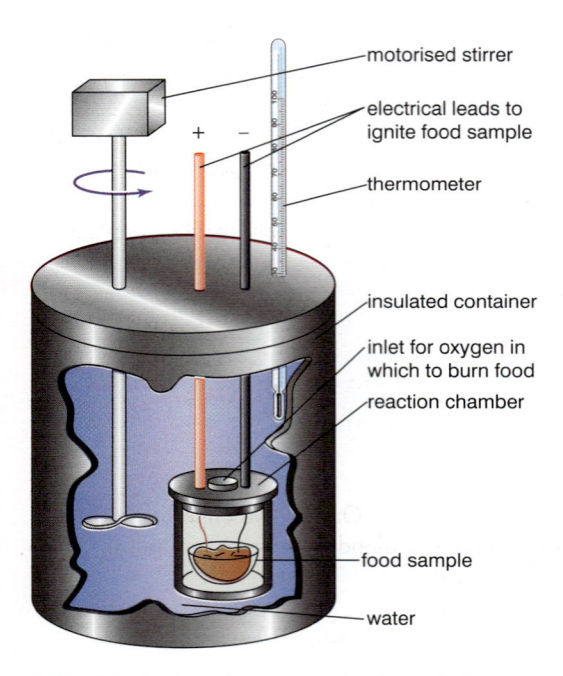

△ Fig. 6.14 Improved apparatus for Example 11.

Glossary

absorption The movement of digested food molecules from the gut into the blood stream.

abstinence Not doing, such as avoiding sexual intercourse.

abundance The number of organisms of a particular species in an area, related to the population size.

accommodation The adjustment of the shape of the lens in the eye to focus images at different distances properly on the retina.

acid rain Rain that contains higher than normal concentrations of dissolved acidic gases, such as sulfur dioxide and nitrogen oxides, which causes the rain to have a lower pH than normal.

active site A 3D-shaped space in the molecule of an enzyme that matches the shape of another molecule that reacts with it (the substrate).

active transport The movement of molecules across a cell membrane using energy from respiration. This movement is often against a concentration gradient.

adaptation The features or characteristics of an organism that makes it suited to its particular environment.

addiction When the body cannot do without a drug.

adrenaline A hormone that stimulates heart rate as preparation for 'fight or flight'.

aerobic respiration Respiration (the breakdown of glucose to release energy) using oxygen.

agitate To mix up, such as in a fermenter to keep the microorganisms in contact with the ideal amounts of nutrients and oxygen.

AIDS (acquired immune deficiency syndrome) The disease caused by the HIV virus.

alimentary canal The tubular part of the digestive system, from mouth to anus.

allele One form of a gene, producing one form of the characteristic that the gene produces.

alveoli The tiny bulges of the air sacs in lungs where gases diffuse between the air in the lungs and the blood.

amino acid The basic unit of a protein.

amniotic fluid Fluid surrounding developing fetus in the uterus, which protects the fetus from mechanical damage.

amniotic sac Membrane sac surrounding developing fetus that contains fluid which protects the fetus.

amylase An enzyme that digests starch.

anaerobic respiration Respiration (the breakdown of glucose to release energy) without oxygen (also called *fermentation*). In animal cells it produces lactic acid; in plant cells it produces ethanol and carbon dioxide.

antenatal care Care of a pregnant woman before the birth of her baby.

anther The male part of a flower that produces pollen.

antibiotic A chemical used to kill pathogens, or stop them from growing, when they have infected the body.

antibiotic resistance Resistance to the effect of an antibiotic by bacteria which are normally killed by that antibiotic.

antibody Chemicals produced by lymphocytes to defend against infection by pathogens. Antibodies are developed to be specific to different pathogens.

artery The vessel that carries blood away from the heart.

artificial insemination Placing sperm artificially in the uterus of a woman, often as part of fertility treatment.

artificial selection Human selection of plants and animals for breeding to improve particular characteristics.

aseptic Precautions involve killing all unwanted microorganisms to prevent their growth, such as in a fermenter.

asexual reproduction The production of new individuals without fertilisation, from division of body cells in the parent.

assimilation The process by which simple food molecules are made into complex molecules in the body.

atrium (plural atria) One of two chambers of the heart that receive blood from veins and pumps it into the ventricles.

auxin Plant growth hormone that controls response of shoot and root tips to light and gravity.

balanced diet Intake of food that supplies all the protein, fat, carbohydrate, vitamins and minerals that the body needs in the right amounts.

base One of four molecules (adenine [A], thymine [T], cytosine [C] and guanine [G]) that join in pairs (A with T, C with G) that link the two strands within DNA.

Benedict's reagent A solution that changes colour in the presence of simple sugars, used to test for their presence in a food sample.

biconcave The shape of a red blood cell in which the middle is pressed inwards, making the cell thinner in the middle than at the edges.
bile A highly alkaline liquid, produced by the liver and stored in the gall bladder, which emulsifies lipids.
bioaccumulation When a toxic substance, such as a pesticide, is stored in living tissue.
biological catalyst Catalysts in living cells that control metabolic reactions – or enzymes.
biological control Using organisms, such as a predator or parasite, to control the numbers of pests.
biological factors Factors that affect the distribution of organisms, caused by other organisms, such as competition for food.
biomass The mass of living material, such as the mass of a living organism.
birth control Methods used to prevent conception (such as condoms, contraceptive pill).
birth rate The number of births in a population over a particular time.
biuret test A test used to indicate the presence of protein in a food sample.
bladder Where urine is held in the body before it is released into the environment.
breed Animals of the same species but with distinct characteristics of their own.
bronchi (single: bronchus) The two divisions of the trachea as it joins to the lungs.
bronchioles The tiny tubes in the lungs that are linked to alveoli.
cancer The uncontrolled division of cells.
canine tooth Teeth with a pointed shape, behind the incisors in the mouth, that hold food while biting and chewing with other teeth.
capillaries Tiny blood vessels in the tissues. Every cell in the body is in very close proximity to a capillary, enabling gaseous exchange.
carbohydrate A molecule, such as starch or glycogen, made of many simple sugars joined together.
carbon cycle The movement of different forms of the element carbon between living organisms and the environment, represented as a diagram.
carcinogenic Something that causes cancer (the uncontrolled division of cells).
carnivore A meat-eater, also called a secondary consumer.
carpel The female structure in flowers which contains one or more ovaries and their stigmas and styles.
carrier protein A protein that sits in cell membranes and carries specific molecules through the membrane, important in active transport.
catalyst A substance that increases the rate of a chemical reaction, such as an enzyme.
cell The 'building blocks' of which tissues are composed; some microorganisms are just one cell.
cell membrane The structure surrounding cells that controls what enters and leaves the cell.
cell wall A cellulose cell wall surrounds plant cells, giving them support and shape.
cellular respiration The release of energy from the chemical bonds in food molecules such as glucose.
cellulose The material that makes up a cell wall.
cement The material that holds teeth firmly in the gums.
central nervous system The part of the nervous system that coordinates and controls responses, consisting of brain and spinal cord.
cervix At the top of the vagina, where sperm is deposited during sexual intercourse.
chemical digestion The breakdown of large molecules into smaller ones using enzymes.
chitin A fibrous carbohydrate that makes up the cell walls of fungi.
chlorophyll The green chemical in chloroplasts that captures light energy for photosynthesis.
chloroplast Organelle found in plant cells and some protoctist cells that can capture energy from light for use in photosynthesis.
chromosome A long DNA molecule that is found in a cell nucleus.
cilia Microscopic hairs that move substances such as mucus across the surface of some cells.
ciliated cell A cell with cilia on the outside of the cell membrane.
cirrhosis Liver damage often caused by excessive long-term alcohol use.
clone A genetically identical copy.
codominance When both alleles for a gene are expressed (visible) in the phenotype.
cohesion Where molecules stick together, such as water molecules in the xylem of a plant.
collecting duct The tube at the end of a nephron, where additional water may be reabsorbed from the urine.
colon The medical name for the large intestine.
combustion Burning, such as the burning of fossil fuels.
community All the populations of organisms that live in an area or ecosystem.
competition The battle for available resources between organisms that live in the same place.
concentration gradient The difference in the amount of a substance between two areas.

Diffusion is usually along the concentration gradient: molecules move from the area of high concentration to the area of low concentration.

condom A rubber sheath placed over the erect penis to act as a mechanical barrier to sperm.

cone cell One of the light-sensitive cells in the retina that respond to the wavelength of light; responsible for colour vision.

conservation The processes of protecting an environment or habitat from change, so that the natural state is maintained and the organisms that live there can flourish.

constipation The abnormally slow movement of digested food through the intestines and rectum as a result of too little fibre in the food for peristalsis to be effective.

consumer (primary/secondary) An animal in a food web. Primary consumers eat plants; secondary consumers eat primary consumers.

continuous variation Variation of features that can have any value, such as height.

contraception Birth control – methods that prevent the fertilisation of an egg during sexual intercourse.

contraceptive pill A pill containing sex hormones that prevents ovulation in women.

convoluted tubules Two sections of the nephron closely associated with a capillary. Selective reabsorption of glucose takes place in the proximal convoluted tubule.

core temperature The base temperature, about 37°C, at which the human body must remain for life processes to go on effectively.

cornea The thick clear part of the eye that bends light rays that enter the eye, focusing them on the retina.

coronary arteries/veins Blood vessels that supply blood directly to the heart muscle, which cause health problems if partly or fully blocked.

coronary heart disease The narrowing of coronary arteries and blood clots that can lead to a heart attack.

cotyledon A food store surrounding the zygote in a plant.

crop rotation Growing a different crop each season in a particular field, alternating leguminous (pea/bean) and non-leguminous crops, to keep the soil fertile.

cuticle A waxy layer that covers leaves, particularly the upper surface, to reduce water loss from the leaf.

cutting A part taken from a plant (such as piece of stem, root or leaf) and treated so that it grows into a new plant, a form of artificial asexual reproduction of plants.

cytoplasm The inside of a cell which contains the organelles and is where many chemical reactions take place.

daughter cell A cell produced by division of a parent cell, as in mitosis.

deamination The removal of the nitrogen-containing part of an amino acid for excretion.

death phase the final phase of a sigmoid growth curve, where some factor in the environment (such as lack of nutrients) prevents reproduction or increases death rate (such as build-up of toxins) so that death rate greatly exceeds birth rate.

death rate The number of deaths in a population over a particular time.

decomposer An organism that causes decay of dead material, such as many fungi and bacteria.

deforestation The destruction of large areas of forest or woodland.

denature To lose shape, such as when a protein denatures at high temperature and is not able to catalyse reactions.

denitrifying bacteria Bacteria that convert nitrates in the soil to nitrogen gas which is released to the atmosphere.

dentine The main layer of tooth below the enamel and softer than enamel.

deoxygenated Lacking in oxygen, as in blood that has passed through body tissues.

depressant A drug that slows down the response of the body.

development Increase in complexity, as in the development of differentiated cells, organs and organ systems in multicellular organisms.

dialysis A technique used to filter the blood artificially if the kidneys do not function.

diaphragm The muscular sheet at the bottom of the lungs that controls breathing.

diaphragm (birth control) A rubber cap placed over a woman's cervix before sexual intercourse to prevent sperm entering the uterus.

dichotomous key A series of pairs of questions based on easily identifiable characteristics that can be used to identify organisms.

dicotyledon A flowering plant, usually broad-leaved, that has two cotyledons in its seeds.

diffusion The net movement of molecules along a concentration gradient; it is a passive process (it does not use energy).

digestion The breakdown of food; mechanical digestion breaks food down into smaller pieces (such as by chewing and the muscular action of the stomach) ready for chemical digestion by enzymes.

digestive enzyme Enzyme in the digestive system that helps to break down large food molecules.

digestive system The organ system that digests food and absorbs nutrients into the body.

diploid A cell that contains two sets of chromosomes.

discontinuous variation Describes characteristics that have only a limited number of possibilities (such as eye colour, fur colour, sex).

disperse Spread out.

distal convoluted tubule Part of the cortex of the kidney, connected to the proximal convoluted tubule by the loop of Henlé.

distribution How organisms are spread out in an area.

DNA (deoxyribonucleic acid) The chemical that forms chromosomes and carries the genetic code.

dominant allele The allele that is expressed in the phenotype (such as spotted coat in leopards).

dormant In a suspended state of non-activity; e.g. a seed that is not active, resting until the conditions are right for the seed to germinate.

double circulation Circulation as in humans where the blood passes through the heart to be pumped first to the lungs, then returned to the heart to be pumped to the rest of the body.

double helix The shape of the DNA molecule, rather like a twisted ladder.

Down's syndrome A condition in humans caused by a chromosome mutation, where three copies of chromosome 21 are inherited instead of the usual two.

drug A substance that, when taken into the body, changes or modifies cell reactions.

dry mass The mass of a body that has had all the water removed.

ecology The study of living organisms and their environment.

ecosystem All the organisms and physical factors in a fairly self-contained area, such as a lake or desert.

effector The part of the body that brings about a response to a stimulus.

egestion The removal of undigested material from the body (faeces).

embryo A developing young, where cell division and differentiation are taking place rapidly. In plants it develops in the seed. In humans, the embryo is the stage between zygote and fetus.

emigration The movement of individuals away from a population.

emulsify To break up the large droplets of a lipid in an aqueous solution into smaller droplets.

enamel The hard outer covering of a tooth that protects the softer layers inside the tooth.

endocrine glands A collection of cells that secrete hormones directly into the blood.

enhanced greenhouse effect An increase in the greenhouse effect, most likely caused by the release of additional greenhouse gases from human activity.

environment The external factors that affect plant and animal life.

enzyme A protein that acts as a catalyst, speeding up reactions.

epidermis (or epidermal tissue in plants) The layer of cells on the outer surface of a body or organ, such as a leaf.

eutrophication The addition of nutrients to water, which may lead to water pollution.

evaporation When particles in a liquid (such as water) gain enough energy to move fast enough and become a gas (as in water vapour).

evolution Adaptation (change) of organisms over time to changes in their environment through the process of natural selection.

excess More than is needed.

excretion removal of waste (often toxic) substances that have been produced from chemical reactions inside the body, such as carbon dioxide and urea in animals.

exhalation Breathing out.

expiration Breathing out.

exponential Where the next value depends on the previous value, such as an exponential population growth, where the number of births depends on the number of mature females that can reproduce.

extinction When an entire species ceases to exist.

faeces The undigested material that remains after food is digested in humans.

family pedigree A diagram that shows the inheritance of different forms of a characteristic through the generations within a family.

famine Starvation of many people in an area.

fatty acid One of the basic units of a lipid. Three fatty acids and one glycerol molecule form a lipid.

female sterilisation Cutting the oviducts of a female to prevent eggs reaching the uterus, and sperm from reaching eggs.

femidom The female equivalent of a condom that prevents sperm entering a woman's body.

fermenter A large vessel for the culture of microorganisms, in which all the conditions can be controlled.

fertilisation Fusion (joining) of male and female sex cells.

fertility drug A drug used to stimulate a woman's ovaries to release eggs to assist her in getting pregnant.

fetus the name given to the developing baby in the uterus.

fibre A plant material that is difficult to digest and keeps the food in the alimentary canal soft and bulky.

fibrin A fibrous protein produced when blood clots.

fibrinogen A protein produced by the liver that is broken down to fibrin by an enzyme released by platelets at the site of damage to a blood vessel.

flaccid Not rigid – what plants become when they lose water by osmosis, and the vacuole shrinks and does not push against the cell wall to give the plant support.

fluoride A soluble chemical that hardens tooth enamel. It may be added to public water supplies to reduce risk of tooth decay.

food additive Any substance added to food during processing to improve its qualities, such as taste or colour, or to reduce the rate of decay.

food chain/web Diagram showing the flow of energy between organisms; a food chain is a simple 'line' (plant, primary consumer, secondary consumer); food webs are a combination of food chains.

fossil fuels Fuels formed from organic material, such as peat, coal and oil.

fruit The structure that develops from the ovary wall and contains the seed or seeds of the plant.

FSH (follicle-stimulating hormone) A pituitary hormone that controls the development of an egg in the ovary during a woman's menstrual cycle.

gamete A sex cell.

gaseous exchange In the tissues, diffusion of oxygen from blood into cells and of carbon dioxide from cells into blood; the reverse occurs in the lungs.

gene Includes an instruction in the form of a code in the DNA.

genetic code The code formed by the order of the bases in DNA that instructs cells how characteristics should be produced.

genetic diagram A diagram that displays how a characteristic may be inherited by offspring from the parents' alleles.

genetic engineering The transfer of a gene from one organism into an organism of a different species.

genetically modified Plants or animals that have had a gene from another species introduced, such as to make a plant resistant to a virus.

genotype The description of the two alleles for a particular characteristic.

geotropism A growth response in plants affected by the direction of the force of gravity.

germination The start of plant growth from a seed, which only occurs when there is the right level of temperature, light and water.

gestation The time when a fetus develops in the uterus, about 40 weeks in humans.

global warming Warming of the Earth's surface and atmosphere, possibly as a result of an enhanced greenhouse effect.

glomerular filtrate The liquid in the renal capsule produced by ultrafiltration.

glomerulus A small knot of capillaries associated with a renal capsule.

glucagon The hormone made by the pancreas that causes liver cells to break down glycogen stores and release glucose into the blood.

glycerol One of the basic units of a lipid. Three fatty acids and one glycerol molecule form a lipid.

glycogen A carbohydrate made from glucose that is stored in liver and muscle cells.

gonorrhoea A sexually transmitted bacterial disease.

graticule scale viewed through a microscope, used to calculate the observed size of a structure.

greenhouse effect The warming effect caused by greenhouse gases in the atmosphere that prevent some of the heat energy radiated by the Earth's surface escaping into space.

greenhouse gases Atmospheric gases, such as carbon dioxide and water vapour, that trap some of the heat radiated from the Earth's surface and prevent it escaping into space.

growth the permanent increase in body size and dry mass of an organism, usually from an increase in cell number or cell size (or both).

habitat A small part of an ecosystem where a species lives.

haemoglobin The red chemical in red blood cells that combines reversibly with oxygen.

haploid A cell that contains only one set of chromosomes, such as gametes.

heart rate The number of heart beats in a given time, for example, beats per minute.

herbicide A chemical that kills plants, used by farmers to control weeds in crop fields.

herbivore A plant-eater or primary consumer.

heroin A dangerously addictive drug with powerful depressant effects.

heterozygous When the two alleles for a gene are different in the genotype.

HIV Human immunodeficiency virus, transmitted during sexual intercourse, by blood, through the placenta or in breast milk. It causes AIDS.

homeostasis The maintenance of a constant internal environment in the body (such as constant temperature, water balance, concentration of oxygen and carbon dioxide).

homozygous When the two alleles for a gene are the same in the genotype.

hormonal system A chemical response system in humans where hormones produced by the endocrine glands are carried in the blood to target organs where they affect the cells.

hormone A chemical that is released from a gland and travels in the blood to another site in the body (its target organ), where it has its effect; hormones are described as chemical messengers.

humidity A measure of the concentration of water molecules in the air.

hypha single thread of fungal mycelium (plural hyphae)

immigration The movement of individuals into a population.

immune system The system that protects the body against infection and includes white blood cells.

incisor The flat-bladed tooth type at the front of the mouth for biting off pieces of food.

ingestion The taking of food into the alimentary canal.

inhalation Breathing in.

inheritance The passing of inherited characteristics from one generation to the next.

inherited A characteristic that is passed on from the parent to the offspring by genes.

insoluble Something that does not dissolve in a solvent such as water.

inspiration Breathing in.

insulin A hormone produced by the pancreas that causes muscle and liver cells to take glucose from the blood.

interspecific predation Predation of one species by another, for example predator (e.g. lion) and prey. (e.g. zebra)

intercostal muscles The muscles between the ribs that can control breathing.

intraspecific predation Predation of individuals in a species by other individuals of the same species.

involuntary Done unconsciously, without thought or control from the central nervous system, such as reflex response.

ionising radiation Radiation, such as gamma rays, X-rays and ultraviolet radiation, that can damage cells and produce mutations in genes.

IUD (intra-uterine device) A mechanical barrier placed in a woman's uterus to prevent the passage of sperm and the implantation of a fertilised egg.

key A system used for identifying an organism by asking a series of linked questions.

labour The stage at the end of a pregnancy when the mother's body prepares to give birth.

lag phase The first phase of a sigmoid growth curve, where there is little change in number as organisms prepare for growth and reproduction.

larynx The 'voice box' at the top of the trachea which produces sounds when air moves through it, as when speaking.

leaching The loss of dissolved mineral nutrients in soil water as it soaks deep into the ground beyond the reach of plant roots.

LH (luteinising hormone) A pituitary hormone that controls when ovulation occurs in the menstrual cycle.

limiting factor A factor (such as light, temperature) that controls the rate of photosynthesis; it is the condition that is least favourable.

line transect A line along which quadrats are placed to sample organisms.

lipase An enzyme that breaks down lipids (fats and oils).

lipid A fat or oil, such as cholesterol, made of the basic units of three fatty acids and one glycerol.

'lock and key' model A model of how an enzyme works, where the substrate fits tightly into the active site of an enzyme.

log phase The second phase of a sigmoid growth curve, where growth is rapid because birth rate greatly exceeds death rate, when conditions are ideal for growth.

lymph Fluid in the lymphatic system.

lymphatic system A system of tubes that collects tissue fluid from the tissues and returns it to the blood system near the heart.

lymphocytes A type of white blood cell that make antibodies to attack a pathogen.

magnification the amount by which a microscope increases the observed size of a structure compared with its actual size: calculated by multiplying the eyepiece magnification with the objective magnification.

malaria A disease in humans caused by a blood parasite transmitted by mosquitoes, which causes many deaths.

malnutrition Not getting the right amounts and balance of nutrients and other essential substances in the diet, including a diet that has too much or too little of any of these.

mechanical digestion The breaking up of food into smaller pieces through biting and chewing by teeth.

medicinal drug A drug used to treat the symptoms or causes of illness, often prescribed by a doctor.

meiosis The form of cell division that produces four genetically different haploid cells from a diploid parent cell, producing gametes.

memory cell A type of blood cell produced by lymphocytes that 'remembers' pathogens that have been destroyed by antibodies, so the same pathogens cannot re-infect the organism.

menstrual cycle The continuous sequence of events in a woman's reproductive organs. Each cycle of ovulation (ripening and release of an egg) and menstruation (shedding of the unwanted uterus lining) takes about 28 days and is controlled by a number of hormones.

mesophyll Tissue in plants that packs the spaces between other tissues.

metabolic reaction The chemical reactions that cause the life processes of organisms to occur.

microbes (microorganisms) Tiny, usually single-celled, organism (such as bacteria) that cause disease.

micropropagation The culture of explants in the lab to produce many clones of a single plant.

microvilli Tiny finger-like extensions of the cell membrane of the surface cells of villi.

minerals (mineral ions) Nutrients that plants and animals need in small amounts, such as nitrates that are needed for making amino acids.

mitochondrion Small structures in the cytoplasm of cells where respiration takes place.

mitosis The form of cell division that produces two identical diploid daughter cells from a diploid body cell, used for growth and repair in the body and in asexual reproduction.

molar The tooth type found at the back of the mouth that has a large grinding surface for chewing.

molecule A particle that makes up substances like water, carbon dioxide and food.

monocotyledon a flowering plant, such as a grass, that has one cotyledon in its seeds.

monohybrid cross The inheritance of a characteristic produced by one gene.

motor neurone A nerve cell that carries impulses from the central nervous system to an effector, such as a muscle cell or gland cell.

mucus Thick liquid produced on the surface of cells.

mutagen A chemical that produces mutations in genes.

mutation A random change in a gene, producing a new allele.

mycelium A mass of hyphae that form the body of a fungus.

natural selection The influence of the environment on survival and/or reproduction, so that organisms with some characteristics are more successful at producing offspring than others.

negative feedback A key mechanism by which homeostasis is achieved; the body monitors its internal environment and reacts by trying to reduce any changes that occur, back to the normal level.

nephron A tiny kidney tubule where ultrafiltration and reabsorption take place to produce urine.

nerves Bundles of cells that connect receptors to the central nervous system and the central nervous system to effectors.

nervous system A response system in humans that uses electrical impulses between receptor cells, nerve cells and effector cells to produce a response to a stimulus.

net movement The movement of molecules in both directions, with more molecules moving from a high concentration to a low concentration. When the concentrations become equal, molecules still move, but now equal numbers are moving in both directions so there is *no* net movement.

neurone A nerve cell, specially adapted for carrying nerve impulses.

nitrifying bacteria Soil bacteria that convert ammonium ions to nitrite ions, and nitrite ions to nitrate ions, important in the nitrogen cycle.

nitrogen cycle The movement of the element nitrogen in different forms between living organisms and the environment, shown as a diagram.

nitrogen-fixing bacteria Bacteria found in the soil and in root nodules of some plants that can convert atmospheric nitrogen into a form of nitrogen that plants can use.

nucleus the organelle in plant and animal cells that contains the genetic material.

nutrition The process by which an organism gets the chemicals and energy it needs (nutrients) from its surroundings.

oestrogen A hormone produced by the ovaries that helps to control the menstrual cycle and produces secondary sexual characteristics in females.

omnivore An animal that eats both plants and animals.

optimum pH The pH at which an enzyme works fastest.

optimum temperature The temperature at which an enzyme works fastest.

organ A group of tissues that work together to carry out a particular function, such as the stomach or the heart.

organ system A group of organs in the body that work together for a particular function, such as the circulatory system or digestive system.

organ transplant Placing an organ from one person into another person, to replace an organ that has failed.

organelle A structure within a cell that carries out a particular function, such as nucleus, chloroplast.

organism A living thing, ranging in size from a single-celled microorganism to an elephant.

osmoregulation The regulation of the concentration of water in the cells, blood and bodily fluids.

osmosis The movement of water molecules across a partially permeable membrane from a lower solute concentration to a region of a higher solute concentration.

ovary (in plants and mammals) A structure that contains immature egg cells.

oviduct Tube that carries the egg from the ovary to the uterus.

ovulation In a woman's menstrual cycle, when an egg is released from the ovary, usually about halfway through the cycle.

ovule The female structure in flowers that contains one egg cell.

oxygen debt The result of anaerobic respiration; extra oxygen is needed after exercise to break down the lactic acid formed during anaerobic respiration.

oxygenated Contains oxygen, as in the blood that has passed through the lungs.

palisade cell One of the cells in the upper part of a leaf that contain the most chloroplasts and carry out most of the photosynthesis.

parasite An entity that can only reproduce inside the cells of another organism. Viruses and some animals are parasites.

partially permeable Describing a membrane (such as the cell membrane) that has tiny pores (holes) which allow some molecules (such as water) but not larger molecules to pass through.

passive The opposite of *active*: something that happens without the need for additional energy.

pathogen An organism that causes disease in another living organism. Examples are found in fungi, bacteria, protoctists and viruses.

peristalsis The rhythmic muscular contractions of the stomach and intestine that move food along the alimentary canal.

pest An organism that causes damage to plants or animals.

pesticide A chemical used to kill pests.

phagocyte A type of white blood cell that engulfs and destroys pathogens.

phenotype The visible characteristics of an organism as a result of its genes.

phloem Tubes formed from many living cells that carry dissolved substances, such as sucrose and amino acids, from the leaves to other parts of a plant.

photosynthesis The chemical process in which plants absorb light/energy, carbon dioxide and water and transfer the energy to chemical energy in glucose, releasing oxygen as a waste product.

phototropism A growth response in plants affected by light.

physical digestion The breakdown of large food pieces, such as by chewing in the mouth.

physical factors Environmental factors such as light, temperature and water that affect the distribution of organisms non-living or even abiotic.

placenta The structure formed by a fetus that attaches to the uterus wall and exchanges essential substances between the mother's blood and the blood of the fetus.

plasma The liquid, watery part of blood which carries dissolved food molecules, urea, hormones, carbon dioxide and other substances around the body and also helps to distribute heat energy.

plasmid A small circle of genetic material found in some bacteria in addition to the circular chromosome.

platelets Fragments of much larger cells that cause blood clots to form at sites of damage in blood vessels.

pleural membranes The two membranes surrounding the lungs.

plumule The part of a plant embryo that will develop into shoots and leaves.

pollen grain A male structure in plants that contains the male gamete.

pollen tube A tube that develops from pollen grain down through the style, carrying the male gamete to the female gamete.

pollination The process in which pollen from one flower is transferred to another flower, before fertilisation can take place.

pollution Damage to the environment, people and other organisms, often as a result of adding chemicals to the air, water or land.

population All the organisms of one species living in the same habitat.

premolar The teeth found behind the canines that help with cutting off food and have a small grinding surface for chewing.

primary consumer An animal that eats plants (also called a herbivore).

producer An organism that produces its own food, such as plants using light energy in photosynthesis to produce glucose.

progesterone A hormone produced in the ovaries that helps to control the menstrual cycle.

protease An enzyme that digests protein.

proteins Large molecules, such as found in muscle tissue, made of many amino acids.

pulp cavity The soft layer in the middle of the tooth containing the blood vessels and nerves.

Punnett square A form of genetic diagram.

pyramid of biomass A diagram that shows the biomass in different trophic levels of a food chain, often a pyramid shape.

pyramid of energy A diagram that shows the energy content of different trophic levels of a food chain, always a pyramid shape.

pyramid of number A diagram that shows the number of individual organisms in different trophic levels of a food chain, often a pyramid shape.

quadrat A square frame used for sampling the abundance and distribution of organisms.

radicle The part of a plant embryo that will develop into roots.

receptor A structure that detects information about the environment (such as heat).

recessive allele The allele that is not expressed in a phenotype unless there are two copies.

recombinant DNA DNA that contains a gene inserted during genetic engineering.

recreational drug A drug not prescribed by a doctor and used for its effects, such as alcohol or heroin.

red blood cell A blood cell that contains red haemoglobin and is adapted for carrying oxygen.

reflex The simplest response to a stimulus, such as blinking.

reflex arc The pathway along which nerve impulses travel during a simple reflex (such as withdrawing your hand from a hot object).

reflex response A rapid response to a stimulus that does not involve the brain, such as blinking.

relay neurone A neurone found in the central nervous system that links other neurones together.

renal capsule A cup-shaped structure at the start of a nephron where filtration occurs.

reproduction (asexual/sexual) The process of creating new members of a species; asexual reproduction does not involve sex cells; sexual reproduction is the combination of sex cells.

respiration the chemical process in which glucose is broken down inside the mitochondria in cells, releasing energy and producing carbon dioxide and water.

respiratory system The body system that includes the lungs, diaphragm and other organs involved in breathing.

retina The part of the human eye, at the back of the eye, containing light-sensitive cells.

rhythm method A natural method of birth control where a couple avoid intercourse, or the penis is removed from the uterus before ejaculation, at the time in the menstrual cycle when the woman is most fertile.

rod cell A type of cell in the retina that is sensitive to light intensity; responsible for black and white images and for night vision.

root hair cells Cells in the epidermis of roots that have a long extension of cytoplasm, where uptake of substances from soil water occurs.

runner A stem that grows along the ground, making it easy to put down new roots and so reproduce asexually.

saprotrophic nutrition The digestion of dead food material outside the body, as in fungi.

scrotum The sac that holds the testes.

secondary consumer An animal that eats primary consumers.

secondary sexual characteristics Physical characteristics that develop at puberty, such as facial hair and deeper voices in boys.

secretion Releasing chemicals that have been made inside the cell into the fluid outside the cell.

seed A hard-shelled structure formed from ovule that contains the plant embryo and food stores.

seed coat The tough outer covering of a seed.

selective breeding Choosing individuals for breeding based on their characteristics.

selective reabsorption The reabsorption of more of some substances (such as glucose from filtrate) than others.

sense organs Organs containing receptor cells adapted for the receiving of a particular type of stimulus.

sensitivity the detection of changes (stimuli) in the surroundings by a living organism, and its responses to those changes.

sensory neurone A nerve cell that carries impulses from a receptor cell to the central nervous system.

sewage Human waste, faeces and urine, which needs to be disposed of.
sex chromosomes Chromosomes that affect the sex of the individual. In humans these are the XX chromosomes in women and XY in men.
sexual reproduction The production of new individuals by the fusion of a male and a female gamete.
sickle cell anaemia A condition in humans caused by the inheritance of two sickle cell alleles.
sigmoid population growth curve A growth curve with a typical shape in four phases, shown by populations living in ideal conditions in a limited area.
simple sugar A basic sugar unit (such as glucose) that can join together with other sugar units to make large carbohydrates such as starch and glycogen.
single cell protein Protein-rich, fibrous food made using fungus grown on carbohydrate, then extracted and dried.
sink Part of a plant where a substance is converted into other substances for growth or storage or used, such as sucrose being broken down to glucose in respiring cells.
social implications Effects on the way people live and work together.
soil erosion The washing away of soil as a result of rainfall when there is little vegetation to hold on to the soil.
soluble Dissolves in a solvent, e.g. salt dissolves in water.
source Where something is produced or enters the body, such as photosynthesising cells for glucose.
specialisation The process by which cells have developed (often in terms of structure) to perform a particular function more efficiently.
specific Describing the action of an enzyme or antibody that is targeted to a particular other substance to produce a particular reaction or result.
sperm The male gamete in humans.
sperm ducts Tubes that carry sperm from the testes to the urethra as they leave the body.
spermicide A chemical that kills sperm.
spongy mesophyll The layer of cells in the lower part of the leaf where there are many air spaces, so increasing the internal surface area to volume ratio.
stamen The male structure in flowers that contains the anther.
starch A form in which carbohydrates are stored as energy in plants.

stationary phase The third phase of a sigmoid growth curve in a population, where some factor becomes limiting, such as a nutrient, and birth rate and death rate are equal.
starvation Eating too little food to supply the body with its need for energy and nutrients.
stigma The female structure in flowers where pollen grains attach in pollination.
stimulus A change in the environment that triggers a response in an organism.
stomata Tiny holes in the surface of a leaf (mostly the lower epidermis) which allow gases to diffuse in and out.
style A structure that supports the stigma in a flower.
substrate A molecule that reacts to a particular enzyme.
succulent A plant with thick, fleshy water-filled tissue, adapted to living in dry conditions such as a desert.
sucrose A form in which carbohydrates are stored as energy in plants.
synthesis The building of larger molecules from smaller ones, such as the formation of proteins from amino acids.
system A group of organs that work together to carry out a particular function, such as the mouth, stomach and intestines in the digestive system.
systemic pesticide A pesticide that is absorbed by a plant and carried through the phloem to all parts, so that pests eating any part of the plant are affected by the pesticide.
target organ An organ of the body containing cells that respond to a particular hormone.
tension A 'pull', such as the tension of the transpiration stream that pulls water through xylem up a plant.
tertiary consumer An animal that eats secondary consumers.
testa In plants, the thick seed coat that protects the seed inside.
testes The site of sperm production in men.
testosterone A hormone produced by the testes that produces secondary sexual characteristics in boys.
thorax The centre part of the body, protected by the ribs, which contains the lungs and heart.
tissue A group of similar cells that have a similar function, such as muscle tissue.
tissue culture Another name for micropropagation.
tissue rejection When the immune system attacks and destroys 'foreign' tissue in the body, such as after an organ transplant.

trachea The tube leading from the mouth to the bronchi, sometimes called the windpipe.

transgenic organism An organism that contains DNA from a different species.

translocation Movement of dissolved sugars and other molecules through a plant.

transpiration Evaporation of water vapour from the surface of a plant.

transplant Tissue taken from an organism and placed elsewhere, e.g. kidney taken from one person and placed in a person suffering from kidney failure.

trophic level A feeding level in a food chain or food web, such as producer, primary consumer.

tropism A growth response to a particular stimulus in plants.

turgid Describing a plant cell that has a full vacuole and the cytoplasm pushes against the cell wall, giving the plant structure and support.

ultrafiltration Filtration on a molecular scale, as happens between the glomerulus and renal capsule.

urea A substance produced in the liver from the breakdown of amino acids not needed in the body.

ureters Tubes that connect the kidneys to the bladder.

urethra A tube that connects the bladder to the outside of the body.

urine A liquid waste produced by kidneys, containing water, urea and salts.

uterus The human female's womb, where a fetus develops before birth.

vaccination Giving a vaccine (to a person or animal) to stimulate the immune system and protect against infection.

vacuole Many plant cells have a large central vacuole that contains cell sap.

vagina The part of the female reproductive system where the penis is placed during sexual intercourse.

valves Flaps in the heart, and in veins, that prevent the flow of blood in the wrong direction.

variation Differences between organisms; continuous variation describes features that can have any value (such as height); discontinuous variation describes characteristics that have only a particular number of possibilities (such as eye colour, fur colour, sex).

variety The range of difference between organisms.

vascular bundle The 'vein' of a plant that contains the conducting tissues of xylem and phloem.

vasectomy Male sterilisation – cutting a man's sperm ducts to prevent sperm leaving the body.

vasoconstriction Narrowing of the blood vessels.

vasodilation Widening of the blood vessels.

vector A mechanism for carrying genetic material into the nucleus of a cell in genetic engineering, such as a virus or bacterial plasmid.

vein (animal) One of the blood vessels that carry blood to the heart.

vein (plant) see vascular bundle

ventilation Another word for breathing.

ventricles The two chambers of the heart that receive blood from the atria and pump it out through arteries.

villus (plural villi) A finger-like projection of the small intestine wall where absorption of digested food molecules occurs.

vitamins Nutrients needed by the body in tiny amounts to remain healthy, for example vitamins A, C and D.

voluntary Done by choice.

waste product A product of a chemical reaction that is not needed, such as oxygen in photosynthesis.

water cycle How water cycles between living organisms and the environment.

water potential The capacity of a plant cell to take up water by osmosis. A turgid cell cannot take up any more water and thus has no water potential.

water potential gradient The difference in water potential between two regions, such as in a plant.

wilting When a plant collapses because its cells are not turgid and so cannot support the plant's structure.

withdrawal symptoms Painful effects on the body caused by not taking a drug that the body has become addicted to.

xylem Tubes, formed from dead cells in the vascular bundles of a plant, which carry water and dissolved substances from the roots to the leaves and other parts of the plant.

yield The amount of food produced from a crop, such as grain from wheat, or a farm animal, such as eggs from chickens.

yoghurt A thick product of milk, made by the action of bacteria.

zygote A fertilised egg, formed from the fusion of a male and a female gamete.

Answers

The answers given in this section have been written by the author and are not taken from examination mark schemes.

SECTION 1 CHARACTERISTICS AND CLASSIFICATION OF LIVING ORGANISMS

Characteristics of living organisms

Page 11

1. a) Any suitable answers for human, such as:

   movement, walking; respiration, combination of oxygen with glucose to release energy, carbon dioxide and water; sensitivity, vision; growth, increase in height; reproduction, having a baby; excretion, producing urine; nutrition, eating food.

   b) Any suitable answers for a specific animal such as:

   movement, crawling; respiration, combination of oxygen with glucose to release energy, carbon dioxide and water; sensitivity, smell; growth, increase in length; reproduction, producing young; excretion, losing carbon dioxide through respiratory surface; nutrition, eating food.

   c) Any suitable answers for a plant, such as:

   movement growing towards light; respiration combination of oxygen with glucose to release energy, carbon dioxide and water; sensitivity detecting direction of light; growth increase in height; reproduction producing seeds; excretion diffusion of waste products out of leaf for photosynthesis (oxygen) and respiration (carbon dioxide); nutrition taking in nutrients from soil and making glucose by photosynthesis.

2. movement – to reach best place to get food or other conditions favourable for growth

   respiration – to release energy from food that can be used for all life processes

   sensitivity – to detect changes in the environment

   growth – to increase in size until large/mature enough for reproduction

   reproduction - to pass genes on to next generation

   excretion – to remove harmful substances from body

   nutrition – to take in substances needed by the body for growth and reproduction.

Page 12 Extended

1. Reproduction.
2. Movement - viruses cannot move themselves. Respiration - viruses do not respire. Sensitivity - viruses do not sense or respond to changes in the environment. Growth - viruses do not build new materials within their structure to produce a permanent increase in size. Excretion - viruses do not carry out metabolism, so they do not have waste products to excrete. Nutrition - viruses do not take in food substances for use in respiration or to build new materials.
3. Any suitable argument with justification, such as: Non-living materials do not reproduce, so viruses must be living.
4. Any suitable argument with justification, such as: Viruses cannot live independently of other organisms and show just one of the seven characteristics of living organisms, so they cannot really be classed as living.

Classification and diversity of living organisms

Page 18

1. a) The unique part is *leo* which is the species name.
   b) The *Panthera* part is the genus name which is shared with other closely related species.
3. Every organism has a two-part name consisting of genus and species that is unique to that organism. This means that people discussing an organism can be certain that they are talking about the same species and not another with the same common name.

Page 19

1. **EXTENDED** A cladogram is constructed by comparing species and grouping them according to how many features they share. It shows the closeness in similarity between the species in the diagram.
2. **EXTENDED** Cladistics shows how similar or different organisms are based on characteristics such as physical features or genetic code sequence, but doesn't attempt to group species into categories as in the five kingdom method.

Page 23

1. a) Fins, tails for swimming, scales, gills.
   b) Soft moist skin, gills in young stages metamorphose to have lungs in adult, external fertilisation.
   c) Scaly dry skin, lungs, internal fertilisation, lay leathery-shelled eggs.
   d) Feathered skin, lungs, internal fertilisation, wings, lay hard-shelled eggs.
   e) Hair or fur, lungs, internal fertilisation, young fed on milk produced from mammary glands.
2. It is a vertebrate (backbone), and a bird, because only birds have feathers.

Page 27

1. Outer protein coat surrounding genetic material.
2. They do not have most of the characteristics of a living cell, and behave like particles until they have infected a cell.
3. Viruses are much smaller than bacteria, in the region of 100 nm (100×10^{-9} m) compared with 2 μm (2×10^{-6} m).
4. Bacteria are single cells, no nucleus, cell wall, single chromosome lying free in cytoplasm; protoctists have a nucleus containing the chromosomes, may contain chloroplasts and are much larger than bacterial cells.
5. a) Cell walls and central vacuole, cannot move around
   b) No chloroplasts, may store carbohydrate as glycogen
6. Toadstools and mushrooms are the reproductive structures of fungi. The main body of the fungus is the mycelium, the mass of tiny thread-like hyphae which is often below the surface of the ground.

Page 31

1. Insects have three parts to the body, six legs and two pairs of wings; crustaceans have a hard jointed exoskeleton, with limbs that may be modified for different purposes; arachnids have a two-part body and eight legs; myriapods have a body of many segments and many legs.
2. Annelids have a body made of many similar segments, while nematodes are round worms with no obvious segments.
3. They have a mantle (large sac-like structure for breathing), no true legs and mostly made of soft tissue.

Page 32

1. Their reproductive structures are found in flowers, and they produce seed that contain food stores called cotyledons.
2. Dicotyledonous plants have two cotyledons in the seed, monocotyledonous plants only have one. Monocotyledons have long strap-like leaves with parallel veins, dicotyledons have broad leaves of many shapes with branching veins.

Simple keys

Page 36

1. The key identifies organisms by using questions, where each question has only two possible answers. So with each question a group of different organisms is divided into two groups. This is rather like the branching of a tree.

2. Any two suitable answers, such as: the key may not include sufficient differences between groups to place an organism at species level accurately; individuals vary within a species, so it may be difficult to decide if an individual does or does not have a particular feature; it might be the wrong time of year to identify particular features e.g. plants don't have flowers at some times of the year.

SECTION 2 ORGANISATION AND MAINTENANCE OF THE ORGANISM

Cell structure and organisation

Page 48 top

1. a) Drawing should be drawn with thin, clear pencil lines, no crossing out, to show the outline of the cell in the photograph and the central shape.
   b) Diagram should be labelled to show nucleus, cytoplasm and cell membrane.
2. Cell wall, large vacuole, chloroplast

Page 48 bottom

1. a) chloroplast
   b) large vacuole
   c) cell wall

Levels of organisation

Page 53

1. a) Any one suitable, such as: muscle tissue, nervous tissue, bone tissue
   b) Any one suitable, such as: heart, liver, brain
   c) Any one suitable, such as: nervous system, digestive system, circulatory system
2. a) Any one suitable, such as: palisade cells, root hair cells
   b) Any one suitable, such as: leaf, root

Page 56

1. a) Ciliated cells line some tubes in animal organs, such as the respiratory tract of humans. The cilia on the outside of the cells help move substances along inside the tubes.
   b) Found in muscles, can contract in length.
   c) Found in blood, carry oxygen around attached to haemoglobin inside the cell.
   d) Near the tips of plant roots, have long cell extensions to increase surface area for absorption of materials into the root.
   e) Found in vascular bundles (veins) of plants, carry water and dissolved substances through the plant and help to support the plant.

Movement into and out of cells

Page 66

1. Any answer that means the same as the following:

    net movement – the sum of movement in all the different directions possible

    diffusion – the sum of the movement of particles from an area of high concentration to an area of lower concentration in a solution or across a partially permeable membrane.

2. Passive, because no energy is provided by the cell for it to happen.

3. Only particles that are small enough to pass through the membrane can diffuse. Larger molecules cannot diffuse through the membrane.

Page 71

1. Any answer that means the same as the following: the net movement of water molecules from a region of their high concentration to a region of their lower concentration.

2. **a)** It is a passive movement of molecules as the result of a concentration gradient.

   b) Osmosis only considers the movement of water molecules, diffusion considers the solute molecules.

3. Diagram should show water molecules leaving the red blood cell as a result of osmosis and entering the solution.

4. The strong cell wall prevents more water entering a plant cell than there is space for in the cell, that is when the cell is turgid. Turgid cells have a specific shape, as a result of the cell wall, and this supports the plant, keeping it upright.

Page 72

1. **EXTENDED** The water potential of the cells inside the plant root is lower than the water potential of the soil water surrounding the root, so water moves down the water potential gradient into the root.

Page 73

1. **EXTENDED** Active transport is the absorption of a substance by a cell against its concentration gradient, using energy.

2. **EXTENDED** Uptake of nitrate ions by root cells in plants because plants need nitrate ions for making amino acids but they are in higher concentration inside plant cells than in soil water. Uptake of glucose from digested food in the small intestine by epithelial cells of the villi in humans, because glucose is essential for use in respiration.

Enzymes

Page 77

1. A catalyst is a substance that speeds up the rate of reaction but remains unchanged at the end of the reaction.

2. A chemical that is found in living organisms that acts as a catalyst.

3. Without enzymes, the metabolic reactions of a cell would happen too slowly for life processes to continue.

Page 80

1. As temperature increases, the rate of the reaction will increase, up to a maximum point (the optimum) after which it decreases rapidly as the enzyme is denatured.

2. The optimum pH for pepsin is around pH 2, which is very acidic as in the contents of the stomach. The optimum for trypsin is around pH 8, which is more alkaline like the contents of the small intestine. Each enzyme has an optimum pH that matches the environment in which they work, so that they work most efficiently there.

Page 81

1. **EXTENDED** It is a model that explains how enzymes work in terms of the close fit of the substrate in the active site of the enzyme.

Page 82

1. **EXTENDED** The shape of the active site matches the shape of the substrate. Different substrate molecules have different 3D shapes, so each enzyme only works on a single specific substrate (or substrates that have a shape that is identical where it fits into the active site).

2. **EXTENDED a)** The cooler molecules are, the slower they move. So the longer it takes for the enzymes and substrate molecules to bump into each other and the substrate to fit into the active site. So the cooler the temperature, the slower the rate of reaction.

   b) As temperature increases the atoms in the enzyme vibrate more. This changes the shape of the active site, making it more difficult for the substrate to fit into the active site and so slowing down the rate of reaction. Eventually, the atoms vibrate so much that the shape of the active site is destroyed and the enzyme is denatured.

3. **EXTENDED** At a pH above and below the optimum of pH 2, the shape of the active site is changed as the interactions between the amino acids in the enzyme are affected by the pH and change. This makes it more difficult for the substrate to fit into the active site, so the rate of reaction slows down.

Page 83

1. **EXTENDED** Enzymes break down the large storage molecules in the seed to smaller substances that can be used to release energy in respiration or be used to make larger molecules needed to make new cells (growth).
2. **EXTENDED** When the seedling has leaves, it can use photosynthesis to make sugars for respiration and growth. And with a root it can absorb mineral nutrients needed to make other substances in new cells, for growth.

Page 86

1. **EXTENDED** Pectinase added to fruit pulp causes more juice to be released and produces a clear juice. This is because the enzyme breaks down substances in the cell walls, releasing more juice and clearing the cloudiness caused by cell wall tissue in the juice.
2. **EXTENDED** Cleans clothes better than without enzymes, saves energy because lower temperatures can be used to produce the same effect.

Page 88

1. **EXTENDED** A fermenter is a large vessel in which microorganisms can be cultured in controlled conditions.
2. **EXTENDED** Any two appropriate answers, such as penicillin and enzymes for biological washing products.
3. **EXTENDED** It produces large quantities of product more quickly than by other means because the microorganisms are kept in ideal conditions. It is easy to extract the product from the liquid produced in the fermenter.
4. **EXTENDED** Ideal nutrient levels, by monitoring concentration continuously and adding more when needed.

    Temperature, which increases due to respiration, is monitored and reduced by circulating cooling water in the jacket around the fermenter.

    pH is monitored continuously and controlled by adding buffer chemicals as needed.

5. **EXTENDED** Aseptic precautions remove any other microorganisms that might compete with the organism grown in the fermenter and reduce the amount of product that is formed.
6. **EXTENDED** *Penicillium* culture is added to the fermenter with the correct amount of nutrients, and the right temperature and pH for maximum rate of growth of the fungus. Air is bubbled through the fermenter to keep conditions aerobic. The liquid contents are drained from the fermenter and the penicillin extracted and purified for use.

Nutrition

Page 95

1. a) fatty acids and glycerol, b) simple sugars, c) amino acids
2. Protein is formed from amino acids, carbohydrates from simple sugars; carbohydrates often made from one kind of simple sugar, proteins from many different kinds of amino acids.

Page 97

1. a) i) A red-brick precipitate would form, because glucose is a reducing sugar.
    ii) The solution wouldn't change colour as there is no starch present.
   b) i) There would be no change in colour because sucrose and the starch in wheat flour are not glucose (reducing sugars).
    ii) The solution would turn blue-black because of the starch in flour.
2. Crush the walnut using a mortar and pestle, then
   a) to test for fat – mix part with ethanol, decant the liquid and add water, if the mixture turns cloudy, then fat is present
   b) to test for protein – mix part with water to form a solution, add a few drops of biuret solution – if protein present, a blue ring forms at the surface, which disappears to form a purple solution.

Page 99

1. $6CO_2 + 6H_2O \rightarrow C_6H_{12}O_6 + 6O_2$
2. CO_2 from air, H_2O from soil water, $C_6H_{12}O_6$ used in cells for respiration or converted to other chemicals for use in cells, O_2 released into air if not needed in respiration
3. Most organisms other than plants get their energy in chemical form from the food that they eat. That energy was originally converted from light energy to chemical energy during photosynthesis in a plant cell and then transferred as chemical energy along the food chain.

Page 100

1. Test the leaf of a variegated plant for starch. Starch is only produced in the green parts of the leaf, where there is chlorophyll, so only the green parts of the leaf photosynthesise.

2. Heat in a water bath, keeping the ethanol away from open flames such as Bunsen burner, because ethanol gives off flammable fumes.

3. Place one de-starched plant in an atmosphere with no (or limited) carbon dioxide (due to potassium hydroxide) and one in an atmosphere high in carbon dioxide (due to carbon dioxide given off in reaction between marble chips and dilute acid). Shine light on the plants. Test one leaf from each plant after several hours. Only the leaf in high carbon dioxide will have produced significant amounts of starch as a result of photosynthesis.

Page 104

1. **EXTENDED** A limiting factor is the factor that is limiting the rate of a reaction because it is the one in shortest supply at that particular time.

2. **EXTENDED** a) As light increases, so rate of photosynthesis increases.
   b) As carbon dioxide increases, so rate of photosynthesis increases.
   c) As temperature increases, the rate of photosynthesis increases up to a maximum, after which it decreases rapidly.

3. **EXTENDED** a) As light increases, more energy is supplied to drive the process of photosynthesis.
   b) As carbon dioxide increases, so there is more reactant for the process.
   c) As temperature increases, up to the maximum the particles in the reaction including enzymes are moving faster and bump into each other more. Above the maximum the rate of photosynthesis decreases because the enzymes that control the process start to become denatured.

Page 105

1. Thin broad leaves, chlorophyll in cells, veins containing xylem tissue that transports water and mineral ions to the leaves and phloem tissue that takes products of photosynthesis to other parts of the plant, transparent epidermal cells, palisade cells tightly packed in a single layer near top of leaf, stomata to allow gases into and out of leaf, spongy mesophyll layer with large internal surface.

2. A large surface area helps to maximise the rate of diffusion, in this case diffusion of carbon dioxide into cells for photosynthesis and oxygen out of cells so that it can be released into the air.

3. It allows as much light as possible to pass through the epidermal cells to reach the palisade cells below where there are chloroplasts.

Page 107 Extended

1. a) The leaves of the plant with limited nitrogen are paler yellow than those of the plant with a lot of nitrogen.
   b) The plant with plenty of nitrogen is larger, bushier and has more leaves than the plant with limited nitrogen.

2. The leaf cells of the plant with limited nitrogen will contain less chlorophyll because nitrogen is needed to make this substance.

3. The plant with plenty of nitrogen is not only able to make more chlorophyll and therefore photosynthesise more and produce more carbohydrate, it also has sufficient nitrogen to convert some of that glucose into proteins. So it can make more new cells more rapidly than the plant with limited nitrogen, and so grow taller and bushier and produce more leaves.

4. When the crop plants grow they take in nitrogen from the soil and convert it to substances such as proteins and chlorophyll in the plant tissues. When the plant is harvested, the nitrogen in the tissues is taken as well. This leaves less nitrogen in the soil for the next crop. With a smaller amount of nitrogen in the soil, the new crop will not grow as well as the previous crop. Additional nitrogen, in the form of nitrogen-containing fertilisers, makes sure the new crop has sufficient nitrogen for rapid and healthy growth.

Page 108

1. Plants make their own foods, and need to convert the carbohydrates made by photosynthesis into other substances, such as proteins, which contain additional elements.

2. a) Nitrogen is an essential element for making substances other than carbohydrates, such as proteins.
   b) Magnesium is needed to make chlorophyll which is the green substance in plants.

3. **EXTENDED** a) Stunted growth: because without proteins, the plant cannot make new cells so the plant will not grow well.
   b) Without enough magnesium the plant will not be able to make enough chlorophyll, so it will lose the green colour and become yellow. Any magnesium in the plant is transported to the new leaves, so that photosynthesis can continue there for making food for growth.

4. **EXTENDED** a) To improve the rate of growth and crop yield from the plants growing in the field.
   b) Any two from: blocking of waterways by rapidly growing plants and algae; toxic compounds in water from algae kill other organisms; death of aquatic plants and animals due to insufficient oxygen in water; problems for young babies that drink the water.

Page 109-110 Extended

1. More intense sunlight, so likely to make sufficient vitamin D naturally in skin and less dependent on vitamin D in diet.
2. Higher latitude so less intense sunlight means less vitamin D produced in skin, and poor diet would mean increased risk of too little vitamin D in diet.
3. Cod liver oil is formed from liver which is a good source of vitamin D.
4. If diets are include more sources of vitamin D, such as fish eggs, cheese and milk, then the risk of vitamin D deficiency is reduced.
5. Lack of light on the skin means little chance of natural vitamin D production, and a vegetarian diet reduces the amount of vitamin D available in the diet. Vitamin D supplements in the diet can help to avoid the deficiency, but care must be taken not to take too much vitamin D over a long period as high levels in the body can be toxic. So supplements during the winter maybe more advisable than supplements all the time.

Page 110

1. Carbohydrates, proteins and fats.
2. Carbohydrates from pasta, rice, potato, bread, wheat flour. Proteins from meat, pulses, milk products, nuts. Fats from vegetable oils, butter, full-fat milk products, red meat.
3. Vitamins, minerals, water and fibre.
4. Vitamins and minerals needed for maintaining health of skin, blood, bones etc. Water needed to maintain water potential of cells. Fibre needed to help digested food to move easily through the alimentary canal.

Page 114

1. Any answer along the lines of: different people need different amounts of energy every day, for example active people need more than people who are seated for much of the day; men have a larger average body mass than women so will need more energy to support that extra tissue; some groups of people need more of a particular group of nutrients than others, e.g. pregnant women need additional iron.
2. Food that contains more energy than the body uses is converted into body fat, leading to obesity which is associated with many health problems. A diet that is too low in energy leads to health problems as a result of low body weight.
3. a) Obesity is caused by a diet that contains too much energy, and is associated with many diseases.
   b) Starvation is a diet too low in energy and/or nutrients, leading to health problems from deficiency diseases or breakdown of muscle tissue for energy.
   c) Constipation is caused by too little fibre in the diet and may lead to diseases such as bowel cancer and diverticulitis.

Page 117

1. Sketch should show the following labels correctly attached to organs shown on the diagram:

   mouth, where food is broken down by physical digestion (chewing) and amylase enzyme starts digestion of starch in food

   oesophagus moves food from mouth to stomach by peristalsis

   stomach, where churning mixes food with protease enzymes and acid to start digestion of protein molecules

   small intestine, where alkaline bile neutralises the acid chyme and enzymes from pancreas complete digestion of proteins, lipids and carbohydrates, and where digested food molecules are absorbed into the body

   large intestine, where water is absorbed from undigested food

   rectum, where faeces are held until they are egested through the anus

   liver, where bile is made and where some food molecules are assimilated

   gall bladder, where bile is stored until needed

   pancreas, where proteases, lipases and amylases are made which pass to the small intestine.

2. Egestion is the removal of undigested food from the alimentary canal – food that has never crossed the intestine wall into the body. Excretion is the removal of waste substances that have been produced inside the body.
3. Peristalsis caused by contraction of the circular muscles of the alimentary canal, followed by relaxation as the longitudinal muscles contract.

Page 120

1. Chemical digestion uses chemicals (enzymes) to help break down large food molecules into smaller ones. Mechanical/physical digestion is the chewing by the teeth to break large pieces of food into smaller ones before swallowing, or the breaking up of large fat droplets into smaller ones by bile.

2. Incisors have a flat blade shape for biting off food. Canines have a sharp point for holding food while other teeth chew off pieces. Premolars and molars have grinding surfaces for chewing.

3. Brushing removes particles of food caught in the teeth. Bacteria growing on these particles can release acids that destroy the tooth enamel leading to tooth decay. So brushing prevents bacterial growth and tooth decay.

4. **EXTENDED** Advantage: strengthens tooth enamel, leading to less tooth decay. Disadvantage, any one from: if in too high a concentration will cause brown mottling of teeth, may cause other health problems, doesn't allow individual choice.

Page 121

1. The digestive enzymes break down food molecules that are too large to cross the wall of the small intestine into smaller ones which can be absorbed across cell membranes and so enter the body. If we did not have enzymes, we would not be able to absorb many nutrients from our food.

2. a) amylase
   b) glucose

3. a) The acid increases the acidity in the stomach, providing the right conditions for enzymes that digest food in the stomach.
   b) Bile neutralises the acidity of food from the stomach, providing the right conditions for enzymes that digest food in the small intestine. It also emulsifies fats, providing a larger surface area for lipase enzymes to work on.

Page 124

1. Absorption is the movement of digested food molecules across the alimentary canal (small intestine) wall into the body. Assimilation is the movement of digested food molecules into cells where they are used to make new molecules or release energy in respiration.

2. There are millions of villi on the surface of the intestine wall projecting into the alimentary canal. This greatly increases the surface area of the wall.

3. The larger the surface area, the more rapidly small molecules can be absorbed across the intestine wall into the body.

4. Glucose absorbed into liver cells is converted to glycogen and stored until needed. Amino acids may be used to make proteins. Excess amino acids are broken down.

5. **EXTENDED** The capillaries provide a large blood supply to remove absorbed food molecules quickly so maintaining a high concentration gradient for diffusion, lacteals in the villi carry absorbed lipid molecules away to the rest of the body.

6. **EXTENDED** To carry blood directly from the small intestine to the liver, for assimilation and processing of many of the absorbed food molecules.

7. **EXTENDED** The nitrogen-containing part of amino acids is alkaline. If excess amino acids are not removed from the body, they could affect the pH of cells and so the effectiveness of enzymes in the cells.

8. **EXTENDED** Small intestine (ileum) and large intestine.

Page 126

1. Any four from: agricultural machinery increases speed at which farming tasks can be completed with fewer people; fertilisers improve yield of crop plants; pesticides reduce damage to crops by pests and so increase crop yield; herbicides kill competing weed plants, and so increase crop yields; artificial selection improves the plant and animals that we grow for food, so producing more food more easily.

2. **EXTENDED** Farmers in poor countries do not have access to modern technologies that increase food production. Drought and flooding can destroy food production. Conflict (wars) can displace people and destroy food production.

3. **EXTENDED** a) *Lactobacillus* culture is added to warm milk that has been heated to destroy other microorganisms. The milk is kept warm to encourage bacterial growth. The bacteria break down a milk protein, causing the milk to thicken and producing lactic acid which gives yoghurt its slightly sour taste.
   b) A fungus is mixed with carbohydrate (as a food source) and kept warm so that the fungus grows rapidly. The fungus is then separated and dried, and made into a food (single cell protein).

Page 128

1. **EXTENDED** Advantage: makes food look more attractive particularly after processing that breaks down natural colour. Disadvantage: some colourings are associated with health problems such as hyperactivity in children.

Transportation

Page 140

1. In vascular bundles that form veins throughout the roots, stems and leaves.

2. Xylem vessels are long continuous tubes formed from dead cells, which allow water and dissolved substances to pass easily through the plant.

3. Phloem cells link together to form continuous phloem tissue in the vascular bundles. They carry dissolved food materials, such as sucrose and

amino acids, from the leaves where they are formed to other parts of the plants that use them for life processes or where they will be stored.

Page 142

1. Enters through root hair cells, moves through root cortical cells to xylem in centre of the root. Moves through xylem up stem and into leaves. In leaves, moves out of xylem into spongy mesophyll cells.

2. Place a stem of a plant in water containing food colouring. The colour will travel through the xylem with the water, and show where the xylem is in the stem, leaves and flowers.

3. **a)** Osmosis, **b)** active transport.

4. **EXTENDED** Diagram should include annotations like the following, at the appropriate point: soil water has higher concentration of water molecules than cytoplasm of cells in the root; water molecules enter root hair cells by osmosis; water molecules pass from cell to neighbouring cell by osmosis until they reach the xylem.

Page 145

1. Transpiration is evaporation from the surfaces of a plant, particularly from the stomata of a leaf into the air.

2. Diagram should include annotations like the following, at the appropriate point: water molecules evaporate from surfaces of spongy mesophyll cells into air spaces; water molecules from air spaces move into and out through stomata into the air – diffusion (net movement) usually from inside leaf to outside; osmosis causes water molecules to move from xylem into neighbouring leaf cells, and then from cell to cell until they reach a photosynthesising cell or a spongy mesophyll cell; transpiration is the evaporation of water from a leaf.

3. Closing stomata reduces diffusion of water molecules out of the leaf. At night, oxygen is not needed for photosynthesis, so keeping stomata open would lose water unnecessarily.

4. **a)** When temperature is higher, particles move faster, so water molecules will diffuse out of the leaf more quickly.

   b) When air humidity is high, there is a high concentration of water molecules in the air. So more water particles will move from the air through the stomata into the leaf while water particles are moving out of the leaf into the air. This means the rate of diffusion will be lower.

5. **EXTENDED** Cohesion of water molecules means they stick to each other. So as water moves out of the xylem in the leaves, down its potential gradient into the spongy mesophyll cells, more water molecules are drawn up the xylem tube through the plant through cohesion of water. This causes a water potential gradient between the root cortical cells and the xylem in the root, drawing more water into the xylem.

Page 147

1. **EXTENDED** Stomata close at night in garden plants to prevent unnecessary loss of water when the plant isn't photosynthesising. Pond plants don't suffer from a shortage of water, so closing stomata isn't essential at night.

2. **EXTENDED** Sketch of cactus with labels to show the following:

   extensive root systems to capture as much water as possible

   reduced or no leaves to reduce water loss by transpiration

   green stems for photosynthesis

   stomata sunk deep in pits in stem to reduce rate of transpiration

   succulent stem (or leaves) to store water

   hairy surface to reduce rate of transpiration from stomata.

Page 148

1. Phloem.

2. Sucrose and amino acids.

3. **EXTENDED** They travel throughout the plant, so will kill any pest eating the plant. They last longer in the plant as they are not washed away with rain, so protecting the plant for longer.

4. **EXTENDED** A source is a part of a plant where a substance is formed or enters the plant, such as a) water from soil, carbon dioxide in air, and b) glucose in photosynthesising cells. A sink is a part of a plant where the substance leaves or is converted into something else, such as a) photosynthesising cells are a sink for water and carbon dioxide, b) respiring cells and storage cells are sinks for glucose.

Page 149

1. To pump blood around the body.

2. Valves in the heart and veins.

3. Blood passes twice through the heart for every once round the body – there are effectively two separate circulations of blood from the heart.

4. The pressure in the two circulations can be different, so the high blood pressure needed to get blood through all of the body doesn't damage the delicate capillaries in lung tissue.

Page 152

1. Vena cava, right atrium, right ventricle, pulmonary artery, pulmonary vein, left atrium, left ventricle, aorta.

2. Resting heart rate varies widely due to many factors, including age, health and fitness, so a single value for the average is too limited.
3. As level of activity increases, so heart rate increases. This is so the blood can circulate faster round the body, delivering oxygen and glucose to muscle cells for the increased rate of respiration to generate the energy needed for contraction. It also removes waste carbon dioxide from muscle tissue more rapidly to prevent it building up and affecting cells.

Page 153

1. To supply the oxygen and glucose needed for the heart muscle cells to respire and to remove waste carbon dioxide.
2. A partial blockage causes heart pains because the heart cells cannot get enough oxygen and glucose to contract properly. A total blockage causes a heart attack because the heart muscle stops contracting.
3. Eat a diet that is relatively low in saturated fat, don't smoke and try to control stress and the effects that it has on your behaviour.

Page 156

1. a) Renal arteries, b) aorta, c) hepatic veins
2. Arteries are large vessels with thick, elastic muscular walls; capillaries are tiny blood vessels with very thin walls that are often only one cell thick; veins are large vessels with a large lumen and valves to prevent backflow of blood.
3. **EXTENDED** The walls stretch as blood enters then, and slowly relax as the blood flows through, balancing out the pressure so that the change in pressure is reduced.

Page 159 Extended

1. Haemoglobin binds with oxygen when it is at high concentration such as in the lungs, and releases oxygen when it is at low concentration, such as in respiring tissues where the oxygen has reacted with glucose to produce carbon dioxide and water.
2. Breathing might become more rapid because with each breath the haemoglobin is combining with less oxygen than the body is used to. So it will deliver less oxygen to cells and the body response will be to increase breathing rate and depth.
3. More red blood cells in a given volume of blood means there will be more haemoglobin in that volume. More haemoglobin can combine with more oxygen, so each cm^3 of blood will carry more oxygen and deliver more oxygen to the body cells.
4. Training at high altitude for several weeks will cause the red blood cell count to increase. This will increase the oxygen-carrying capacity of the blood. When the athlete then competes at low altitude, their blood will be delivering more oxygen to their muscle cells than if they had trained at low altitude. So their muscles will be able to work harder aerobically than after low-altitude training.

Page 161

1.

Blood component	Function
plasma	carries dissolved substances, such as carbon dioxide, glucose, urea and hormones
	also transfers heat energy from warmer to cooler parts of the body
red blood cell	carries oxygen
white blood cell	protects against infection
platelet	cause blood clots to form when a blood vessel is damaged

2. The biconcave disc shape increases surface area to volume ratio, so rate of diffusion of oxygen into and out of cell is maximised. Haemoglobin inside cell binds with oxygen when oxygen concentration is high and releases oxygen when oxygen concentration is low. Cell has no nucleus, so there is as much room as possible for haemoglobin. Cell has flexible shape so can squeeze through the smallest capillaries and reach all tissues.
3. Phagocytes engulf pathogens inside the body and destroy them. Lymphocytes produce antibodies that attack pathogens.
4. Damage to a blood vessel can create an easy route of infection into the body. So forming a blood clot where there is damage, as quickly as possible, helps to reduce the risk of infection.

Page 163

1. **EXTENDED** Antibodies attach to pathogen cells in the body and either cause the cell to die or attract phagocyte cells to engulf the pathogen.
2. **EXTENDED** Antibodies are produced against any cell that isn't from the body, including cells from another person. So the antibodies attack and will destroy the transplanted tissue unless the immune system is suppressed with drugs.
3. **EXTENDED** The lymphatic system returns fluid to the blood that left the capillaries in tissue and did not return directly to the capillaries.

Respiration

Page 171

1. **a), b)** and **c)** glucose (from digested food from alimentary canal) + oxygen (from air via lungs) → carbon dioxide (excreted through lungs) + water (used in cells or excreted through kidneys) (+ energy (transferred to other chemicals in cell processes))

    d) Glucose replaced by fats from hump, and very little water excreted through kidneys.

2. Inside cells.
3. Any three from: muscles cells for contraction, synthesis of new molecules, such as proteins, for growth, active transport across cell membranes, passage of nerve impulses, maintenance of core body temperature.
4. $C_6H_{12}O_6 + 6O_2 \rightarrow 6CO_2 + 6H_2O$

Page 173

1. During vigorous exercise, they may not be able to get enough oxygen from the blood for all the energy they need for contracting. So the additional energy comes from anaerobic respiration.
2. Similarities: use glucose as substrate, produce energy, don't need oxygen. Differences: animals produce lactic acid, plants produce ethanol and carbon dioxide.
3. The amount of energy released from a mole of glucose molecules is much greater during aerobic respiration (around 2900 kJ) than in anaerobic respiration (around 150 kJ) in muscle cells.
4. Ethanol produced during brewing, carbon dioxide makes bread dough rise in baking.

Page 176

1. Exchange of gases between the body and the environment is by diffusion. Organisms need plenty of oxygen for respiration to provide energy for all life processes, and need to get rid of the waste carbon dioxide. So a rapid rate of diffusion supports a higher rate of respiration and all the other processes in the body.
2. Trachea carries air from mouth down to lungs; bronchi - the two large divisions of the trachea as it reaches the lungs, supported with rings of cartilage to prevent collapse during breathing; bronchioles – the fine tubes in the lungs that carry air to alveoli; alveoli, the bulges of the air sac, which have large surface area and are very thin for efficient diffusion of gases; ribs and intercostal muscles, protect the lungs but also help expand the volume of the thorax during forced or deep breathing; diaphragm, muscular sheet below lungs which control relaxed breathing.
3. Sketch similar to Fig. 2.108, with annotations showing: thin lining of alveolar wall and wall of capillary allows rapid diffusion; high concentration gradients for gases between blood and air in alveolus due to continuous blood flow through capillary and ventilation of alveolus (lungs); large area of contact between capillary and alveolus, maximising area over which diffusion can occur.
4. **EXTENDED** The mucus traps particles and microorganisms that are in the air breathed in, and the cilia moves the mucus and anything trapped in it up out of the lungs to the throat where it can be swallowed. This protects the lungs from damage and infection.

Page 181

1. **EXTENDED** **a)** The diaphragm contracts and flattens, pulling downwards on the thorax; the intercostal muscles contract lifting the ribs out and up; these actions increase the volume of the thorax, causing the volume of the lungs to increase and so drawing air into the lungs.

    b) The diaphragm muscle relaxes, and is pushed upwards by the organs below it; the intercostal muscles relax, so the ribs fall back and down; both actions reduce the volume of the thorax, so reducing the volume of the lungs which pushes air out of the lungs.

2. Percentage of oxygen is less in exhaled air than inhaled air, because oxygen in the body is used for respiration. Percentage of carbon dioxide is greater in exhaled air than inhaled because the body produces carbon dioxide in respiration. Percentage of water vapour is higher in exhaled than inhaled air because water molecules evaporate from the surface of the alveoli due to warmth of the body. (Percentage of nitrogen and other gases don't change because they aren't used by the body.)

3. **a)** As level of exercise increases, rate and depth of breathing increase.

    b) **EXTENDED** More exercise means more carbon dioxide is produced from an increased rate of respiration. Carbon dioxide is a soluble acidic gas so causes the body tissues and blood to become more acidic. A change in pH can affect many enzymes and so affect the rate at which life processes are carried out in the body. Slowing down the rate of life processes may harm the body.

Excretion in humans

Page 186

1. Excretion is the removal from the body of waste substances from metabolic reactions in cells.
2. They excrete carbon dioxide, which is the waste product of respiration.

3. Excess amino acids, that the body doesn't need, are broken down to form urea; hormones broken down; toxic substances such alcohol and drugs broken down.

Page 187

1.

Structure	Function
kidneys	produce urine by filtering waste substances from the blood
ureters	carry urine from kidneys to bladder
bladder	stores urine until it is released to the environment
urethra	short tube linking bladder to environment

2. Filtration and reabsorption
3. Water, urea, mineral salts

Page 190

1. **EXTENDED** Blood in glomerulus, nephron (renal capsule, proximal convoluted tubule, loop of Henlé, distal convoluted tubule, collecting duct), pelvis of kidney, ureter, bladder, urethra.
2. a) **EXTENDED** renal capsule (in conjunction with glomerulus of blood capillaries), (b) proximal convoluted tubule for most substances, also collecting duct for additional water.
3. a) **EXTENDED** Glucose
   b) Selective reabsorption using active transport.

Page 192

1. **EXTENDED** Kidney dialysis cleans the blood artificially every few days. Kidney transplant places a healthy kidney into the patient's body and attaches it to the blood system to carry our natural blood cleaning.
2. **EXTENDED** You only need one kidney to keep blood cleaned sufficiently to remain healthy.
3. **EXTENDED** Normal kidney function: blood filtered in renal capsule – large amounts of water and soluble substances pass into the tubule; reabsorption of water, glucose and some ions, restores the correct balance of substances in the blood. Dialysis: substances exchanged only by diffusion between the blood and dialysis fluid.
4. **EXTENDED** Getting a kidney transplant depends on making a good tissue match between the patient and new kidney. The patient must wait until a suitable kidney becomes available, which may be a matter of luck (unless they have a relative willing to donate one of their kidneys).
5. **EXTENDED** Any two from: risk of tissue rejection meaning another transplant would be needed, need to take drugs to suppress immune response for life, which increases risk of other infections; (risk of having an operation).

Coordination and response

Page 198

1. The ability to detect changes and respond to them
2. a) A change in the environment that causes a response
   b) An organ that carries out the response to a stimulus, such as a muscle or gland in a human.
   c) A nerve cell, adapted for carrying nerve impulses
   d) The brain and spinal cord that coordinate responses
3. The nervous system

Page 200

1. Sensory neurones have long dendrons and axons that link the sense organ with the central nervous system.
   Relay neurones are short neurones with many dendrites, found in the central nervous system, that link sensory neurones to motor neurones or other relay neurones.
   Motor neurones have many dendrites to link with relay neurones and end on the effector, such as a muscle and a long axon.
2. A reflex response is a simple response of receptor>nerve>spinal cord>nerve>effector, that does not usually include the brain. This makes it possible to respond to a stimulus very quickly. Reflex responses are usually important in survival, e.g. to protect you from touching something dangerous, or blinking to protect the eye if something comes toward it.
3. **EXTENDED** A voluntary action is one that is consciously chosen (i.e. thought about), such as choosing what to eat. An involuntary reaction is one that is done unconsciously, such as blinking when something comes toward the eye.

Page 203

1. a) The cornea is transparent so light passes through it easily into the eye.
   b) The pupil is a hole surrounded by the iris, that lets light pass through to the back of the eye.
   c) The retina contains the light-sensitive cells that respond to light. The retina is also very dark, to absorb as much light as possible.

2. As light intensity increases, the pupil gets smaller, reducing the amount of light that can enter the eye. This change happens because the radial muscle in the iris relaxes and circular muscle contracts. As light intensity decreases, the pupil gets larger, increasing the amount of light that can enter the eye. This change happens because the radial muscle in the iris contracts and circular muscle relaxes.

3. Light entering the eye from a near object needs to be refracted more than light from a distant object in order to focus it on the retina. The ciliary muscles contract, which reduces the tension on the ligaments that are attached to the lens. This allows the lens to become thicker and more rounded, which increases its focusing power.

4. **EXTENDED** Rod cells are found more around the periphery of the retina, and respond to light intensity, so we use them most when light levels are low. Cone cells are found more in the centre of the retina and respond to different colours of light, so we use them to distinguish colour when light levels are high.

Page 204

1. Antagonistic muscles are a pair of muscles that work together to cause movement of a bone.

2. Muscles can only contract actively; they cannot extend actively. So one muscle of the pair contracts to extend the other antagonistic muscle, and vice versa.

Page 205

1. a) A chemical messenger in the body that produces a change in the way some cells work
   b) Glands that secrete hormones
   c) Organs that contain cells which are affected by hormones

2. When faced with attack, or when suddenly frightened.

3. It prepares the body for action by increasing the amount of oxygen and glucose delivered to muscle cells for rapid respiration, and improves vision.

4. **EXTENDED** Advantage: makes animals grow faster and develop more muscle (which we eat). Disadvantage: may affect human health, and may affect other organisms if it gets into the environment.

Page 208

1. A tropism is a growth response of a plant to a stimulus.

2. a) Shoots grow towards light. b) Roots grow in the direction of the force of gravity.

3. **EXTENDED** Auxin is produced in the tip of the growing shoot and diffuses down the shoot. Auxin on the bright/light side of the shoot moves across the shoot to the darker side as it diffuses down the shoot. Cells on the dark side of the shoot elongate more than the cells on the light side of the shoot, so the shoot starts to bend as it grows so that the tip is pointing towards the light.

4. **EXTENDED** Auxins are absorbed into the leaves of broad-leaved weed plants more than into the leaves of grass-type crop plants. They cause the weed plants to grow so rapidly that they cannot sustain growth and die, leaving more space for the crop plants.

Page 212

1. Homeostasis is the maintenance of conditions inside the body within limits that allow cells to work efficiently.

2. Control of core body temperature. (Other answers possible.)

3. Blood vessels near the skin surface dilate when the core body temperature is too high. This allows heat energy carried by the blood to reach the skin surface more easily and so be transferred to the environment more rapidly. Blood vessels near the skin surface constrict when the core body temperature falls too low. This reduces blood flow to near the skin's surface, so heat energy cannot be transferred as easily to the skin surface and so cannot be transferred to the environment as quickly. This keep more heat energy within the body.

4. **EXTENDED** Where a change in a stimulus causes a control centre to trigger the opposite change in response, so keeping a condition within limits.

5. **EXTENDED**

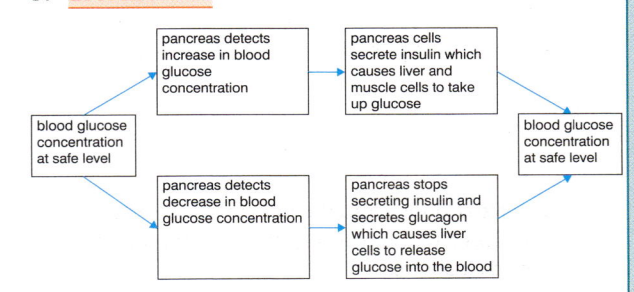

```
blood glucose         pancreas detects         pancreas cells secrete insulin which
concentration   →    increase in blood    →   causes liver and              →   blood glucose
at safe level         glucose                  muscle cells to take              concentration
                      concentration            up glucose                        at safe level
                ↘                                                           ↗
                      pancreas detects         pancreas stops secreting
                      decrease in blood   →    insulin and secretes glucagon
                      glucose concentration    which causes liver
                                               cells to release
                                               glucose into the blood
```

Page 215

1. A drug is a chemical that, when taken into the body, affects the way that the body works.

   An antibiotic is a chemical that kills bacteria or prevents them from growing.

2. Heroin is a powerful depressant, very addictive so difficult to give up, needs increasing amounts to produce the same effect as the body gets used to it, generally leads to crime to pay for the drug, may lead to infection by blood-borne diseases such as HIV as a result of sharing needles for injection.

3. Causes reaction time to lengthen which is dangerous when fast responses needed (e.g. in driving), in

large amounts causes loss of control which can result in violence, vomiting, unconsciousness, and liver damage.

4. Any four from: nicotine is addictive, so smoking difficult to give up; tar damages cilia so more difficult to remove mucus, causing more lung infections; tar and smoke particles cause coughing which breaks down alveoli, reducing area for gas exchange; carbon monoxide replaces oxygen on haemoglobin, reducing the amount of oxygen that blood can carry; other chemicals cause cancers.

5. **EXTENDED** Antibiotics only work on the structures in bacterial cells, so they have no effect on viruses that do not have a cell structure.

SECTION 3 DEVELOPMENT OF THE ORGANISM AND THE CONTINUITY OF LIFE

Reproduction

Page 243

1. Reproduction without the fusion of gametes, using a cell from only one parent.
2. a) Binary fission, where the genetic material is copied and the cell splits in half. Only one cell is involved and there is no fusion of parent cells before division.
   b) Production of spores by fungi such as *Mucor*, where spores are produced directly from cells in the hyphae. So there is no fusion of cells from different individuals before spores are produced.
   c) Growth of new plants from tubers of old plants, such as potatoes. There is no fusion of gametes before this growth occurs, so the new plants are genetically identical to the old plant.

Page 244

1.

	Sexual reproduction	Asexual reproduction
Similarities	• produce new individuals	
Differences	• fusion of male gamete with female gamete • offspring genetically different from parents and from each other • two parents • slower because male and female need to find each other and mate	• new individuals produced from division of body cell of parent • offspring genetically identical to parent • only one parent • faster because no search for mate

2. a) A sex cell or gamete such as a sperm or an egg cell.
   b) A zygote, or fertilised egg cell.

3. **EXTENDED** Any suitable example that refers to changing environment where genetic variability in offspring increases chance of survival of offspring that are genetically different to parent.

4. **EXTENDED** Answer should include:
   - summer not long so asexual reproduction increases numbers more rapidly
   - summer short period, so variability of environment not as big a problem, so genetic variability not as important
   - if parent is feeding well on food plant, the offspring from that parent are equally likely to survive and grow well on that food plant, so variability would be a disadvantage.

Page 247

1. It produces the pollen.
2. In pollen grains.
3. Stigma where pollen grains attach. Style which supports the stigma. Ovary which surrounds and protects the ovule, inside which is the female gamete.

Page 250

1. Wind-pollinated flowers are usually small, no colour (white), make masses of lightweight pollen.
2. The lightweight pollen can be carried far in wind, stamens hang outside the flower so pollen more likely to be caught by wind.
3. Insect-pollinated flowers are usually large, may be brightly coloured, produce nectar and sometimes scent, make small amounts of larger pollen grains.
4. Scent, nectar (as food), large petals with bright colours all help to attract insects to the flower. As the insect feeds, they pick up pollen which are then carried to other flowers.
5. **EXTENDED** Anthers and stigma ripe at different times; plant only has male or female flowers; pollen tubes of own pollen grow more slowly than from other plants.

Page 251

1. Pollination is the transfer of pollen from one flower to another. Fertilisation is the fusion of a male gamete with a female gamete to form a zygote.
2. Pollination (pollen grain lands on stigma); pollen tube grows down through style to ovule; pollen tube delivers male gamete to egg cell; nucleus of male gamete and nucleus of female gamete fuse to form zygote.
3. Female gamete develops into embryo plant; ovule wall forms hard outer shell of seed; ovary forms fruit.

Page 252

1. So offspring grow at a distance from parent, avoiding competition with parent plant and other offspring. Also increases chance of growing in a new habitat.
2. Any two from: seeds that stick to animal's coat, e.g. cleavers; berries or other soft fruits eaten by animals, where seed is not damaged by digestion and is deposited with animal's faeces; nuts, buried by squirrels for eating later.
3. Any two from: dandelion seeds with 'parachute' of fine hairs; sycamore seeds with wings.

Page 253

1. Sketch as diagram on page 253 of this Student Book.

2/3. Labels and annotations as follows:

- testes, where sperm (male gametes) produced
- sperm duct, carry sperm to urethra
- prostate gland and seminal vesicles, produce liquid in which sperm swim
- penis, when erect delivers sperm into vagina of female
- urethra, tube that carries sperm from sperm ducts to outside the body.

Page 254

1. Sketch as diagram on page 254 of the Student Book.

2/3. Labels and annotations as follows:

- ovaries, where egg cells form
- oviducts, carry the eggs to the uterus and where fertilisation by sperm takes place
- uterus, where embryo implants into lining and fetus develops;
- cervix, base of uterus where sperm are deposited during sexual intercourse
- vagina, where penis is inserted during sexual intercourse.

4. **EXTENDED**

	Egg cell	Sperm cell
size	very large, 0.2 mm diameter	very small, 45 μm long
numbers	thousands in ovary but usually only one released each month	>100 million produced each day
mobility	unable to move on its own	self-propelling with tail

Page 256–257

1. Testosterone.
2. Oestrogen and progesterone.
3. To make sexual reproduction possible, and to show that the individual is sexually mature.
4. Circle numbered 1 to 28 around the circle; **a)** ovulation at around day 14; **b)** increase in oestrogen days 8–12, decrease days 12–14, increase again day 15, decrease again around day 18; **c)** increase in progesterone days 14 to around 18, decrease in progesterone around days 23–28.
5. **EXTENDED** a) Pituitary gland in the brain
   b) LH triggers ovulation. FSH stimulates egg development in the ovary.

Page 259

1. a) Zygote is the cell produced by fusion of a male gamete and female gamete.
   b) Formed from the division of cells in the zygote – until distinctive structures are obvious, such as limbs when it becomes a fetus.
   c) Developing baby in the uterus (womb), from about three months after fertilisation.
2. In an oviduct.
3. In early stages, rapid cell division, and differentiation of cells to produce the main structures; later, development of nervous system and movement; increase in size and weight.
4. Provides nutrients from mother's blood and carries waste to mother's blood to be excreted.

Page 262

1. Care of the mother during her pregnancy to make sure she and the fetus are healthy.
2. Eating a healthy diet providing all the nutrients needed to keep her healthy and for the healthy development of the fetus. Limiting of alcohol intake, to minimise effect of alcohol on fetus. Additional nutrients such as folic acid to prevent problems in the fetus, such as spina bifida.
3. Any two from: strong contractions of the uterus; blood-tinged mucus from plug lost from cervix; breaking of waters.
4. **EXTENDED** a) It usually provides the best balance of nutrients for the baby's growth, including antibodies that protect it from disease.
   b) If the mother was sick, or unable to produce sufficient milk for the baby.

Page 264

1. To control when they have children and the number of children they have.

2. Abstinence, chemical methods (pill and IUD), surgical methods.
3. Because it depends on getting the timing right, and sperm survive for more than a day, so it is easy to misjudge the time when the woman is fertile.
4. Provide protection against transmission of STDs.
5. They prevent the sex cells getting to a place where they can meet.
6. **EXTENDED** Can help where a man is producing few sperm, by collecting and concentrating them before placing them in the woman's uterus. Or using a sperm donor, when the man's has no useful sperm.
7. **EXTENDED** Advantage: couple more likely to have children; disadvantage: risk of multiple births which is more dangerous for the mother and for the babies.

Page 266
1. A disease that is transmitted in body fluids during sexual intercourse.
2. Many people do not show symptoms of the infection, but will later suffer from long-term health problems and possibly infertility.
3. Through sexual body fluids during sexual intercourse; via blood through cuts or sharing of needles for injecting drugs; across the placenta from mother to fetus; through milk from mother to baby when breastfeeding.
4. **EXTENDED** The virus attacks the immune system, reducing the ability of the body to fight off other infections. This leads to AIDS.
5. **EXTENDED** Only have intercourse with a partner who is not infected with HIV; use barrier methods such as condoms or femidoms during intercourse.

Growth and development

Page 273
1. a) A permanent increase in size and dry mass as a result of an increase in cell number, cell size or both.
   b) The mass of all the tissues without the water.
   c) An increase in complexity of cells (by differentiation), tissues and organs.
2. An increase in mass may be just an increase in the amount of water in the cells, so this is not a permanent increase in tissue and therefore not growth.

Page 276
1. Germination is when the embryo in a seed starts to grow, splitting the seed coat and increasing in size and complexity.
2. a) Seeds need a supply of oxygen for growth, although they may be able to start germination using anaerobic respiration.
   b) Seeds need water for germination and will not germinate in dry soil.
   c) Seeds need warmth for germination, though the amount of warmth they need may depend on where they naturally grow. Seeds from plants that live in colder areas may need a period of deep cold before they will germinate. Seeds from plants that live in areas prone to fire may not germinate until after a fire.

Inheritance

Page 282
1. Gene, chromosome, nucleus, cell.
2. A gene codes for a protein or characteristic, an allele is one form of the gene coding for a variation in the protein or characteristic. Any suitable example, e.g. gene for eye colour, allele for blue eye colour or brown eye colour.
3. Men are XY and women are XX.
4. Inheritance is the passing on of inherited characteristics from one generation to the next due to the passing on of genes which code for those characteristics.

Page 283–284
1. a) The characteristic expressed in the phenotype when the organism has only one allele of that form.
   b) The characteristic expressed in the phenotype when both genes in the genotype are the alleles for this form.
   c) Having two identical copies of that allele for a particular gene.
   d) Having different alleles for a particular gene.
2. a) 2   b) 1   c) 2

Page 286
1. The inheritance of a characteristic produced by one gene.
2. Genotype (the alleles in the chromosomes) BB, phenotype (what the organism looks like) brown; genotype Bb, phenotype brown (because the brown allele is dominant); genotype bb, phenotype black (because the organism doesn't have the brown allele).
3. a) Answer may be presented as a full layout diagram or a Punnett square, showing the adult genotypes and phenotypes (male BB brown and female bb black), the possible gametes produced (male B and B, female b and b), genotypes and phenotypes of possible offspring (BB brown, Bb brown, Bb brown, bb black).
   b) This cross produces a theoretical probability of one black rabbit for every three brown rabbits, a ratio of 1:3, probability of 1 in 4 or 25%.

Page 290 top

1. So that, when the plants were bred together, the results in the offspring were not confused by a mix of alleles in one or both of the parents.
2. Random variation is possible in the results. So the larger the sample, the more likely that any random variation will be averaged out.
3. He removed the stamens from every flower, so that pollen could not be transferred by insect. He also covered each flower after he had hand-pollinated it, so that other pollen could not get to the stigma.
4. Any characteristic may be used, with alleles appropriately designated with capital letter for dominant and lower-case letter for recessive allele. Parents used should show one with phenotype of dominant allele, homozygous, e.g. BB, and one parent with phenotype of recessive allele, i.e. bb. First cross will produce all individuals with phenotype of dominant allele but heterozygous in genotype, i.e. Bb. Crossing of these individuals will produce characteristic 1 BB : 2 Bb : 1 bb in genotype and 3 dominant characteristic to 1 recessive characteristic in next generation.
5. If Mendel had not been as thorough about his method, then his results would not have been as clear and predictable. So he would not have been able to have drawn clear and repeatable conclusions about the way characteristics are inherited in pea plants.

Page 290 bottom

1. **EXTENDED** When both alleles are expressed in the phenotype, and there is no dominance of one allele over the other.
2. **EXTENDED**

Phenotype	Genotypes
A	$I_A I_A$ or $I_A I_o$
B	$I_B I_B$ or $I_B I_o$
AB	$I_A I_B$
O	$I_o I_o$

3. **EXTENDED** Genetic layout diagram or Punnett square with following outcomes:

		father's gametes	
		I_A	I_B
mother's gametes	I_o	$I_A I_o$ blood group A	$I_B I_o$ blood group B
	I_o	$I_A I_o$ blood group A	$I_B I_o$ blood group B

Page 291

1. XX
2. XY
3. At each fertilisation there is a 50% chance that the X egg will be fertilised by an X sperm or a Y sperm. So the chance of the child being born male is 50%.

Page 292

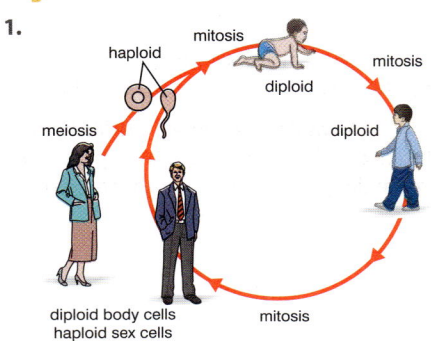

1. [diagram of life cycle showing meiosis producing haploid sex cells, fertilisation producing diploid body cells, and mitosis through growth stages]

2. a) Mitosis b) It produces cells that are identical and are diploid – so all the cells of the body contain all the genes for making all parts of the body.
3. a) Meiosis   b) Mitosis
4. Meiosis produces non-identical cells, so there is variety in the gamete cells. When the gamete cells fuse, this will mean that the offspring will vary from each other.

Page 295

1. a) Genes, e.g. eye colour; environment, e.g. weight as a result of diet. (Other suitable examples are acceptable.)
   b) Any suitable example showing a combination of genes and environment, e.g. human height which depends on genes and a healthy diet to achieve the potential of the genes.
2. A change in a gene or chromosome that produces a different form of a characteristic.
3. Three copies of chromosome 21 are inherited instead of the usual two copies.
4. Ionising radiation, such as ultraviolet radiation, X rays or gamma rays; chemical mutagens such as the chemicals in tobacco smoke.

Page 297

1. People who want to develop new varieties of plants and animals that have economic importance (will earn more money for the breeder).
2. Any two suitable examples, e.g. increased meat production of beef cattle, large eggs produced by chickens, unusual flower colours, crops that produce large amounts of grain.

3. Parent organisms with desirable characteristics are selected and bred together. Offspring with the best combination of characteristics are selected and then bred together. The process is repeated over many generations to produce a number of individuals that all exhibit the desired characteristics.

Page 298 Extended

1. In sexual reproduction, the offspring inherit half their alleles from one parent and half from the other. In selective breeding, only a few parent types are used for breeding. So this limits the range of variation in the alleles in the parents. So the offspring can only inherit from this limited range.
2. This means that when they cross-breed in the field, the seed produced is more likely to contain the characteristics that they have been bred for. Some of that seed planted in the following year will produce plants that still show the required characteristics.
3. The environment is continually changing, and may change a lot over the next century or so as a result of climate change. This means that characteristics that are useful now, may be less important than others in the future, such as the ability to withstand drought (if the climate gets hotter and drier in some places), or resistance to particular pests or diseases that are not common now by may become more common as climate changes. If the varieties we are growing do not have these characteristics, then crop yield will be reduced as growth of the crop plants is reduced.
4. Any suitable answer with appropriate justification: e.g. in the short-term this doesn't seem to make sense because we need to grow enough food for everyone to eat, but in the long-term it does make sense because the environment changes and we cannot predict how it will change and what characteristics we will need for crops in the future.

Page 302

1. Answer along the lines of: the influence of the environment on a characteristic, such that some variations of the characteristic are more successful at producing offspring than others.
2. **EXTENDED** a) If the individuals in a population are all the same, natural selection will favour (or disadvantage) them all equally.
    b) If there is competition, only those that are better adapted will be fit and healthy enough to produce offspring and pass on their genes.
    c) If individuals with a particular variation of a characteristic have a greater survival advantage, they are more likely to produce offspring that carry their genes, and so their genes will become more common in the next generation.
3. **EXTENDED** If conditions within the environment change continuously, or a mutation produces a new variation that is more successful than existing variations, the characteristics of the population will change so much that a new species will have evolved.
4. **EXTENDED** Diagrams should show the following: person infected with bacteria > bacteria grow in number inside patient > treatment of patient with antibiotics kills off least resistant bacteria but most-resistant bacteria survive > some of these bacteria escape into the environment from the patient and infect another person > the same antibiotic cannot be used on that patient as the bacteria are resistant.

Page 303

1. Taking a gene out of one organism and putting it into the DNA of an organism of a different species, so that the gene produces its protein or characteristic in the new organism.
2. Any two suitable examples, such as: female sheep that produce human hormones, crops with a gene for a bacterial toxin that is a natural insecticide, crop plants that contain a gene that makes them resistant to a herbicide, bacteria containing the human insulin gene.
3. **EXTENDED** The human insulin gene is cut out of a human chromosome > the gene is inserted into a bacterial plasmid > the plasmid is inserted back into a bacterium > the insulin gene causes the bacterium to make human insulin > when the bacterium divides, all the cells it produces contain the insulin gene.

SECTION 4 RELATIONSHIPS OF ORGANISMS WITH ONE ANOTHER AND WITH THEIR ENVIRONMENT

Energy flow, food chains and webs

Page 322

1. Any description that means the same as:
    a) All the living organisms that interact with each other and the physical environment with which they interact in an area
    b) Organism that produces its own food from simpler materials, e.g. plants making carbohydrates in photosynthesis
    c) Animal that gets its food from eating other organisms
    d) Organism that gets its food from dead plants and animals or waste material, such as some fungi and bacteria
2. The Sun provides light energy, transferred as chemical energy to build plant tissue, which is then

transferred as chemical energy through all other organisms in the ecosystem.

3. A food chain shows the relationship between one producer, one herbivore, the carnivore that eats the herbivore etc.

   A food web shows the feeding relationships between all the organisms living in an area.

4. Food webs help us to understand the relationship between organisms in an area, and can help us predict what might happen to the organisms as a result of a change to the ecosystem.

   It can be difficult to organise the information in a food web because some organisms feed at many trophic levels, and it may not be possible to include all organisms (e.g. decomposers) on a food web because of space for the drawing.

Page 327

1. a) A diagram showing the numbers of organisms at each trophic level in a food chain or food web in an area
   b) A diagram showing the biomass of organisms at each trophic level in a food chain or food web in an area
   c) A diagram showing the energy in the organisms at each trophic level in a food chain or food web in an area

2. Any suitable example that includes producers, primary consumers and secondary consumers from a reasonable food chain. Count the number of individuals feeding at each level within the area. Draw a pyramid of three layers, starting with producers at the bottom and ending with secondary consumers at the top, with the bar for each level drawn to scale.

3. A pyramid of biomass only shows the mass at a particular time in an area. If some trophic levels have a shorter life-span than others, they will be under-represented in the pyramid, which may cause an inverted shape.

Page 329

1. Light energy from Sun (gain) > some reflected, some passes straight through, some wrong wavelength (losses) > light energy converted to chemical energy during photosynthesis > heat energy transferred to environment from photosynthetic reactions and from respiration (losses) > chemical energy in plant biomass.

2. Chemical energy in food (gain) > chemical energy in undigested food lost as faeces (loss) > chemical energy in absorbed food molecules converted to chemical energy in waste products such as urea lost in urine (loss) > heat energy from respiration transferred to environment (loss) > chemical energy in animal biomass

3. **EXTENDED** The pyramid for plant/animal/human should show three layers, widest at the bottom for plant and shortest at the top for human. The pyramid for plant/human should show two layers, the bottom one the same width as in the other pyramid, and the top one for humans may be wider than in the other. Explanation should indicate that eating the plants ourselves means more food available, as energy is not lost to the environment from an intermediary animal level.

Nutrient cycles

Page 334

1. a) Evaporation is the conversion of liquid water on the Earth's surface to water vapour in the atmosphere.
   b) Transpiration is the evaporation of water from the surface of leaves.
   c) Condensation is the conversion of water vapour in the atmosphere into liquid water droplets in clouds.
   d) Precipitation is the falling of liquid (or solid) water to the Earth's surface from clouds.

2. Biotic route: drunk by an animal, or absorbed through a plant > water transpired by plant or evaporates from animal's skin > water vapour in atmosphere > condenses in cloud > falls as rain or snow to ground/pond

   Non-biotic route: water evaporates from pond > water vapour in atmosphere > condenses in cloud > falls as rain or snow to ground/pond.

3. All organisms need water in their bodies, to transport soluble materials around their bodies, and to allow cell reactions to take place. Plants also need water for photosynthesis.

Page 335

1. a) Respiration releases carbon dioxide into the atmosphere from the breakdown of complex carbon compounds inside organisms.
   b) Photosynthesis is the fixing/conversion of carbon dioxide from the atmosphere into complex carbon compounds in plant tissue.
   c) Decomposition is the decay/breakdown of dead plant and animal tissue by decomposers, releasing carbon dioxide into the atmosphere during respiration.

2. a) carbon dioxide, b) complex carbon compounds, c) complex carbon compounds

Page 337

1. **EXTENDED** Combustion releases carbon dioxide into the atmosphere from complex carbon compounds where the carbon has been locked away for millions of years. This greatly increases the rate at which carbon dioxide is being returned to the atmosphere.

2. **EXTENDED** Combustion increases the carbon dioxide concentration in the atmosphere more rapidly than natural processes such as respiration. Deforestation removes trees so this reduces the amount of oxygen taken from the atmosphere for photosynthesis and increases the amount of carbon dioxide released if the forest is burnt. So this can rapidly change the oxygen/carbon dioxide balance in the atmosphere near the forest.

Page 339

1. **EXTENDED** a) Bacteria that increase the amount of nitrate ions in the soil by converting ammonium ions to nitrite ions and then to nitrate ions.
   b) Bacteria that convert atmospheric nitrogen gas directly into nitrates.
   c) Bacteria that reduce the amount of nitrate ions in soil by converting them to nitrogen gas.

2. **EXTENDED** Nitrifying bacteria increase the fertility of soils because plants can only take in nitrogen in the form of nitrates dissolved in soil water. Without nitrogen the plants will not grow well, and become stunted.

3. **EXTENDED** Decomposers break down complex nitrogen compounds in dead plant and animal tissues and animal waste. This releases ammonium ions that nitrifying bacteria convert to nitrate ions that plants need. Without decomposers, the bacteria would have nothing to work on, and the concentration of nitrate ions in the soil would decrease.

Population size

Page 348

1. A population is a group of organisms of the same species living in the same place at the same time.
2. Births and immigration increase population size; deaths and emigration decrease population size.
3. Food supply can increase population growth because it can increase birth rate and survival, reducing death rate. It can also cause an increase in immigration and decrease in emigration.
4. Predation and disease can decrease population growth because they increase the death rate.

Page 350

1. Lag phase, exponential (lag) phase, stationary phase, death phase.
2. Microorganisms in a fermenter (or other suitable example) because conditions for growth are ideal and there is no other organism to predate on the populations.
3. **EXTENDED** Lag phase is when the individuals in the population are preparing for growth and reproduction but population size is increasing very slowly.

   Exponential (lag) phase is where growth in population size is rapid because birth rate is fast due to ideal conditions.

   Stationary phase is where growth levels off, birth rate and death rate are equal, due to a limiting factor such as limited nutrients.

   Death phase is where population size falls because death rate is greater than birth rate, due to lack of a nutrient or increase in toxic conditions.

Page 353

1. Rate of growth shows exponential growth/same shape as the log and exponential phases of a sigmoid growth curve.
2. Because different estimates include predictions of different birth rates and death rates over this period.
3. Any two of: different birth rates, immigration, death rates or emigration.
4. Any positive from: improved food availability, improved chances for work, improved health care (or similar).

   Any negative from: slum developments because city growing too rapidly, unrest leading to violence, increased risk of transmission of infectious disease.

Human influences on the ecosystem

Page 358

1. a) The destruction/cutting down of large areas of forest and woodland.
   b) The washing away of soil by heavy rainfall
   c) When soluble nutrients dissolve in soil water and soak away deep into the ground

2. a) Trees take up water from the ground and release it to the atmosphere through transpiration, which affects the moisture in the air above the forest and affects the water cycle.
   b) Soil erosion because there are no tree roots to hold the soil, and increased leaching because there are few plant roots to absorb the nutrients,

both remove soluble mineral nutrients from the soil, which will reduce the rate of plant growth.

c) Burning or rotting of trees releases the carbon stored in the wood as carbon dioxide to the atmosphere at a much faster rate than normal, increasing atmosphere carbon dioxide.

3. a) Biodiversity is the number of organisms, and the different species, in an area.

b) Cutting down large areas of trees destroys the habitats of many species that can only live in those habitats. So deforestation greatly reduces biodiversity in an area.

Page 360

1. The addition of nutrients to water.
2. Runoff of fertiliser into water as a result of heavy rainfall, leaching of soluble nutrients in fertiliser through soil into water systems.
3. Eutrophication leads to the rapid growth of algae and other microorganisms that remove large amounts of oxygen from the water for respiration. This leaves not enough oxygen in the water for the fish, so they die.

Page 363

1. Sewage added to water > adds nutrients to water = eutrophication > plant and microorganism growth rate increases > respiration rate of microorganisms increases removing dissolved oxygen from water > less dissolved oxygen for other organisms which die = water pollution.
2. Poisoning of the water by toxic chemicals.

Page 376

1. It is produced by human activities in large enough amounts to harm organisms and the environment.
2. **EXTENDED** a) Smoke/emissions from factories contains acidic gases, such as sulfur dioxide, which dissolve in water droplets in clouds that then fall as acid rain.

    b) The clouds containing the acidic water droplets can be blown over great distances away from the industrial areas by wind.

3. **EXTENDED** Damage direct to delicate tissues in lungs, of soft-skinned organisms such as fish and amphibians, and of single-celled organisms. Indirect damage by changing the acidity of the soil, affecting its fertility due to leaching of minerals, or making poisonous minerals more soluble. Effects as a result to changes in food web may affect other organisms due to interdependency.

Page 376

1. a) Natural, respiration; human, combustion of fossil fuels.

    b) Natural, soil bacteria in the nitrogen cycle; human, addition of nitrogen-containing fertilisers to soil.

    c) Natural, digestion of food in animal guts and decay of waterlogged vegetation; human, increase in herd animals and artificial water-logged vegetation in rice paddy fields.

2. Any from: increase in number and intensity of storms, more drought, more flooding, change to summer/winter temperatures and precipitation.

3. **EXTENDED** The greenhouse effect is a natural process that warms the Earth's surface when greenhouse gases in the atmosphere prevent longer wavelength radiation escaping into space. The enhanced greenhouse effect is the additional warming caused by the addition of greenhouse gases to the atmosphere as a result of human activity.

Page 379

1. A pesticide is a chemical used to kill organisms that we don't want, including insects that eat our crops or weed plants that compete with our crops.
2. It increases the yield of the crop by reducing the damage done by pests or making more water and nutrients available to a crop so that the crop plants grow better.
3. a) Any suitable answer, such as: kills other non-pest organisms so damages food webs; may kill off predators of the pest so that numbers of pests can increase even more rapidly; increasing resistance of pests to pesticides so farmers use even more of the chemicals.

    b) Any suitable answer, such as: removes food and shelter plants for other insect species which may include predators of the pests, so pest numbers will increase further; some herbicides can be toxic to other insects and soil organisms.

Page 381

1. The radioactive dust that falls out of the air after release from a power station, processing plant or bomb.
2. Accidental leakage or an explosion.
3. The radioactive dust was sent high into the atmosphere. Winds spread the dust across the continent. The areas with highest fallout are those that are mountainous and nearest to the explosion.
4. Any suitable answer such as: burning by contact, causing cancers.

Page 376

1. They take a very long time to break down and they may leak poisonous chemicals into ground water which can leak away into water systems.
2. The land cannot be used for many purposes for many years after the plastic has been dumped.
3. The plastics cause problems for wildlife in the oceans that eat them by accident, or entangle them.

Page 377

1. Protecting species and habitats from damage.
2. Any suitable example: such as breeding a tiger species in a zoo, the Hawaiian goose breeding programme at Slimbridge.
3. If the habitat is not conserved, then the organism cannot ever be returned to the wild. Some species are important for the effect on an ecosystem, some species might be worth money and some are important for their aesthetic value.

Page 381

1. Humans use a large proportion of the fresh water resources that are available, and our population size is still growing. We also need to avoid pollution of water by our activities.
2. They are non-renewable, and we use them for much of our energy needs and as a raw material for products such as plastics.
3. **EXTENDED** It is cleaned of large material, the organic material is then broken down by microorganisms in treatment beds; it is then treated with chemicals to kill any microorganisms and to remove toxic dissolved minerals.
4. **EXTENDED** It uses less energy than making paper from wood, and fewer chemicals, so reduces pollution also.

Index

A

absorption 94, 114, 122–3
   reabsorption 189–90
abstinence 262
accommodation 203
accuracy 413, 418, 420–2
acid rain 363–5
active site 81, 82
active transport 72–3, 106, 141, 189
addiction 213, 214
adrenaline 205
aerobic respiration 169–70
age structure of a population 351–2, 394–5
agriculture 357–60
   bird populations 384
   see also fertilisers; pesticides
AIDS 24, 213, 265–6, 395
air pollution 363–6
   see also global warming; greenhouse gases
alcohol 213–14
algae 324, 326, 360
alimentary canal 114, 116
alleles 281, 282
   codominant 288–90
   dominant and recessive 283–5, 287–8, 308, 310
alveoli 175–6, 214
amino acids 95, 123, 147
amniotic fluid 259
amphibians 20–1
amylases 77, 78, 80, 84, 120, 121
anaemia 109
   sickle cell 294–5
anaerobic respiration 171–3
angiosperms 32
animal cells 46–8
   osmosis in 68
animals
   classification 16
   energy losses 327–9
   in the food chain 320
   gas exchange in 173–4
   nutrition 94
   in the water cycle 334
   see also carnivores; herbivores; human; vertebrates
annelids 30
anomalous results 403–4, 405, 419, 420
antagonistic muscles 204
antenatal care 260
anthers 245, 246, 247, 248
antibiotic resistance 26, 89, 212, 301, 313
antibiotics 87, 212
antibodies 159, 161, 162, 261
anus 115, 116
arachnids 29
arteries 155
arthropods 43
artificial cells 48
artificial insemination 264
artificial respiration 178
artificial selection 125, 296–8
artificial system 16
asexual reproduction 240, 242–3, 291
assimilation 94, 114, 123–4
athletes
   diet schedule 134–5
   respiration in 172
atria 150
auxins 207–8, 236

B

bacteria 25–6
   antibiotic resistance 26, 89, 212, 301, 313
   binary fission 242
   classification 16
   in the food chain 320, 321
   in the nitrogen cycle 338–9, 341
balanced diet 110–11, 134–5
bar chart 406–7, 408
Benedict's solution 96–7
bile 115, 120, 121
binary fission 242
binomial system 16–17, 42
bioaccumulation 372
biodiversity 358
bioindicators 361
biological washing products 85, 86
biomass 323–4
   pyramid of 325, 326, 392
birds 22, 384
birth 260–1, 264, 346–7
birth control 262–3
birth rate 347, 348, 352
biuret test 97
bladder 187
blood 156–7, 229
   carbon dioxide in 180, 181
   oxygenated and deoxygenated 149–50
   see also red blood cells; white blood cells
blood glucose levels 211, 303
blood groups 288–9, 314
blood pressure 148–9, 155, 156
blood tests 160, 260
blood vessels 136, 154–6
breastfeeding 261
breathing in and out 176–81
bronchi 174, 177

C

calcium 109
cancer 295, 374
canine teeth 117, 118
capillaries 155, 156
carbohydrates 94, 95, 106, 108
   see also starch; sugars
carbon cycle 335–7, 357
carbon dioxide
   atmospheric 335, 336–7, 343, 357, 367, 385–6
   in the blood 180, 181
   excretion 186
   in photosynthesis 98, 100, 101, 102, 236
   test for 179
   see also gas exchange
carbon monoxide 215
carnivores 320, 326
carpel 245, 247
carrier proteins 72–3
catalysts 77
cell division 281, 291–2
cell membrane 46, 47, 48
   movement through 63, 65, 70
cell respiration 73, 168–9
cell wall 39, 46, 47, 48, 56
   breaking down 84
   and osmosis 68
cells
   movement in and out of 63–75
   organisation 46–51, 52–3
   specialisation for function 53–6
   structures 46–51, 281, 310
   see also specific cells
central nervous system 197, 198, 200

cervix 254, 260
CFCs 386
characteristics 17, 18
   inheriting 282–3, 287
   living organisms 10–14
chemical digestion 117, 120–1
chemical energy 98, 325, 328, 333
Chernobyl 374
chlorophyll 48, 98, 100
chloroplasts 46, 47, 48, 105
cholesterol 113, 153
chromosomes 48, 280–1, 282, 283, 290–1
cilia 174
ciliated cells 54
circulatory system 148–60
cirrhosis 214
cladistics 18
cladograms 18–19
codominance 288–90, 313
colon 115, 123
colostrum 261
combustion 336, 337
   food 112
concentration gradient 64, 65, 70, 71, 139
   active transport 72
   in diffusion 174, 175
   and transpiration 142
concentrations in solutions 67, 68, 69, 416
condensation 333, 334
condoms 263, 266
cone cells 201, 202
conservation 376–80
constipation 113
consumers 320, 321, 322, 323
continuous variation 293
contraception 262–3, 266
contraceptive pill 262
control variables 413, 417
coordination and response 195–219
core temperature 209–11
coronary heart disease 153
cortex 187
cotyledons 32, 251
cross-pollination 249–50
crustaceans 28–9
cuticle 104
cystic fibrosis 310
cytoplasm 46, 47, 48
   and osmosis 68, 70

D

daily energy requirements 111
daughter cells 291
DDT 372
deamination 123
death 346–7
death phase 349
death rate 347, 348
decomposers 320, 335
deforestation 336–7, 357–8, 388
denatured 77, 82
denitrifying bacteria 338–9
deoxyribonucleic acid (DNA) 280–1
   mutations 293–5
   recombinant DNA 302
dependent variables 413, 416
depressants 212, 213
desert plants 146
development 273
diabetes 303
dialysis 190–1, 235
   haemodialysis 66
diaphragm (body) 176–7
diaphragm (contraceptive) 263
dichotomous key 35, 36, 43
dicotyledons 32
diffusion 64–6, 70, 141
   gas exchange 173–4, 175
   surface area-to-volume ratio 220–2
digestion 114, 117–21
   liver role in 123–4
digestive system 114–22
dinosaurs 301, 326
diploid cells 244, 292
discontinuous variation 293
dominant allele 283, 284–5, 287, 288, 308
double circulation 148
Down's syndrome 294
drink-driving 213
drugs 212–13, 230
dry mass 272, 324
duodenum 115, 121

E

ecosystems 319
effector 196, 197, 199, 200
egestion 114
egg cell 254, 255, 256, 257
embryo (human) 257
embryo (plant) 245, 251, 273
emigration 346–7
emphysema 214, 215

emulsification 120
endocrine glands 196, 204
endocrine system 196–7
energy input 319
energy transfer losses 327–9
enhanced greenhouse effect 369, 387
environment, and natural selection 298–300
enzymes 76–91
   digestive 120–1
   in germination 83
   and pH 78, 80, 82, 86, 232
   in plant growth 102
   and temperature 77–9, 82, 86, 232
   useful 84–7
epidermis 104
errors 419–20
eutrophication 107, 359–60, 361, 362
evaporation 142, 209, 333, 334
evolution 300–1
examinations 395–8
excretion 11, 13, 185–94, 227
exercise
   effect on breathing 179–80
   and heart rate 151–2
exoskeleton 27, 28
expiration 176
expired air 178–9
exponential phase 349, 355
extinction 376, 392
eyes 201–3

F

faeces 114, 115
famine 125–6
fats 95, 108
   digestion 120
   saturated 113, 153
   test for 97
fatty acids 95
female reproductive system 254, 311
female sterilisation 263
femidom 263, 266
fermenters 86–7, 230, 348–9
fertilisation 244
   in humans 257
   in plants 250–2
fertilisers 107, 124, 340
   overuse of 359–60
   from waste water 379

fertility treatment 263–4
fetus 257–9
fibre 108, 113, 116
fibrinogen 160
filtration, kidneys 189, 234
fish 20
Fleming, Alexander 89
flight or fight 205
flower structure 245–7, 312
flowering plants 32
　　sexual reproduction 32, 245–52
　　transport in 137–48
fluoride 119
focusing light 203
food
　　additives 127–8
　　colouring 127, 128, 141
　　combustion 112
　　consumption 318
　　production 84, 124–6, 205
　　supply 347, 348
　　see also nutrients; nutrition
food chains 320–1, 392
　　human 329
　　insecticide effects on 372
　　and radioactivity 374
food molecules, tests for 96–7
food webs 321–3
　　acid rain effects on 364
　　on a rocky shore 389
formula milks 261
fossil fuels 335, 336, 337, 340, 363, 366, 380
fossil record 301
fresh water 377–8
fruit 36
FSH (follicle-stimulating hormone) 255
fungi 26–7, 39–40
　　classification 16
　　in the food chain 320, 321
　　penicillin from 86, 87, 89
　　spore formation 242

G

gall bladder 115, 116
gas exchange
　　in animals 173–4
　　in humans 174–6
genes 281, 282, 283
　　mutations 293–5
　　see also alleles
genetic diagram 284, 289
genetic engineering 26, 302–3
genotype 284, 286, 287, 289, 309

genus 16, 17
geotropism 206, 207
germination 83, 170, 273–6, 365
gestation 257–8
global warming 337, 367–8, 385–7
global warming potential 386
glomerulus 187, 189
glucagon 211
glucose 72
　　absorption 123
　　assimilation 123
　　blood glucose levels 211, 303
　　from photosynthesis 98
　　in respiration 169, 171–2
　　test for 96–7
　　in translocation 148
glycerol 95
glycogen 123, 211
gonorrhoea 265
Gram-negative/positive bacteria 25
graphs 406–11
graticule 60
greenhouse gases 366–9, 385–7
greenhouses 102
growth 11, 13, 272
growth rings 139
guard cells 70

H

habitat conservation 376–7
haemodialysis 66
haemoglobin 55, 158, 215, 294
haemophilia 314
haploid cells 244, 245, 292
heart 52–3, 149–50, 229
heart rate 151–2
heat energy 82, 112, 209, 328, 333
herbicides 124, 208, 370
　　using 373
herbivores 320, 326
heroin 212–13
heterozygous 283, 285–9
HIV virus 24, 212, 265, 266
homeostasis 209
homozygous 283, 285–8
hormonal system 196–7
hormones 186, 204–5
　　sex hormones 254–5, 256
human
　　circulatory system 148–60
　　classification 17

digestive system 114–22
excretion 185–94
food chains 329
gas exchange in 174–6
immune system 159, 161–3
nutrition 108–13
organ systems 52
organisation at cell level 53
population 346–7
sexual reproduction 253–66
human population size 350–3, 355, 391
　　and AIDS 395
　　and species extinctions 392
　　and water requirements 378
humidity 143
hyphae 26, 27, 39, 242, 321

I

ileum 115, 122
immigration 346–7
immune system 159, 161–3
　　damage to 266
　　and transplants 162, 191–2
incisors 117, 118
independent variables 413, 416, 417
influenza viruses 24
ingestion 114
inheritance 280
　　characteristics 282–3, 287
　　and sex determination 290–1, 314
insect-pollination 247, 248, 249, 312
insecticides 370–2
insects 28
inspiration 176
inspired air 178–9
insulin 211, 302–3
intercostal muscles 176–7
invertebrates 19, 27–31
iodine solution 78, 80, 96, 99
ionising radiation 295
iris 202
iron 109, 111
IUCN Red List 376
IUD (intrauterine device) 263

K

keys 35–6, 43
kidneys 185, 186–90, 234
　　failure 66, 190–2
　　see also dialysis
kingdom 16, 17
kwashiorkor 110

L

labour 260–1
lacteals 123
lactic acid 172
lactose 122
lag phase 349, 350
landfill sites 375
large intestine 115, 116, 123
larynx 174
leaching 358, 363
leaf
    section through 58, 228
    stomata 45, 70, 105, 142, 143
    structure 104–5
    see also photosynthesis
lens 203
levels of organisation 52–8
LH (luteinising hormone) 255
life processes, seven 10–11
light intensity
    and carbon dioxide
        concentration 343
    and photosynthesis 98, 100–1,
        102, 103, 235
    and transpiration 143
lightning 338
limiting factors 100–1
line graph 406–7, 408–11
line of best fit 404
lipases 77, 120, 121
lipids 95, 108
liver 115, 116
    cell 231
    cirrhosis 214
    in digestive process 123–4
    excretion role of 186
living organisms
    characteristics 10–14
    classification and diversity
        15–34, 42–3
'lock and key' model 81
locusts 371
log phase 349, 350
lungs 136, 174–6
    and smoking 214
lymphatic system 162–3
lymphocytes 159, 161, 162, 163

M

magnification 59–60, 62
malaria 295
male reproductive system 253
malnutrition 113
mammals 22–3
manure 340, 361
markers 192
Mars 369
mechanical and physical digestion 117
medulla 187
meiosis 246, 291–2
memory cells 161, 162
menstrual cycle 255, 256
metabolic reactions 77
methane 367, 379, 385–6
microorganisms
    and fermenters 86–7, 230, 348–9
    in food production 126
    see also bacteria; viruses
micropyle 250
microscope 49, 59–60
migration 345
milk 122
millipedes 30
Minamata Bay 362
mineral deficiencies 106
mineral ions 106, 141, 233, 363
minerals 108, 109
mitosis 291
molars 117, 118
molecules 64
molluscs 31
monocotyledons 32
monohybrid crosses 284–7, 309
motor neurones 198, 199, 200
mould 26, 242
    see also fungi
mouth 115, 116
movement 11, 13
    in and out of cells 63–75
Mucor 26, 39–40, 242
mucus 54, 174
multiple births 264
muscle cells 54–5, 172
mushrooms 27
mutations 293–5
mycelium 26, 27, 40
myriapods 30

N

natural selection 298–300, 301
natural system 16
negative feedback 211
nematodes 30
nephrons 187, 188
nerves 197, 198, 199
nervous system 196, 197
    CNS 197, 198, 200
net movement 64, 70
neurones 198, 199, 200
nitrate ions 72, 106, 338, 340
nitrifying bacteria 338–9
nitrogen cycle 337–9, 341
nitrogen fertilisers 107
nitrogen-fixing bacteria 338–9, 341
nitrogen oxides 363, 385–6
non-biodegradable 375
nuclear fall-out 373–4
nucleus 46, 47, 48
nutrients 94
    in a balanced diet 111
    cycles 333–44
nutrition 10, 13, 92–135
    definition 94
    in humans 108–13
    saprotrophic 26, 40
    see also food

O

obesity 111, 113
oesophagus 115, 116
oestrogen 255, 262
omnivores 329
order 17, 19
organ systems 52
organ transplant 162, 191–2, 214
organisation, levels of 52–8
organs 52
    sense organs 197, 198, 200
    target organs 204
osmosis 66–71, 98, 140–1, 233
ovary 247, 251, 254
oviduct 254, 257
ovulation 255, 262
oxygen
    concentration in water 361, 394
    diffusion 65
    and germination 274
    from photosynthesis 98, 102, 103
    see also gas exchange
oxyhaemoglobin 158

P

palisade cells 47, 58, 104, 105
pancreas 115, 116, 121
paper recycling 380–1
parasites 30
partially permeable 65, 67, 72
passive process 64, 66
pathogens 24, 25, 26, 159, 161, 162
pectinase 84, 231, 232
penicillin 86, 87, 89

penis 253, 257
peristalsis 116
pesticides 124, 147
   pollution from 370–3
pH 78, 86, 225, 232, 363
   optimum 80, 82
   and seed germination 365
phagocytes 159, 161
phenotype 284–9, 308
phloem 105, 137, 138, 139, 236
photosynthesis 48, 98
   investigating 99–100, 102, 103, 235–6
   leaf adaptations for 104
   limiting factors 100–1
phototropism 206, 207
phylum 17, 19
placenta 259, 261
plant cells 46–8
   osmosis in 68–71
   water potential gradient 72
plants
   classification 16
   energy losses 327
   in the food chain 320
   in greenhouses 102
   mineral requirements 106–7
   natural selection 299–300
   nutrition 94, 98–9
   tropic responses in 206–8
   tuber formation 242–3
   water loss 146
   water uptake 70–1, 140–1, 334
   see also flowering plants; leaf; root hair cells
plasma 157, 234
plasmids 25, 26, 302–3
plastics 375
platelets 159, 160
pollen tube 250
pollination 245, 247–50, 312
pollution 360
   air pollution 363–6
   greenhouse gases 366–9
   from nuclear fall-out 373–4
   from pesticides 370–3
   from plastics 375
   water pollution 361–2, 375, 394
pond plants 146
population 345–7
population pyramids 351–2, 394–5
population size 345–55
   birds on farmland 384
   see also human population size

precipitation 333, 334
precision 413, 418
predators 320, 347–8
predictions 285, 287
premolars 117, 118
prey 320, 347–8
primary consumers 320, 321, 322, 323
probabilities 285, 287
producers 320, 321, 322, 323
progesterone 255, 256, 262
proteases 77, 79, 80, 84–5, 120–1, 224–5
protein fibres 54–5
proteins 95, 108
   carrier proteins 72–3
   nitrogen in 338
   single-cell protein 126
   test for 97
   see also enzymes
protocists 16
Punnett square 284–5, 291, 309
pupil (eye) 202
pyramid of biomass 323–5, 326, 392
pyramid of energy 325–7
pyramid of numbers 323, 325, 326, 332

R

radioactivity 373–4
random error 419–20
reabsorption, kidneys 189–90
receptor 196, 197, 200
recessive allele 283, 284–5, 287, 288, 310
recombinant DNA 302
rectum 115, 116
recycling
   paper 380–1
   water 378–9
red blood cells 51, 55, 63, 158, 294, 295
   osmosis 67
reflex 199–200
reflex arc 199
relay neurones 198, 199
renal capsule 187, 189
reproduction 11, 13, 240–71
   asexual 240, 242–3, 291
   see also sexual reproduction
reptiles 21
respiration 11, 13, 167–84
   aerobic 169–70

   anaerobic 171–3
   artificial 178
   cell respiration 73, 168–9
respiratory system 175
retina 201, 202, 203
rhythm method 262
ribs 177
rickets 109
rod cells 202
root hair cells 55, 70–1, 72, 106, 140, 141

S

safety precautions 400–2, 413, 418–19
saliva 115
salivary glands 115, 121
saprotrophic nutrition 26, 40
scattergraphs 406, 409
scorpions 29
seawater 377
secondary consumers 320, 321, 323
seed dispersal 32, 252
seed formation 250–2
seed germination 83, 170, 273–6, 365
selection 296–302
selective breeding 296–8
self-pollination 249–50
semen 253
sense organs 197, 198, 200
sensitivity 11, 13, 196–205
sensory neurones 198, 199, 200
sewage 361–2, 394
sewage treatment 378–9
sex chromosomes 281, 282, 283
sex determination 290–1, 314
sex hormones 254–5, 256
sexual intercourse 257
sexual reproduction 240, 244–5
   in flowering plants 32, 245–52
   in humans 253–66
   meiosis 246, 291–2
sexually transmitted diseases 265–6
sickle cell anaemia 294–5
sigmoid population growth curve 348–9
single-cell protein 126
sinks 147–8
skin 209, 210
small intestine 115, 116, 122–3

smoking 153, 214–15
soil erosion 357
soil fertility 340
solar energy 319, 333
solutes 64, 67, 68
solutions concentrations 67, 68, 69, 416
solvents 65
sources 147–8
specialisation 53–6
species 8, 16, 17, 42
   conservation 376–7
   extinction 376, 392
specimen size 59–62
sperm 253, 257, 264
spermicide 262
spiders 29
spinal cord 199
spirometer 179
spongy mesophyll cells 105, 145
spores 242
stamens 245, 246
standing crop 324
starch 76, 95
   from photosynthesis 98, 99, 100
   test for 78, 80, 96
starvation 113, 125
stationary phase 349
stem cells 54
stigma 246, 247, 248, 249
stimulus 196, 199, 200
stomach 115, 116, 121
stomata 45, 70, 105, 142, 143
substrate 77, 81, 82
sucrose 94, 99, 138, 147
sugar solution 64, 66–7, 226, 233
sugars 94, 95, 97
   see also glucose
sulfur dioxide 363, 366
sulfur hexafluoride 386–7
surface area-to-volume ratio 167, 220–2
symbol equation 98, 169, 172, 173
systematic error 420
systemic pesticides 147

T
tables 404–5, 406
target organs 204
taxonomy 15
teeth 117–19
temperature
   core 209–11
   and enzymes 77–9, 82, 86, 232
   and germination 274, 275, 276
   global surface 367–8
   on Mars and Venus 369
   and photosynthesis 101, 102
   and transpiration 142–3
test crosses 288
testes 253
testosterone 254
tissue rejection 162, 192
tissues 52
tobacco mosaic virus 24
toxins 124, 362, 372
trachea 174, 177
transgenic organism 302, 303
translocation 138, 147, 148
transpiration 71, 142, 334
   rate of 142–4
   sources and sinks 147–8
   water potential 145
transplant 162, 191–2, 214
transportation 136–66
   in flowering plants 137–48
   in humans 148–63
trophic levels 320–7
   energy transfer losses 328–9
tropic responses 206–8
tubers 242–3
turgid cells 68, 70

U
ultrasound scans 258
umbilical cord 259
urea 124, 186, 190
urinary system 186–7, 227
urine 187, 188
   composition 227, 234
   test 185, 260
uterus 254, 255, 256
   fetal development 257, 258

V
vaccination 161, 162
vacuole 39, 46, 47, 48
vagina 254, 257
valves 150, 155, 156
variables 413, 416, 417
variation 293–5
   and artificial selection 296–8
   and natural selection 298–300
vascular bundles 56, 105, 138
vasectomy 263
vasoconstriction 210
vasodilation 209
vectors 26
veins 155, 156
ventilation 168, 174
ventricles 150
Venus 369
vertebrates 19
   features and adaptations 19–23, 42–3
villi 122–3
viruses 12, 23–4, 212
vitamin C 109
vitamin D 109, 110
vitamins 108, 109

W
waste products 11, 66, 98, 169
   removal 186, 191
water 108
   absorption 123
   conservation 377–80
   eutrophication effects 359–60, 361, 362
   and germination 274, 275
   osmosis 66–71, 140–1
   reabsorption 189–90
   from respiration 170
water cycle 333–4
   and deforestation 357
water pollution 361–2, 375, 394
water potential gradient 72, 141, 145, 233
weeds 69, 373
white blood cells 159, 266
wilt 68, 142
wind-pollination 247, 248
withdrawal symptoms 213
womb see uterus
word equation 98, 169, 173
worms 30

X
xylem 56, 71, 72, 105, 137–8, 236
   in transpiration 145
   in transport 139

Y
yeast 26, 40, 231
   respiration in 173
yield 107
yoghurt 126

Z
zygote 244, 245, 250–1, 257
   division 279, 280